Applications of Adaptive Control

ACADEMIC PRESS RAPID MANUSCRIPT REPRODUCTION

Papers included in this book were presented at
the *International Workshop on Applications of Adaptive Control*
held at Yale University, August 23–25, 1979.

APPLICATIONS OF ADAPTIVE CONTROL

edited by

Kumpati S. Narendra
Department of Engineering and Applied Science
Yale University
New Haven, Connecticut

Richard V. Monopoli
Electrical and Computer Engineering Department
University of Massachusetts
Amherst, Massachusetts

1980
ACADEMIC PRESS
A Subsidiary of Harcourt Brace Jovanovich, Publishers
New York London Toronto Sydney San Francisco

ACADEMIC PRESS, INC.
111 Fifth Avenue, New York, New York 10003

United Kingdom Edition published by
ACADEMIC PRESS, INC. (LONDON) LTD.
24/28 Oval Road, London NW1 7DX

Library of Congress Cataloging in Publication Data

International Workshop on Applications of Adaptive Control, Yale University, 1979.
Applications of adaptive control.

1. Adaptive control systems—Congresses. I. Narendra, Kumpati S. II. Monopoli, Richard Vito, Date III. Title.
TJ217.I59 1979 629.8'36 80-18879
ISBN 0-12-514060-6

PRINTED IN THE UNITED STATES OF AMERICA

80 81 82 83 9 8 7 6 5 4 3 2 1

Contents

Contributors vii
Preface ix
Acknowledgments xiii

TUTORIAL

Self-Tuning Regulators—Design Principles and Applications 1
K. J. Aström
Design of Stable Model Reference Adaptive Controllers 69
Kumpati S. Narendra and Yuan-Hao Lin

SIGNAL PROCESSING

An Overview of Optimum-Adaptive Control in Sonar Array Processing 131
N. L. Owsley
On the Application of Recursive Least Squares Methods to Adaptive Processing 165
E. H. Satorius and M. J. Shensa

POWER SYSTEMS

Adaptive Power Plant Control: Problems and Prospects 193
Raman K. Mehra, John S. Eterno, and Sadashiva S. Godbole
New Developments in Improving Power Network Stability with Adaptive Generator Control 221
E. Irving

AIRCRAFT SYSTEMS

Adaptive Control Application to Aircraft 245
E. G. Rynaski
Perspectives on the Application of Adaptive Control to Aircraft Systems 269
Gerhard Kreisselmeier

Adaptive Flight Control: A Pragmatic View 291
Gunter Stein
Toward Adaptive Control of Large Structures in Space 313
Mark J. Balas and C. Richard Johnson, Jr.

PROCESS CONTROL

Adaptive Control in the Process Industries 345
Pierre R. Bélanger
Adaptive Level Control in a Glass Furnace 367
R. Haber, J. Hetthéssy, L. Keviczky, I. Vajk, A. Kovács, Z. Csaszar, N. Czeiner, A. Feher, and A. Turi
Experimental Evaluation of Hyperstable Model Reference Adaptive Control 379
R. M. Johnstone, S. L. Shah, D. G. Fisher, and R. Wan
Discrete Adaptive Control of a Radio Telescope 397
S. I. Haque and R. V. Monopoli
Adaptive Control of a Laser Pointing and Tracking System: A Feasibility Study 423
R. L. Gutmann, R. V. Monopoli, and R. Van Allen
Classical Control Interpretation and Design of Microcomputer Adaptive Controllers 453
C. P. Neuman and R. L. Morris
Discrete Time Modelling of the Torque Response of a Spark-Ignited Fuel-Injected Engine 491
H. G. Hopkins and R. H. Borcherts
Application of Adaptive Systems in Process Control 509
H. Unbehauen and Chr. Schmid

Contributors

Numbers in parentheses indicate the pages on which the authors' contributions begin.

K. J. Åström (1), Department of Automatic Control, Lund Institute of Technology, Lund Sweden

Mark J. Balas, (313), Electrical and Systems Engineering Department, Rensselaer Polytechnic Institute, Troy, New York

Pierre R. Bélanger (345), Department of Electrical Engineering, McGill University, Montreal, Quebec, Canada

R. H. Borcherts (491), Engineering and Research Staff, Ford Motor Company, Dearborn, Michigan

Z. Csaszar (367), Research Group for Automation Control, Research and Design Institute for Silicate Industry, Budapest, Hungary

N. Czeiner (367), Research Group for Automation Control, Research and Design Institute for Silicate Industry, Budapest, Hungary

John S. Eterno (193), Scientific Systems, Inc., Cambridge, Massachusetts

A. Feher (367), Research Group for Automation Control, Research and Design Institute for Silicate Industry, Budapest, Hungary

D. G. Fisher (379), Department of Chemical Engineering, University of Alberta, Edmonton, Alberta, Canada

Sadashiva S. Godbole (193), The Babcock and Wilcox Company, Lynchburg Research Center, Lynchburg, Virginia

R. L. Gutmann (423), Boeing Company, Seattle, Washington

R. Haber (367), Department of Atuomation, Technical University of Budapest, Hungary

S. I. Haque (397), Intermetrics, Inc., Cambridge, Massachusetts

J. Hetthéssy (367), Department of Automation, Technical University of Budapest, Hungary

H. G. Hopkins (491), Engineering and Research Staff, Ford Motor Company, Dearborn, Michigan

E. Irving (221), Automatic Control Division, Direction des Etudes et Recherches, Electricité de France, Clamart, France

C. Richard Johnson, Jr. (313), Department of Electrical Engineering, Virginia Polytechnic Institute and State University, Blacksburg, Virginia

R. M. Johnstone (379), Department of Chemical Engineering, University of Alberta, Edmonton, Alberta, Canada

L. Keviczky (367), Department of Automation, Technical University of Budapest, Hungary

A. Kovács (367), Department of Automation, Technical University of Budapest, Hungary

Gerhard Kreisselmeier (269), DFVLR-Institut fuer Dynamik der Flugsysteme, Oberpfaffenhofen, Federal Republic of Germany

Yuan-Hao Lin (69), Department of Engineering and Applied Science, Yale University, New Haven, Connecticut

Raman K. Mehra (193), Scientific Systems, Inc., Cambridge, Massachusetts

R. V. Monopoli (397, 423), Department of Electrical and Computer Engineering, University of Massachusetts, Amherst, Massachusetts

R. L. Morris[1] (453), Carnegie Institute of Technology, Carnegie-Mellon University, Pittsburgh, Pennsylvania

Kumpati S. Narendra (69), Department of Engineering and Applied Science, Yale University, New Haven, Connecticut

C. P. Neuman (453), Carnegie Institute of Technology, Carnegie-Mellon University, Pittsburgh, Pennsylvania

N. L. Owsley (131), Naval Underwater Systems Center, New London, Connecticut

E. G. Rynaski (245), Advanced Technology Center, Calspan Corporation, Buffalo, New York

E. H. Satorius[2] (165), Naval Ocean Systems Center, San Diego, California

Chr. Schmid (509), Department of Electrical Engineering, Ruhr-University Bochum, Germany

S. L. Shah (379), Department of Chemical Engineering, University of Alberta, Edmonton, Alberta, Canada

M. J. Shensa (165), Hydrotronics, Inc., San Diego, California

Gunter Stein (291), Honeywell Systems and Research Center and Massachusetts Institute of Technology, Cambridge, Massachusetts

A. Turi (367), Research Group for Automation Control, Research and Design Institute for Silicate Industry, Budapest, Hungary

H. Unbehauen (509), Department of Electrical Engineering, Ruhr-University Bochum, Germany

I. Vajk (367), Department of Automation, Technical University of Budapest, Hungary

R. Van Allen (423), USAF, Kirtland Air Force Base, Albuquerque, New Mexico

R. Wan (379), Department of Chemical Engineering, University of Alberta, Edmonton, Alberta, Canada

[1]Present address: Control Systems Department, Scientific Research Laboratory, Ford Motor Company, Dearborn, Michigan

[2]Present address: Dynamics Technology, Inc., Torrance, California

Preface

The adaptive control of systems has been a dream of control system theorists and practitioners for at least a quarter of a century. It arose from a desire and need for improved performance of increasingly complex engineering systems with large uncertainties. This is especially important in systems with many unknown parameters that are changing with time. Roughly speaking, an adaptive system is one that continually monitors these changes and adjusts its control parameters automatically to maintain good performance. In mathematical terms, such a system is generally a nonlinear nonautonomous dynamic system. Hence, adaptive systems can be considered as a class of control systems that have been made deliberately nonlinear.

We have engaged in many hours of discussion throughout the past decade as to why the promise of adaptive control has been so slow in coming to fruition. From these discussions several key reasons emerged. First, early schemes for adaptive control, though ingeneous, did not guarantee global stability of the adaptive system. Those who tried to apply these schemes either had great difficulty in showing even local stability of the resulting control systems and, therefore, were reluctant to use them, or, worse, implemented them on systems with unfortunate, and perhaps even disastrous, results. Second, globally stable adaptive control algorithms have been developed only recently, and these are not yet known to most practitioners. Third, researchers in adaptive control, for the most part, are not well versed in the potential applications of the theory. Finally, the technology of the past decade was not adequate to the task.

With recent advances in microprocessor technology, it has become feasible to implement adaptive algorithms efficiently in real time at reasonable cost. This fact together with recent developments in theory and the keen interest shown by industry in adaptive control prompted us to organize the Workshop on Applications of Adaptive Control, which was held in August 1979 at Yale University. Of the one hundred participants at the workshop, approximately sixty percent were potential users of adaptive control from industry and forty percent were researchers from universities. The objectives of the workshop were to :

(i) make potential users aware of recent advances in control theory;
(ii) make theorists aware of the potential applications and especially aspects of applications which do not satisfy the assumptions being made by the theorists;
(iii) determine if the reasons for the concept of adaptive control falling into disfavor in the flight control community are still valid.

Workshop participants agree that the objectives were met reasonably well. This motivated us to share, via this book, the workshop experience with a broader audience. Papers included here represent a selection of those presented at the workshop and do illustrate that it is becoming increasingly realistic to consider using adaptive control in many practical problems.

The first two papers, "Self-tuning regulators—design principles and applications" by Åström and "Design of stable model reference adaptive controllers" by Narendra and Lin are tutorials that bring together important recent results in two of the most studied approaches to adaptive control—self-tuning regulators (STR) and model reference adaptive control (MRAC). In addition to presenting a clear exposition of the theory of STR and its relationship to MRAC, Åström also gives numerous examples of successful applications. The importance of error models in the stability analysis of MRAC is stressed in the paper by Narendra and Lin, and it is shown that the different independent proofs of stability given recently are equivalent. Although STR and MRAC were developed separately by different research groups, current research is beginning to reveal the similarities between these two approaches.

The section on signal processing has papers contributed by N. L. Owsley and E. H. Satorius and M. J. Shensa. Though not dealing with adaptive control per se, it was felt essential to include these papers because algorithms used for adaptive signal processing are very similar to those used in adaptive control. For this reason, control theorists and practitioners can benefit from an understanding of work in this area.

In the section on power systems, the paper by R. K. Mehra and J. S. Eterno presents a good description of the types of power systems problems that could benefit from application of adaptive control. Electricité du France has had a research and development effort in adaptive controls for several years now, and E. Irving describes how they are attempting to apply adaptive control algorithms for controlling large electric generators.

Adaptive control in aircraft systems is discussed in the papers by E. G. Rynaski, G. Kreisselmeier, G. Stein, and M. J. Balas and C. R. Johnson. G. Stein makes the point that many flight control problems are solvable by gain scheduling based on data from air sensors. However, he does point out some areas in flight control where an adaptive controller may be necessary. In his paper, Rynaski describes how adaptive control fell into disfavor in the flight control community, and points out that a residual negative bias still exists. He also describes the utility of the model following approach in flight control, where, in the examples he considers, the adaptive feature is not required. Kreisselmeier dicusses the desirability of eliminating the cost of air data sensors in less-sophisticated flight control systems. This may be made possible by use of adaptive control techniques. In the paper by Balas and Johnson, adaptive control of distributed parameter systems is considered. The problem they address, control of large structures in space, is one where adaptive control may be the only method to deal with ignorance of system structure, order, and parameters.

The set of papers on process control includes applications not only to chemical processes, but to electromechanical systems as well. P. Bélanger, in his paper "Adaptive control in the process industries" offers a number of interesting reasons as to why adaptive control, so far, seems to have found wider applicability in the area of chemical process control than in electromechanical systems. One of these, which strikes us as particularly important, is the fact that in process control the dynamics are almost always well damped, whereas electromechanical systems often have several highly underdamped modes usually related to bending or vibration. Also, parameters of electromechanical systems can change rapidly, whereas changes in many chemical processes take place slowly enough to allow time for parameter identification to take place. Harber *et al.*, in their paper on "Adaptive level control in a glass furnace" show experimental results which clearly demonstrate the improvement possible with adaptive control. Their self-tuning regulator ensures a much smaller variance of the output around the reference value than does the traditional controller. Johnstone *et al.* show that model reference adaptive control applied to liquid level control is robust and has superior tracking and regulation properties as compared to the classical PI control. It is important to note that their results were achieved in spite of the fact that their pilot plant had nonlinearities and rapid changes of parameters which were not covered by the theory. Haque and Monopoli, in "Discrete adaptive control of a radio telescope," show the effectiveness of the discrete model reference control algorithms in removing the undesirable effects of unknown Coloumb friction level and unknown system inertia. Three adaptive gains in the telescope control system resulted in a decrease of tracking error by a factor of two. Gutmann, Monopoli, and Van Allen show the potential for using model reference adaptive control to improve the pointing accuracy of a high power laser pointing and tracking system. In their paper, "Adaptive control of a laser pointing and tracking system: A feasibility study," it is pointed out that good results are obtainable even though the design is based on a dominant second order model of a sixth order plant. In "Classical control interpretation and design of microcomputer adaptive controllers," Neuman and Morris compare three classes of model reference adaptive controllers in a common framework from the viewpoint of microcomputer implementation. This is a particularly important contribution because of the blending of the latest technology and theory. Hopkins and Borcherts, in "Discrete time modelling of the torque response of a spark-ignited fuel-injected engine," give experimental results of a model reference identification technique on which to base control strategies generated by linear quadratic theory. In the paper "Application of adaptive systems in process control" by Unbehauen an excellent survey of successful applications of adaptive control in various process control industries is given. Also, Unbehauen makes the important point that further successful applications of adaptive control will depend on a close and continuous collaboration between control design engineers and researchers in adaptive control theory.

Sound theory provides the basis for good practice, even though in real world problems the assumptions made for theoretical analysis are almost never strictly satisfied. The applications given in this volume are no exceptions since, in one way or another, they fail to satisfy the assumptions that have to be made to prove theoretically global stability. However, they appear to be robust, and to achieve the desired objective of improved performance over nonadaptive systems. These successful applications provide motivation for researchers to re-examine current theories and develop new ones to explain their behavior. This continuing interaction between theory and practice should lead to systematic design procedures for adaptive control systems.

Acknowledgments

We wish to express out appreciation to all participants in the Workshop on Applications of Adaptive Control held in August 23- 25, 1979 at Yale University. Our special thanks go to Ben Peterson who attended to many of the details of organizing the workshop and to Ian Gemmell for his help in running it.

Papers appearing in this book were presented at the workshop, and we are grateful to the authors for their continued cooperation. Carol Phelps's assistance in making the final drawings is also very much appreciated.

It is a great pleasure to acknowledge the invaluable assistance provided by Jean Gemmell at every stage in the preparation of this book: in organizing the workshop, in the arrangement and typing of the final manuscript, and by her attention to many other details with her usual thoroughness. The publication of the book on schedule is due in large measure to her efforts.

SELF-TUNING REGULATORS - DESIGN PRINCIPLES AND APPLICATIONS[1]

K. J. Åström

Department of Automatic Control
Lund Institute of Technology
Lund, Sweden

The basic principles of self-tuning regulators are discussed. The regulators are motivated from the viewpoint of nonlinear stochastic control theory. Self-tuning regulators based on pole-zero placement, minimum variance control and linear quadratic gaussian control are described in a common framework. Relations between self-tuning regulators and model reference adaptive control are discussed. Applications to different industrial process control problems are described. Practical and theoretical issues of relevance to design of self-tuning regulators are also treated.

[1]This research was supported by the Swedish Board of Technical Development (STU).

ISBN 0-12-514060-6

1. INTRODUCTION

The tuning problem is one reason for using adaptive control. It is a well-known fact that many processes can be regulated satisfactorily with PI or PID regulators. It is fairly easy to tune a PI regulator which only has two parameters to adjust. However, for an installation which has several hundred regulators it is a substantial task to keep all the regulators well tuned. A PID regulator which has three or four parameters is not always easy to tune, particularly if the process dynamics is slow. The derivative action is, therefore, frequently switched off in industrial controllers.

Since many control loops are not critical three term controllers will undoubtedly be used extensively in the future too. With an increasing demand for efficiency in the use of energy and raw material there are, however, an increasing number of control problems where it is reasonable to use regulators which are more complicated than PID regulators. Such regulators, which may include feedforward, state feedback, and observers, can often have more than 10 adjustable parameters. It is not possible to adjust so many parameters without a systematic procedure. The lack of a suitable tuning procedure is one reason why modern control theory has not been used more extensively.

One possibility to tune a regulator is to develop a mathematical model for the process, and its disturbances, and to derive the regulator parameters from some control design procedure. The appropriate mathematical models can be obtained from physical modeling or from system identification. The drawback with such a procedure is that it may be fairly time consuming, and that it requires personnel with skills in modeling, system identification, and control design. The self-tuning regulator can be regarded as a convenient way to combine system identification and control design. Its name does, in fact, derive from such applications.

One reason for using adaptive control is thus to avoid the tuning problem. Another motivation for using adaptive control is that the characteristics of the process and its disturbances may change with time. If the changes are not too rapid a properly designed self-tuning regulator may be used for continuous tuning to obtain close to optimal performance.

Adaptive control has been a challenge to control engineers for a long time [1]. Many different schemes have been proposed [2-4]. In spite of this, progress in the field has been comparatively slow. One reason for this is that adaptive systems are difficult to understand because they are inherently nonlinear. The field is also fairly immature as a scientific discipline. Many ideas that are basically the same are derived and presented using very different approaches. There is a wide divergence in notations. A fundamental conceptual framework is also lacking. Recently there has been an increased interest in adaptive control. One reason for this is the availability of microprocessors which make it possible to implement adaptive controllers conveniently and cheaply [5-7]. Another reason is the success of adaptive control in pilot installations in industry [8-16]. A third reason is that some progress has recently been made in the theory of adaptive control [17-23].

The purpose of this paper is to give an overview of self-tuning regulators, which is <u>one</u> approach to adaptive control. The focus will be on concepts, theory, and applications. The paper is organized as follows.

A brief review of nonlinear stochastic control theory is given in Section 2. This theory gives a conceptual framework and a general structure of an adaptive regulator with many interesting features. The regulator obtained from nonlinear stochastic control theory is, however, so complicated that it can only be computed numerically in almost trivial cases. To obtain something useful it is thus necessary to make approximations. Self-tuning regulators are one approximation. The principles for design of self-tuning regulators are also discussed in Section 2. The basic idea can be described as follows. Start with a design method that will give adequate results if the parameters of models for the dynamics of the process and its environment are known. When the parameters are unknown, they are replaced by estimates obtained from a recursive parameter estimator. Since there are many available methods for designing control systems there are at least as many ways to design self-tuning regulators.

The general principles are illustrated in Section 3 by giving some details for self-tuning regulators based on assignment of poles and zeros. In this way, it is also possible to bring up the notion of algorithms with implicit and explicit identification. Self-tuning regulators based on minimum variance control and on linear quadratic control theory are briefly discussed in Section 4. In Section 5 self-tuning regulators are compared with model reference adaptive systems, which is another approach to adaptive control. It is shown that the two approaches are closely related. Applications of self-tuning regulators to real processes are discussed in Section 6. This includes control of papermachines, heat exchangers, ore crushers, and autopilots for ship steering. Some aspects on the available theory and practice of self-tuning regulators are given in Section 7. The paper ends with conclusions and references.

2. DESIGN PRINCIPLES

The principles for designing self-tuning regulators (STR) are presented in this section. A brief review of nonlinear stochastic control theory is first given. This provides a convenient conceptual framework. Self-tuning regulators are then introduced as approximations to the general nonlinear stochastic problem. It is also shown that the self-tuning regulators can be formulated for many problems which are not stochastic control problems.

Nonlinear Stochastic Control

In discrete time nonlinear stochastic control theory, the model of the system and its environment are specified by giving the conditional probability distribution of the state x(t+1) at time t + 1 given the state x(t) and the control u(t) at time t. The observation process is similarly specified by giving the conditional probability distribution of the observation y(t) at time t given the state x(t) and the control u(t) at time t. The criterion is formulated as to minimize the expected value of a loss function which is a scalar function of states and controls.

The problem of finding a control which minimizes the expected loss function is difficult. Useful explicit conditions for existence are not known, in general. Under the assumption that a solution exists a functional equation can, however, be derived using standard dynamic programming [24]. This functional equation, which is called the _Bellman equation_, can be solved numerically only in very simple cases. The approach is, nevertheless, of interest because it gives insight into the structure of the optimal controller. This structure is shown in Fig. 1. The controller can be thought of as composed of two parts, an estimator and a feedback regulator. The estimator generates the conditional probability distribution of the state of the process given all past data. This distribution is sometimes called the _hyperstate_ of the problem. The feedback regulator is simply a nonlinear static function which maps the hyperstate into the space of controls.

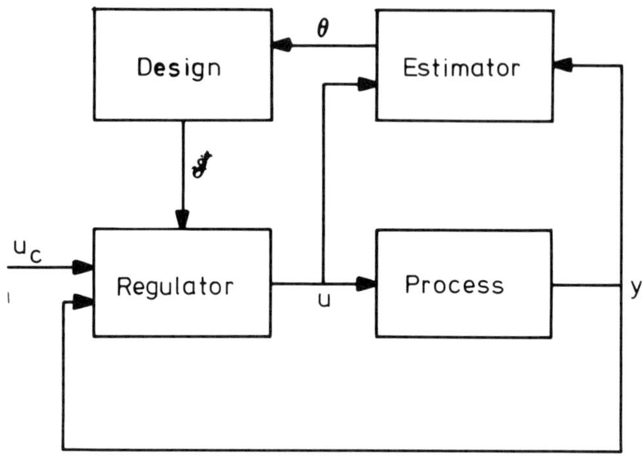

FIGURE 1. Structure of the regulator obtained from nonlinear stochastic control theory. The hyperstate w is generated from a dynamical system using u and y as inputs. The regulator is a static nonlinearity which gives the control variable u(t) as a function of the hyperstate w(t).

To solve a nonlinear stochastic control problem it is necessary to determine the estimator, which updates the hyperstate, and to solve the functional equation which gives the feedback regulator. The structural simplicity of the solution is obtained at the price of introducing the hyperstate which is a quantity of very high dimension. For example, a problem whose state space is R^4 has a hyperstate which is a distribution over R^4.

Nonlinear stochastic control is of substantial interest for adaptive control, because many of the aspects of adaptive control are captured by the nonlinear stochastic control formulation. In a problem with unknown but constant parameters the parameters are simply introduced as auxiliary state variables. Let a be a constant parameter, the associated state equation is then

$$a(t+1) = a(t). \tag{2.1}$$

If it can be assumed that the parameter is drifting like a random walk the following is an alternative to (2.1):

$$a(t+1) = a(t) + v(t), \tag{2.2}$$

where $\{v(t)\}$ is a sequence of uncorrelated random variables. For a system with constant but unknown parameters or drifting parameters it is natural to separate the state variables into two groups, the original state variables and the state variables associated with the parameters of the process. It is often assumed that the 'parameters' are changing at much slower rates than the other state variables. It is, however, not necessary to make such a distinction. Nonlinear stochastic control theory thus makes it possible conceptually to deal with rapid parameter variations.

Approximations

Since it is difficult to solve the Bellman equation, approximate solutions are of considerable interest. A simple example will be used to illustrate some common approximations. Consider a process described by

$$y(t+1) = y(t) + bu(t) + e(t) \tag{2.3}$$

where u is the control, y the output, e white noise and b a constant parameter. Equation (2.3) is a sampled data model of an integrator with unknown gain. Let the criterion be to minimize

$$\lim_{N\to\infty} E \frac{1}{N} \sum_{1}^{N} y^2(t). \tag{2.4}$$

If the parameter b is known, the control law which minimizes (2.4) is given by

$$u(t) = -\frac{1}{b}\, y(t). \tag{2.5}$$

If the parameter b has a gaussian prior distribution, it follows that the conditional distribution of b, given inputs u, and outputs y up to time t is normal with mean $\hat{b}(t)$ and covariance P(t). The hyperstate of the problem can then be characterized by the triple $[y(t),\hat{b}(t),P(t)]$. In this simple case the Bellman equation can be solved numerically [25-28]. The control law obtained can not be characterized in a simple way. It has, however, a very interesting property. The control signal will not only attempt to bring the output close to zero. When the parameters are uncertain the regulator will also inject signals into the system to reduce the uncertainty of the parameter estimates. This is referred to as <u>probing</u> [29]. The optimal control law will give the right balance between the tasks of keeping the control errors and the estimation errors small. This is called <u>dual control</u> [30].

The following control law

$$u(t) = -\frac{1}{\hat{b}(t)}\, y(t) \tag{2.6}$$

is an approximation which is called the <u>certainty equivalence control</u>. It is obtained simply by solving the control problem in the case of known parameters and substituting the known parameters with their estimates.

In the control law (2.6) the control variable u(t) is obtained by dividing the measured variable y(t) by the estimate $\hat{b}(t)$. This implies that there will be difficulties if the estimate $\hat{b}(t)$ is zero or very small. When using a certainty equivalence control like (2.6) it is therefore necessary to limit the controller gain. One simple way to do this, which works well in practice, is simply to put a hard bound on u(t) or its increment u(t) - u(t-1).

The difficulty associated with division by $\hat{b}$ in (2.6) is avoided in the control law

$$u(t) = -\frac{\hat{b}(t)}{\hat{b}^2(t) + P(t)}\, y(t) = -\frac{1}{\hat{b}(t)}\, \frac{\hat{b}^2(t)}{\hat{b}^2(t) + P(t)}\, y(t) \qquad (2.7)$$

which is called _cautious control_ because it hedges and uses lower gain when the estimates are uncertain. Notice that the cautious control law minimizes the criterion

$$E[y^2(t)\,|\,y(t-1), y(t-2), \ldots] \qquad (2.8)$$

where $E[\cdot\,|\underline{D}]$ denotes the conditional expectation with respect to $\underline{D}$. The control law (2.7) does, however, not minimize (2.5). The minimization of (2.8) can be taken as the definition of cautious control in the general case.

There is empirical evidence that the cautious controller hedges too much. It has been observed in many simulations that the control signal will occasionally be zero over certain periods when the cautious controller is used. The mechanism can intuitively be explained as follows. If the estimate $\hat{b}$ is uncertain, i.e. $\hat{b}^2 \ll P$, then it follows from (2.7) that the control signal will be small. The uncertainty P may then increase and the control signal will decrease further until it finally becomes zero. When applied to stable systems the control signal will be zero for a period and normal operation will then be resumed. This phenomenon is called _turn off_. When the cautious controller is applied to an unstable system the output may become so large that the system can no longer be brought back into normal operation. The phenomenon is then called _escape_. The possibilities of turn off and escape makes the cautious control law less useful.

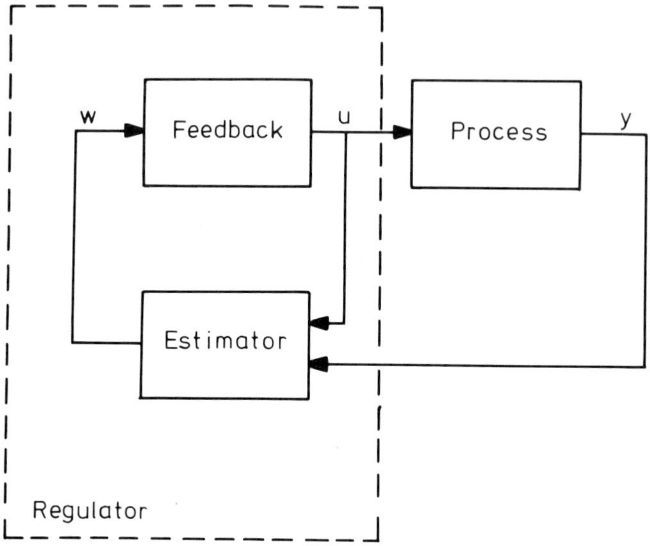

FIGURE 2. Schematic diagram of a self-tuning regulator.

Self-Tuning Regulators

A block diagram of a self-tuning regulator is shown in Fig. 2. The regulator can be thought of as composed of three parts, a recursive parameter estimator, a design calculator, and a regulator with adjustable parameters. The design method is chosen so that it gives the desired results when the parameters characterizing the process and its environment are known. When the parameters are unknown they are simply replaced by estimates. From the viewpoint of nonlinear stochastic control a self-tuning regulator is clearly a certainty equivalence control. Notice that the state is separated into two parts corresponding to the original process states and the parameters and that there are different signal paths for the two parts. Also notice that although the STR can be viewed as a certainty equivalence control the principle of STR can be applied to many problems which can not be formulated as stochastic control problems.

Since there are many different ways to do control design and to estimate parameters, there are many varieties of self-tuning regulators. Because the self-tuning regulators are certainty equivalence controls, it is also clear that they could be modified by introducing approximations of the effects of caution and probing if needed.

Disregarding the special structure of the regulator in Fig. 2 it can clearly be regarded as a nonlinear regulator which accepts an input y and generates a control u. The special structure is, however, very useful because it helps to understand how the regulator works and how it should be designed.

3. SELF-TUNERS BASED ON POLE-ZERO ASSIGNMENT

Design of self-tuning regulators based on pole-zero assignment will now be worked out in detail. A pole-zero assignment method for systems with known parameters is first described. It is then shown how this method is used to design self-tuners. Only discrete time systems are discussed.

Formulation

Consider a process characterized by

$$Ay = Bu \tag{3.1}$$

where A and B are polynomials in the forward shift operator. It is assumed that A and B are coprime and that

$$\deg B < \deg A,$$

where deg A denotes the degree of the polynomial A. It is desired to find a controller such that the closed loop is stable and that the transfer function from the command input u_c to the output is given by

$$G_M = \frac{Q}{P} \tag{3.2}$$

where P and Q are coprime and

$$\deg P - \deg Q >= \deg A - \deg B. \tag{3.3}$$

Condition (3.3) is necessary for the problem to be solved using a causal regulator.

Design Procedure

A general linear two-degree-of-freedom [31] regulator can be described by

$$Ru = Tu_c - Sy. \tag{3.4}$$

The closed loop transfer function relating y to u_c is given by

$$\frac{TB}{AR + BS} = \frac{Q}{P}, \tag{3.5}$$

where the right hand side is the desired closed loop transfer function G_M given by (3.2). The design problem is thus equivalent to the algebraic problem of finding polynomials R,S, and T such that (3.5) holds. It follows from (3.5) that factors of B, which are not also factors of Q, must divide R. Since factors of B correspond to open loop zeros it means that open loop zeros, which are not desired closed loop zeros, must be canceled. Factor B as

$$B = B^+ B^- \tag{3.6}$$

where all the zeros of B^+ are in a region $\underline{Z}$ of the complex plane and all zeros of B^- outside $\underline{Z}$. The region $\underline{Z}$ corresponds to modes that are sufficiently well damped. This means that all zeros of B^+ correspond to well damped modes and all zeros of B^- correspond to unstable or poorly damped modes.

A necessary condition for solvability of the servo problem is that the specifications are such that

$$Q = Q_1 B^- . \tag{3.7}$$

Since deg P is normally less than deg (AR + BS) it is clear that there are factors in (3.5) which cancel. In state space theory it can be shown that the regulator (3.4) corresponds to a combination of an observer and a state feedback. See [32]. It is natural to assume that the observer is designed in such a way that changes in command signals do not generate errors in the observer. This means that the factor which cancels in the right hand side of (3.5) can be interpreted as the observer polynomial T_1.

The design procedure can be formulated as follows.

Data: Given the desired response specified by the polynomials P and Q, subject to the conditions (3.3) and (3.7), and the desired observer polynomials T_1. It is assumed that P and T_1 have all their zeros in $\underline{Z}$.

Step 1: Solve the equation

$$AR_1 + B^- S = PT_1 \tag{3.8}$$

with respect to R_1 and S.

Step 2: The regulator which gives the desired closed loop response is given by (3.4) with

$$R = R_1 B^+ \tag{3.9}$$

and

$$T = T_1 Q_1 . \tag{3.10}$$

□

The equation (3.8) can always be solved because it was assumed that A and B were coprime. This implies that A and B^- are also coprime. Equation (3.8) has infinitely many solutions. If R_1^0 and S^0 is one solution then another solution is given by

$$R_1 = R_1^0 + B^- F$$

$$S = S^0 - AF$$

where F is an arbitrary polynomial. All solutions give closed loop systems with the same transfer function (3.2) from command signal u_c to the output y. The response of the closed loop system to disturbances will, however, depend on the particular solution which is chosen.

The lack of uniqueness can, for example, be used to ensure that the regulator has integral action, that it is causal and that it is insensitive to measurement noise. Since the following discussion will not depend on the particular solution which is chosen, it is unnecessary to go into further details here. Let it suffice to mention [33] that there will always be a causal regulator if

$$\deg T_1 = \deg A - \deg B^+ - 1.$$

It is often reasonable to choose a solution such that

$$\deg R = \deg S = \deg T,$$

which means that there will be direct terms both in the feedback and feedforward paths.

Least Squares Parameter Estimation

A recursive parameter estimator is an important part of a self-tuning regulator. See Fig. 2. There are many different methods which could be used for parameter estimation, for example stochastic approximation, least squares, extended least squares, generalized least squares, multistage least squares, instrumental variables, and maximum likelihood. A review of recursive estimation methods is given in [34] where also many references are given. There is unfortunately no recursive parameter estimator which is uniformly best. Consequently there are many different possibilities. For simplicity only recursive least squares estimation is used in this paper. Least squares is one of the simplest recursive estimation procedures. The method unfortunately gives biased estimates if the disturbances are correlated.

In recursive least squares estimation the criterion

$$V = \sum_{k=1}^{t} \lambda^{t-k} \varepsilon^2(k) \tag{3.11}$$

where

$$\varepsilon(t + \deg A) = Ay(t) - Bu(t) \tag{3.12}$$

is minimized. The parameter λ is a weighting factor which gives lower weight to old measurements. To describe the algorithm the model (3.1) is written explicitly as

$$y(t) + a_1 y(t-1) + \ldots + a_n y(t-n) =$$

$$= b_0 u(t-k) + \ldots + b_m u(t-m-k). \tag{3.13}$$

Introduce a vector of parameter estimates

$$\theta = [\hat{a}_1 \ldots \hat{a}_n \quad \hat{b}_0 \ldots \hat{b}_m]^T \tag{3.14}$$

and a vector of regressors

$$\phi(t) = [-y(t-1) \ldots -y(t-n) \quad u(t-k) \ldots u(t-m-k)]^T. \tag{3.15}$$

The recursive least squares estimate is then given by

$$\theta(t+1) = \theta(t) + P(t+1)\phi(t+1)\varepsilon(t+1), \tag{3.16}$$

where

$$\varepsilon(t+1) = y(t+1) - \theta^T(t)\phi(t+1), \tag{3.17}$$

and

$$P(t+1) = [P(t) - P(t)\phi(t)R(t)\phi^T(t)P(t)]/\lambda, \tag{3.18}$$

where

$$R(t) = [\lambda + \phi^T(t)P(t)\phi(t)]^{-1}.$$

There are other ways to perform the least square calculations. Square root algorithms, [35-38], are useful if the problem is poorly conditioned. Fast algorithms, [39-41], can be used if many parameters have to be estimated and if computing time is crucial.

Interpretations

Equation (3.16) can be interpreted as a quasi-Newton iteration for minimizing ε^2. It follows from (3.17) that

$$\phi = -\operatorname{grad}_\theta \varepsilon. \tag{3.19}$$

The term $\phi\varepsilon$ in (3.16) can thus be interpreted as the gradient of $\varepsilon^2/2$. The matrix P in (3.16) modifies the gradient direction and determines the step length. The matrix P can thus be interpreted as a <u>gain factor</u> which determines the rate of change of the estimate. The matrix P satisfies the equation

$$P^{-1}(t+1) = \lambda P^{-1}(t) + \phi(t+1)\phi^T(t+1). \tag{3.20}$$

When λ is equal to one, the matrix P will thus decrease monotonically. This means that the gain of the estimator will decrease to zero. Recall (3.11) which shows that λ can be interpreted as a factor which discounts past measurements. When $\lambda < 1$ the matrix P will not go to zero as time increases. The magnitude of the gain factor will depend on λ. The gain P will be larger if λ is smaller. The gain of the estimator will thus depend on the rate in which past data is discounted. There are also other possibilities to obtain an estimator with nondecreasing gain. If it is assumed that the parameters are drifting as in (2.2) and $\lambda = 1$ Equation (3.18) is replaced by

$$P(t+1) = P(t) + P(t)\phi(t)R(t)\phi^T(t)P(t) + R_1, \tag{3.21}$$

where

$$R_1 = \operatorname{cov}[v(t),v(t)].$$

Under stationary conditions when R_1 is different from zero the gain factor P will settle on a level which is determined by R_1. The larger R_1 is, i.e. the larger the assumed drift rate is, the larger will the gain factor be. The model (3.21) is more flexible than (3.18) because the rate of change of the parameters can be set individually. Different drift rates for different parameters can, however, also be incorporated in (3.18) by replacing λ in (3.18) by a diagonal matrix.

There are proposals to adjust the forgetting factor automatically [42]. The principle is to choose λ as

$$\lambda = 1 - \alpha\varepsilon^2/\overline{\varepsilon^2} \tag{3.22}$$

where $\overline{\varepsilon^2}$ is the mean value of ε^2 over a certain period. It has been found empirically [43] that a forgetting factor λ between 0.95 and 0.99 works well if there are continuous disturbances, provided that the problem is not overparametrized. If there are too many parameters the matrix

$$\Sigma\phi(t)\phi^T(t)$$

will be poorly conditioned. With a forgetting factor the matrix P can then be very large in certain directions.

In a servo problem with few disturbances the major excitation comes from the changes in the command signal. Such changes may be irregular and it has been found that there may be bursts in the process output if Eq. (3.18) is used with λ less than one. The presence of bursts can be understood intuitively as follows. The negative term in (3.18) represents the reduction in parameter uncertainty due to the last measurement. When there are no changes in the set point the vector $P(t)\phi(t)$ will be zero. There will not be any changes in the parameter estimate and the negative term in the right hand side of (3.18) will be zero. The equation (3.18) then reduces to

$$P(t+1) = \frac{1}{\lambda} P(t)$$

and the matrix P will thus grow exponentially if $\lambda < 1$. If there are no changes for a long time the matrix P may thus become very large. A change in the command signal may then lead to large changes in the parameter estimates and in the process output. The large values of the matrix P may also lead to numerical problems. Examples which illustrate this behaviour are found e.g. in [42] and [44]. These results are partly due to bad numerics.

There are many ways to eliminate bursts. Perturbation signals may be added to ensure that the process is properly excited. The estimation algorithm may be modified. One possibility is to stop the updating of the matrix P(t) when the signal $P(t)\phi(t)$ is smaller than a given value. Another possibility is to replace P(t) by

$$[\alpha I + \sum_{k=1}^{t} \lambda^{t-k} \phi(k)\phi(k)]^{-1}$$

where α is a small number. This ensures that the matrix P stays bounded.

An Explicit Self-Tuning Regulator

A self-tuning controller based on pole-placement design and least squares estimation can be obtained by implementing the scheme of Fig. 2 directly. The following algorithm is obtained.

ALGORITHM E1. (Basic explicit algorithm)

Data: The polynomials P, T_1, and Q_1 are given.

Step 1: Estimate the parameters of the model

$$Ay(t) = Bu(t) \tag{3.1}$$

by least squares.

Step 2: Factor the estimated polynomial $\hat{B}$ into $\hat{B}^+$ and $\hat{B}^-$ where $\hat{B}^+$ has all its zeros in the region $\underline{Z}$ and $\hat{B}^-$ has all its zeros outside $\underline{Z}$.

Step 3: Solve the linear equation

$$\hat{A}R_1 + \hat{B}^- S = PT_1. \tag{3.8}$$

Step 4: Calculate the control variable u from

$$Ru = Tu_c - Sy \tag{3.4}$$

where

$$R = R_1\hat{B}^+$$

and

$$T = T_1Q_1.$$

(Notice that there are many solutions and that a choice has to be made [50].)

The steps 1,2,3, and 4 are repeated at each sampling period. □

This algorithm is called an algorithm based on <u>estimation of process</u> parameters or an algorithm with <u>explicit identification</u>, because the parameters of the process model (3.1) in the standard form are estimated. Using the terminology of model reference adaptive systems, the algorithm is also called <u>indirect</u> [45], because the parameters of the regulator are updated indirectly via estimation of the process parameters (Step 1) and the design calculations (Steps 2,3, and 4).

If the parameter estimates converge the closed loop system obtained will have the transfer function

$$G = \frac{Q_1\hat{B}^-}{P}$$

from the command signal to the output. The polynomial $\hat{B}^-$ is the polynomial that corresponds to the unstable or poorly damped process zeros. Notice that the closed loop response will change if $\hat{B}^-$ changes.

There are two difficulties with Algorithm E1. The factorization may be difficult and time consuming. Equation (3.8) is poorly conditioned if the polynomials $\hat{A}$ and $\hat{B}$ have factors which are close. It is, therefore, of interest to consider cases where the factorization problem can be avoided. Such schemes are discussed in [46],[47] and [50]. When the polynomial $\hat{A}$ has zeros which are close to the zeros of the polynomial $\hat{B}$ it is reasonable to cancel such factors before Step 3. An example of this is discussed in [48].

A third difficulty with the algorithm E1 is that the least squares parameter estimation method will give biased estimates if there are correlated disturbances. One possibility to avoid this is to use another estimation method. The first step in the algorithm can also be replaced by:

Step 1*: Estimate the parameters of the model

$$A\bar{y}(t) = B\bar{u}(t),$$

where

$$\bar{y} = \frac{1}{T_1}\,y, \quad \bar{u} = \frac{1}{T_1}\,u$$

by least squares.

This requires, of course, that the polynomial T_1 is known. If the polynomial T_1 is not known the following algorithm can be used.

ALGORITHM ESR. (Explicit algorithm for combined servo and regulation problem)

Data: The polynomials P and Q_1 are given.

Step 1: Estimate the parameters of the model

$$Ay(t) = Bu(t) + Ce(t)$$

by recursive maximum likelihood.

Step 2: Factor the estimated polynomial $\hat{B}$ into $\hat{B}^+$ and $\hat{B}^-$ where $\hat{B}^+$ has all its zeros in the region $\underline{Z}$ and $\hat{B}^-$ has all its zeros outside $\underline{Z}$.

Step 3: Solve the linear equation

$$\hat{A}R_1 + \hat{B}^-S = PT_1,$$

where $T_1 = \hat{C}$.

Step 4: Calculate the control variable u from

$$Ru = Tu_c - Sy,$$

where

$$R = R_1\hat{B}^+$$

and

$$T = T_1Q_1.$$

The steps 1,2,3, and 4 are repeated at each sampling period. □

An Implicit Self-Tuning Regulator

The design calculations required for the explicit algorithms may be time consuming. It is possible to obtain different algorithms where the design calculations are simplified considerably. The basic self-tuning regulator [59] is a prototype for algorithms of this type. The basic idea is to rewrite the process model in such a way that the design step is trivial. By a proper choice of model structure the regulator parameters are updated directly and the design calculations are thus eliminated. Algorithms of this type are called algorithms based on <u>implicit identification</u> of a process model. In the terminology of model reference adaptive systems the corresponding algorithms are also called <u>direct methods</u> [45] because the parameters of the regulator are updated directly. An example of an explicit algorithm will now be given.

Consider a process described by (3.1) with $B^- = 1$, which means that all zeros are well damped. Assume that it is desired to find a feedback such that the transfer function from the reference value to the output is

$$\frac{z^{\deg B}}{P}.$$

Equation (3.5) gives

$$\frac{TB}{AR + BS} = \frac{z^{\deg B}}{P}.$$

The polynomial B thus divides R. Introduce

$$R = R_1 B.$$

Then

$$PT = z^{\deg B}(AR_1 + S). \tag{3.23}$$

Hence

$$PTy = z^{\deg B}(AR_1y + Sy) = z^{\deg B}(R_1Bu + Sy) =$$

$$= z^{\deg B}(R_1Bu + Sy) = z^{\deg B}(Ru + Sy), \tag{3.24}$$

where the second equality is obtained from (3.1). The process can thus be represented either by (3.1) or by (3.23). The representation (3.23) has the advantage that the polynomials R and S required by the polynomial design occur explicitly in the model. If the model (3.23) is available the pole-placement design is thus trivial because the regulator (3.4) is obtained from the model (3.23) by inspection. The following self-tuning control algorithm can now be obtained.

ALGORITHM I2. (Implicit algorithm with all process zeros cancelled)

Data: Given the polynomials P and T, where P is normalized such that $P(1) = 1$.

Step 1: Estimate the parameters of the polynomials R and S in the model

$$PTy = z^{\deg B}(Ru + Sy) \tag{3.25}$$

by least squares.

Step 2: Calculate the control signal using

$$\hat{R}u = Tu_c - \hat{S}y, \tag{3.26}$$

where $\hat{R}$ and $\hat{S}$ are the polynomials estimated in Step 1. The steps 1 and 2 are repeated at each sampling period. □

This algorithm was originally proposed in [49]. Since the specifications require that all process zeros are cancelled, they must be sufficiently well damped for the algorithm to function. The algorithm will thus not work for nonminimum-phase systems.

To write the equation (3.25) it is necessary to know the pole excess

$$k = \deg A - \deg B$$

of the sampled system. The number k can often be estimated based on physical knowledge. There are, however, cases where it can be difficult to determine k apriori. It has been found empirically that it is not dangerous to overestimate k. If the pole excess of the sampled model is underestimated the leading coefficient r_0 of the polynomial R will be zero. The control law (3.26) is then not causal. If the estimate of r_0 is small the gain of the controller will be very large. Compare with the discussion of Equation (2.6).

Since the least squares method gives biased estimates if the disturbances are correlated, the following variation of the algorithm I2 is useful.

ALGORITHM I3.

Data: Given the polynomials P and T where P is normalized such that P(1) = 1.

Step 1: Estimate the parameters of the polynomials

$$Py = z^{\deg B}(R\bar{u} + S\bar{y}),$$

where

$$\bar{u} = \frac{1}{T}\, y, \; \bar{y} = \frac{1}{T}\, y$$

by least squares.

Step 2: Calculate the control variable using

$$\hat{R}u = Tu_c - \hat{S}y,$$

where $\hat{R}$ and $\hat{S}$ are the polynomials estimated in Step 1. The steps 1 and 2 are repeated at each sampling period. □

Further discussions on self-tuners based on pole-zero assignment are given in [46], [50-55], where many additional references are given.

4. SELF-TUNERS BASED ON STOCHASTIC CONTROL THEORY

Self-tuners based on minimum variance control and on linear quadratic gaussian control theory (LQG) will be discussed in this section. The discussion is limited to single input single output systems. It is assumed that the process to be controlled is governed by the model

$$Ay(t) = Bu(t) + Ce(t) \tag{4.1}$$

where u is the input, e white noise and y the output. It is assumed that the reader is familiar with the design problem for systems with known parameters [56],[57].

Minimum Variance Control

In this case the criterion is to minimize

$$E[y^2(t)] \tag{4.2}$$

in steady state. A typical application is minimization of fluctuations in quality variables in process control. If the parameters of the model (4.1) are known the control strategy which minimizes (4.2) is obtained by solving (3.23) with

$$P = z^{\deg A}$$

$$T = C.$$

The equation (3.23) then becomes

$$z^{\deg A - \deg B} C = AR_1 + S.$$

The particular solution such that

$$\deg R_1 < \deg A - \deg B = k$$

$$\deg S < \deg A$$

is chosen. The minimum variance control law is then given by (3.4) with $R = R_1 B$, S and $T = C$. Notice that B divides R, which means that all process zeros are cancelled. This means that the controller can not be expected to work well unless all process zeros are well damped. A minimum variance controller for non-minimum phase systems is given in [57]. The minimum variance controller attempts to bring the predicted output equal to the desired output after a time interval of length

$$h\ (\deg A - \deg B) = hk$$

where h is the sampling period. This means that the sampling period is the major design variable which determines the closed loop bandwidth.

For the case $C = z^{\deg A}$ it is easy to obtain a minimum variance self-tuner. The algorithm is a special case of the implicit pole placement Algorithm I2. It is easy to derive it directly. Equation (4.3) gives

$$\begin{aligned} y(t + 2 \deg A - \deg B) &= AR_1 y(t) + Sy(t) = \\ &= R_1 Bu(t) + Sy(t) + R_1 e(t + \deg A) = \\ &= Ru(t) + Sy(t) + R_1 e(t + \deg A), \end{aligned} \tag{4.4}$$

where Equation (4.1) was used to obtain the second equality. The transformed model (4.4) is a convenient starting point for an implicit self-tuner because it contains the controller parameters explicitly. A self-tuning minimum variance regulator is now obtained as follows.

ALGORITHM IMV. (Implicit minimum variance controller)

Step 1: Estimate the parameters in the model

$$y(t + k + \deg R) = Ru(t) + Sy(t) \tag{4.5}$$

$$k = \deg A - \deg B$$

by least squares.

Step 2: Calculate the control variable from

$$\hat{R}u(t) = y_r(t + \deg \hat{R}) - \hat{S}y(t) \tag{4.6}$$

where the coefficients of the polynomials $\hat{R}$ and $\hat{S}$ are the estimates obtained in Step 1.

The steps 1 and 2 are repeated in each sampling period. □

This is the basic self-tuner discussed in [58-60]. It is natural that the algorithm will work well when applied to a process (4.1) such that $C = z^{\deg A}$. In this case, the parameter estimates will converge to the corresponding minimum variance regulator. It is more surprising that the regulator will work well also when applied to a model (4.1) with $C \neq z^{\deg A}$. In this case, the least squares parameter estimates will be biased. In spite of this, the regulator will, however, converge to the minimum variance controller if it converges at all [58],[59]. Notice, however, that the parameter estimates may not necessarily converge [17], [85].

When using minimum variance control the variable y is often introduced as the control error. With this choice of variables the reference value y_r is zero. The control law (4.6) then has one redundant parameter. The redundant parameter can be eliminated by reparametrizing the estimation model (4.5) as

$$y(t+k) = r_0[u(t) + r_1'u(t-1) + \ldots + r_{n_R}u(t-n_R)] + \\ + s_0y(t) + s_1y(t-1) + \ldots + s_{n_S}y(t-n_S). \quad (4.7)$$

The control law (4.6) then becomes

$$u(t) = -\frac{1}{\hat{r}_0}[\hat{s}_0y(t) + \ldots + \hat{s}_{n_S}y(t-n_S)] - \\ - \hat{r}_1'u(t-1) - \ldots - \hat{r}_{n_R}'u(t-n_R). \quad (4.8)$$

It is shown in [59] that the estimate $\hat{r}_0$ can be fixed apriori provided that

$$0.5 \leq \hat{r}_0/r_0 < \infty. \quad (4.9)$$

If the algorithm converges with $\hat{r}_0 = r_0$ it will still converge if (4.9) holds. The convergence rate of the parameters is, however, influenced by $\hat{r}_0$. The fastest convergence is obtained for $r_0 = \hat{r}_0$.

Linear Quadratic Control

The self-tuning regulators based on minimum variance control suffer the same drawback as minimum variance control for systems with known parameters, namely that all process zeros are cancelled in the design. This is of little importance if the zeros are well inside the unit disc. The cancellation is, however, disastrous if the system is not minimum-phase. Another drawback with minimum variance control is that the control signals may be excessively large. This can be overcome by a suitable choice of the sampling period. Another possibility is, however, to base the design on linear quadratic gaussian control theory [56]. For single input single output systems the model (4.1) is thus used and the criterion is

$$\lim_{N\to\infty} E \frac{1}{N} \sum_{t=1}^{N} [y^2(t) + \zeta u^2(t)]. \tag{4.10}$$

For the explicit algorithm it is necessary to find the steady state solution of a Riccati-equation [32] or to solve a spectral factorization problem [33]. It is thus necessary to perform a reasonable amount of calculations at each step. With the increasing computing power of micro-processors such calculations are, however, feasible in many situations. Further discussions of explicit self-tuners based on stochastic control theory are given in [60-62], where many additional references are given. It is an interesting problem to find the corresponding implicit algorithms.

5. RELATIONS TO MODEL REFERENCE ADAPTIVE SYSTEMS

Model reference adaptive systems (MRAS) is another approach to adaptive control. The original MRAS concept was developed by Prof. H. P. Whitaker at MIT [63] in connection with work on adaptive flight control systems. The main design goal was to obtain an adaptive servo for a continuous time problem. Superficially MRAS appears to be very different from the STR which was originally developed for stochastic regulation of a discrete time system. Disregarding the types of problems to which STR and MRAS were first developed, it turns out that the algorithms are very similar. The purpose of this section is to explore the connections between STR and MRAS in some detail. The basic principles of MRAS are first presented. A servo design problem is then considered. For systems with known parameters it is shown that the system obtained by applying the pole-placement design can be interpreted as model following. It is then shown that a self-tuning regulator based on implicit identification is equivalent to a model reference system.

Model Reference Adaptive Systems

In MRAS the specifications are given in terms of a reference model which tells how the process output should respond to the command signal. A schematic diagram of an MRAS is shown in Fig. 3. Notice that the reference model is part of the control system. The regulator can be thought of as having two loops. The inner loop is an ordinary control loop composed of the process and a regulator. The parameters of the regulator are adjusted in an outer loop in such a way that the error e between the process

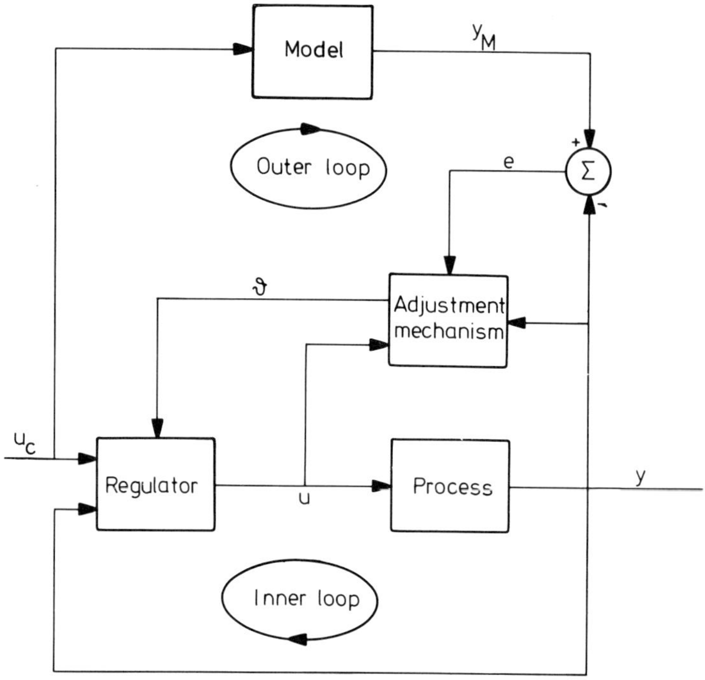

FIGURE 3. Block diagram of model reference adaptive system (MRAS).

output y and the model output y_M becomes small. The key problem is to determine the adjustment mechanism so that a stable system, where the error goes to zero, is obtained.

The following parameter adjustment mechanism, called the 'MIT-rule', was used in the original MRAS [63].

$$\frac{d\nu_i}{dt} = -k \frac{\partial e}{\partial \nu_i} e, \qquad i = 1,\ldots,n. \tag{5.1}$$

The variables $\nu_1,\ldots,\nu_n$ are the adjustable regulator parameters. The quantity e denotes the model error

$$e = y - y_M \tag{5.2}$$

and $\partial e/\partial \nu_i$, $i = 1,\ldots,n$, are the sensitivity derivatives. The constant k is a parameter which determines the adjustment rate.

The MIT rule can be interpreted as an algorithm for minimizing e^2. It was found that the adjustment rule could easily give unstable closed loop systems. A major step forward in the design of MRAS was when Parks [64] showed that Lyapunov stability theory could be used to obtain a modified adjustment rule which guaranteed stability. The modification consisted in replacing the model error and the sensitivity derivative in (5.1) by filtered signals. When the pole excess of the system is greater than one, the filtering proposed by Parks did, however, amount to taking derivatives of the error. The number of derivatives increased with the pole excess. Much work was devoted to the search for modified adjustment rules. The major tools were Lyapunov stability theory [65], [66] and Popov's hyperstability theory [67], [68]. A major innovation was made by Monopoli [69]. He showed that the derivatives that appeared when filtering the error could be eliminated by augmenting the model error e by terms which were different from zero only when the parameter estimates changed. In the adjustment rule proposed by Monopoli the error e in (5.1) was thus replaced by an *augmented error*. Both the error and the associated sensitivity derivatives were also filtered. The adjustment rules based on the augmented error have been generalized to the discrete time case [70],[71]. It is, however, only recently that correct stability results have been obtained [19],[20],[22],[23]. These stability results are dealt with in detail in [72].

Interpretation of Pole Placement as Model Following

The case of systems with known parameters will first be considered. It was shown in [33] that the regulator obtained by the pole-placement design method discussed in Section 3 can be interpreted as model following control. This is natural because the regulator aims at making the transfer function from the command signal u_c to the output equal to the model (3.2). To see this explicitly the feedback law (3.4) will be rewritten. It follows from (3.8), (3.9), and (3.10) that

$$\frac{T}{R} = \frac{T_1 Q_1}{B^+ R_1} = \frac{(AR_1 + B^- S)Q_1}{PB^+ R_1} = \frac{AQ_1}{B^+ P} + \frac{SB^- Q_1}{B^+ R_1 P} = \frac{A}{B}\frac{Q}{P} + \frac{S}{R}\frac{Q}{P} .$$

The feedback law (3.4) can thus be rewritten as

$$u = \frac{A}{B} y_M + \frac{S}{R} (y_M - y), \tag{5.3}$$

where

$$y_M = \frac{Q}{P} u_c. \tag{5.4}$$

Notice that the signal y_M can be interpreted as the model output. It follows from (4.2) that the control law can be thought of as composed of two parts. There is one feedforward term

$$\frac{A}{B} y_M = \frac{AQ}{BP} u_c \tag{5.5}$$

and one feedback term

$$\frac{S}{R} (y_M - y). \tag{5.6}$$

The feedforward term is generated by feeding the command signal through a cascade combination of the desired model $G_M = Q/P$ and an inverse A/B of the process model. Notice that the transfer function of the inverse process model A/B is not causal. The combination AQ/(BP) is, however, causal because of (3.3). The feedback term (5.6) is obtained by feeding the error e through a system with the transfer function S/R.

Implicit STR and MRAS

It will now be shown that the implicit self-tuning pole placement algorithm I1 is equivalent to an MRAS. For this purpose it is necessary to go into some details of the algorithm.

Introduce

$$\phi(t+1) = [y(t+n_S)\ldots y(t)u(t+n_R)\ldots u(t)]^T, \tag{5.7}$$

where

$$n_S = \deg S$$

$$n_R = \deg R.$$

Since the regulator is causal it follows that $n_R \geqq n_S$. In the implicit algorithm the estimated parameters are equal to the regulator parameters. Hence

$$\theta = \nu = [s_0\ldots s_{n_S}\ r_0\ldots r_{n_R}]. \tag{5.8}$$

The residual ε can then be written as

$$\varepsilon(t+1) = PT_1y(t) - Ru(t) - Sy(t) = PT_1 - \phi^T(t+1)\nu \tag{5.9}$$

where

$$1 = \deg PT_1.$$

If recursive least squares parameter estimation is used the formula for updating the parameter estimates is given by (3.16). Since $\nu = \theta$ it follows that

$$\nu(t+1) = \nu(t) + P(t+1)\phi(t+1)\varepsilon(t+1) \tag{5.10}$$

where ϕ is given by (5.7), ε by (5.0), and P by (3.18). Since the estimated parameters θ equals the regulator parameters ν the equation (5.10) can clearly be interpreted as an adjustment rule for the regulator parameters. Notice that it follows from (5.9) that

$$\phi(t) = - \text{grad}_{\nu}\ \varepsilon(t). \tag{5.11}$$

The vector ϕ can thus be interpreted as a sensitivity derivative. A comparison with (5.1) now shows that (5.12) is a discrete time version of the MIT rule. The main difference is that the model error $e = y - y_M$ in the MIT rule is replaced by the residual ε in the recursive least squares given by (5.9). Another difference is that the parameter k in (5.1) is replaced by the matrix P which is given by (3.18). The matrix P serves two purposes: it modifies the gradient direction and it gives an appropriate step length. While (5.1) can be viewed as a gradient algorithm to minimize e^2 (5.10) can be considered as a quasi-Newton method for minimizing ε^2. To explore the relations between MRAS and STR further it is of interest to find the relations between the model error e and the residual ε.

An Interpretation of the Residual ε

Equation (5.9) gives

$$\varepsilon = PT_1 y - Ru - Sy.$$

It follows from (5.9) that the prediction of y(t+1) based on data available at time $t + n_R$ is given by

$$PT_1\hat{y} = \hat{R}u + \hat{S}y.$$

It thus follows that

$$\varepsilon = PT_1(y-\hat{y}) = PT_1[(y-y_n) - (\hat{y}-y_M)]$$

or

$$\varepsilon = PT_1[e - \hat{e}] \tag{5.12}$$

where

$$e = y - y_M$$

is the model error.

It thus follows that the residual ε is obtained by forming the difference $e - \hat{e}$, where $\hat{e}$ is the prediction of e, and filtering the difference with PT_1.

The residual ε is identical to the augmented error introduced by Monopoli [69]. This equivalence was shown in [73-75]. Monopoli was led to introducing the augmented error by very different arguments. It is interesting to see that the augmented error falls out automatically from the STR formulation. Notice that the error estimate $\hat{e}$ can be zero in certain cases. If $\theta(t+1)$ goes to $\theta(t)$ then $\hat{e}$ will be zero. Some control laws are designed so that $\hat{y} = y_M$ which also implies $\hat{e} = 0$. In these cases the residual is simply a filtered model error.

Another Interpretation of the Residual ε

It is also possible to give still another interpretation of the residual ε. Equation (5.9) gives

$$\varepsilon(t+1) = PT_1 y(t) - \hat{R}_{t+1-1} u(t) - \hat{S}_{t+1-1} y(t), \tag{5.13}$$

where the subscripts on R and S have been introduced to denote that the coefficients of the polynomials are based on estimates obtained at time t + 1 - 1. The control signal u is however generated from

$$\hat{R}_{t+n_R} u(t) = Tu_c(t) - \hat{S}_{t+n_R} y(t).$$

It follows from (3.10) that

$$Tu_c = T_1 Q_1 u_c(t) = T_1 P y_M(t).$$

Hence

$$\hat{R}_{t+n_R} u(t) = T_1 P y_M(t) - \hat{S}_{t+n_R} y(t).$$

Subtracting this equation from Equation (5.13) gives

$$\begin{aligned}\varepsilon(t+1) &= PT_1[y(t) - y_M(t)] + [\hat{R}_{t+n_R} - \hat{R}_{t+1-1}]u(t) + \\ &\quad + [\hat{S}_{t+n_R} - \hat{S}_{t+1-1}]y(t) \\ &= PT_1 e(t) + [\theta(t+n_R) - \theta(t+1-1)]\phi(t+n_R).\end{aligned} \tag{5.14}$$

The residual ε can thus be generated by filtering the model error $e = y - y_M$ and adding a correction term which is proportional to the differences between parameter estimates at times which differ by deg P + deg T_1 - deg R - 1. The correction term will vanish if the parameter estimates converge. The error form (5.14) was used by Landau [71], [86]. The correction term was introduced in order to apply the hyperstability theorem. The notions of 'apriori' and 'aposteriori' reference models were introduced to give an interpretation of the correction term. Again it is interesting to observe that the correction term is obtained directly from the STR formulation.

6. APPLICATIONS

This section gives an overview of different applications of self-tuning regulators.

Control of Basis Weight and Moisture Content on Paper Machines

A schematic drawing of a paper machine is shown in Fig. 4. Basis weight and moisture content are important variables which characterize the quality of the finished product. Control of these quality variables can be conveniently described by stochastic control theory. Disturbances arise from many different sources.

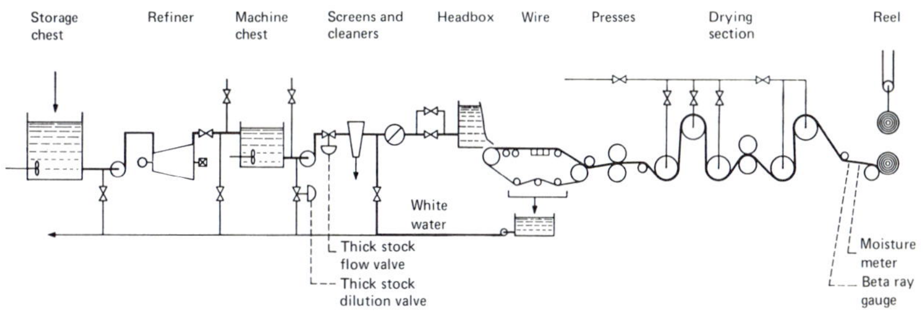

FIGURE 4. Schematic drawing of a paper machine. From [56].

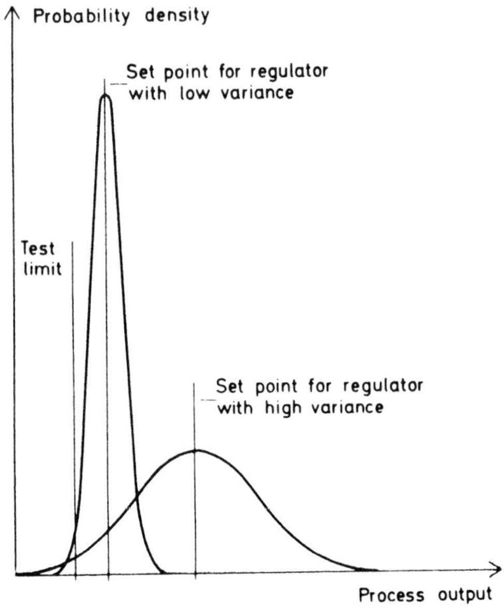

FIGURE 5. Benefits of minimum variance control. By reducing the variance of the output signal, the mean value can be moved closer to the target. From [56].

Their net effect on the output can be modeled as stochastic processes. The purpose of the control is to reduce the fluctuations in the quality variables. This goal is well described mathematically as to minimize the variances of the process outputs. See Fig. 5. There is no natural way to introduce any penalty on the control variables. The thick stock flow valve is the control variable for the basis weight loop and the steam pressure is the control variable for the moisture loop. The process dynamics of both loops are characterized by time delays and low order dynamics. There are couplings in the sense that changes in the thick stock flow valve will influence both basis weight and moisture content. Changes in the steam flow to the drying section influences the moisture content only. There are also interactions in the measuring devices because the beta ray gauge measures both basis weight and moisture content. These couplings can, however, easily be eliminated. Simple processing of the measured signals gives dry basis weight and moisture content. The basis weight is then controlled by a feedback from the dry basis weight signal to the thick stock flow valve and the moisture content is controlled by a feedback from the moisture content signal to the steam flow to the drying section. A feedforward signal e.g. from the thick stock flow measurement to the steam flow to the dryers, is introduced to compensate for the interaction in the process. A more detailed description of the control problem is given in [56], [76], and [77].

It has been verified by experiments on many plants that the dynamics of the plant and the disturbances can be adequately described by an ARMAX process of low order. See [56]. By identification of process dynamics and disturbance characteristics based on plant experiments it was demonstrated that substantial improvements over PID control could be achieved by using a basis weight regulator having the form

$$u(t) = -\frac{s_0 + s_1 z^{-1}}{1 + r_1 z^{-1} + r_2 z^{-2} + r_3 z^{-3}} y(t). \qquad (6.1)$$

The parameters of a regulator like (5.1) can not be conveniently tuned by hand. Instead the parameters were determined by system identification based on plant experiments and control design as described in [56]. Such a procedure is comparatively costly because it requires appropriate identification software and skilled personnel. To obtain a reasonably good model it is necessary to experiment on the plant for at least 2 hours. It was, therefore, attempted to tune the parameters with a self-tuning regulator. Since the criterion was well defined and since the sampling period and the regulator structure could be chosen based on prior experience it was very straightforward to apply the self-tuning regulator.

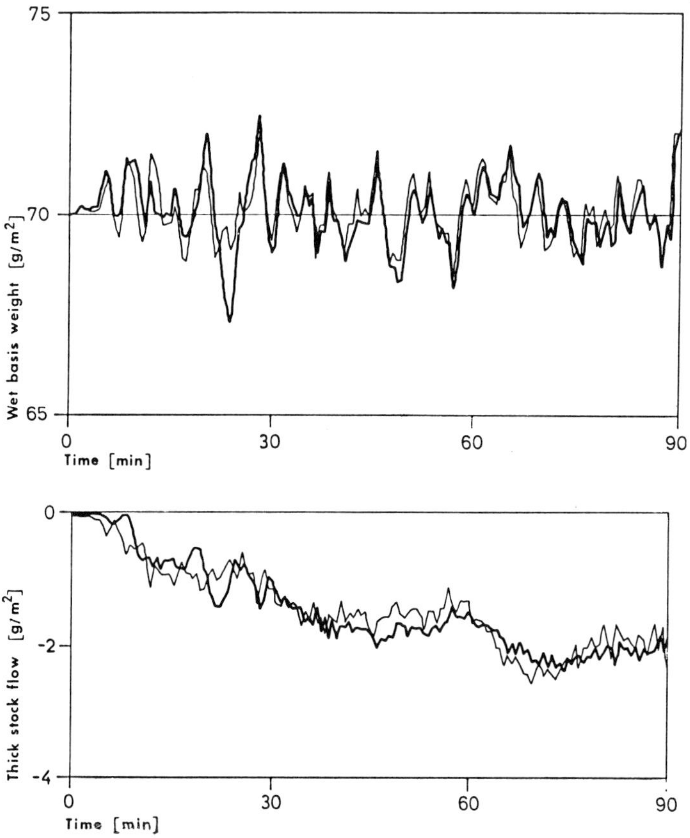

FIGURE 6. Simulation of minimum variance control (thin lines) and self-tuning control (thick lines) of basis weight of a paper machine. From [59].

Process control computers were available at the plant. The code for the simple self-tuning regulator based on least squares identification and minimum variance control (Algorithm IMV) was simply introduced as a special control algorithm, and experiments were started. A typical result is illustrated by the simulation shown in Fig. 6. This simulation is based on measured plant disturbances and models for the process dynamics estimated from plant experiments. In Fig. 6 the self-tuning regulator was initialized with all estimates equal to zero. It is seen from Fig. 6 that the fluctuations in basis weight obtained from the self-tuning regulator are larger than those obtained from the minimum variance regulator for the first 30 minutes. After 30 minutes there are, however, very small differences between the outputs of the two regulators. It is perhaps even more instructive to look at the control signals generated by the different regulators. It is seen from Fig. 6 that the self-tuner is fairly sluggish in the initial period. After 30 minutes there are, however, only minor differences in the control signal. Results similar to those shown in Fig. 6 were obtained when controlling the actual plants. On typical basis weight and moisture control loops the self-tuner will give close to optimal performance after a tuning period of 15 minutes to 2 hours.

In the example shown in Fig. 6 no apriori information about the parameters was assumed. In many applications in the paper mill it is possible to start the algorithm with reasonable estimates. The convergence time will then be shorter. The tuning time should be compared with the time required to make a good identification experiment. This time is between 2 and 5 hours. In the paper machine applications there have not been any problems with phenomena like 'turn off' or 'covariance blow up'. One reason is that the disturbances are persistent and fairly stationary.

In the paper mill the self-tuning regulator was used as a tuner, since the chief instrument engineer did not like the parameters of important control loops to be changed in his absence. The simple self-tuning regulator has been applied to many simple flow and level loops, to basis weight and moisture loops [78] on several different paper machines and also to recovery boilers in the paper mill.

It is much easier to implement a simple self-tuning regulator than to go through the procedure of process experiments, system identification and control design. It is thus clear that the self-tuning regulator gives a considerable saving of engineering work compared to previously used methods.

Control of a Heat Exchanger

This application is described in [79]. A schematic drawing of the heat exchanger is shown in Fig. 7. This type of heat exchanger is a common component in heating and ventilation systems. The warm air leaving a room gives away some of its enthalpy to the fresh air supplied to the room. The warm and cold air streams pass through the rotor which has axial channels. The rotor is made of a material which can absorb heat and moisture. The enthalpy

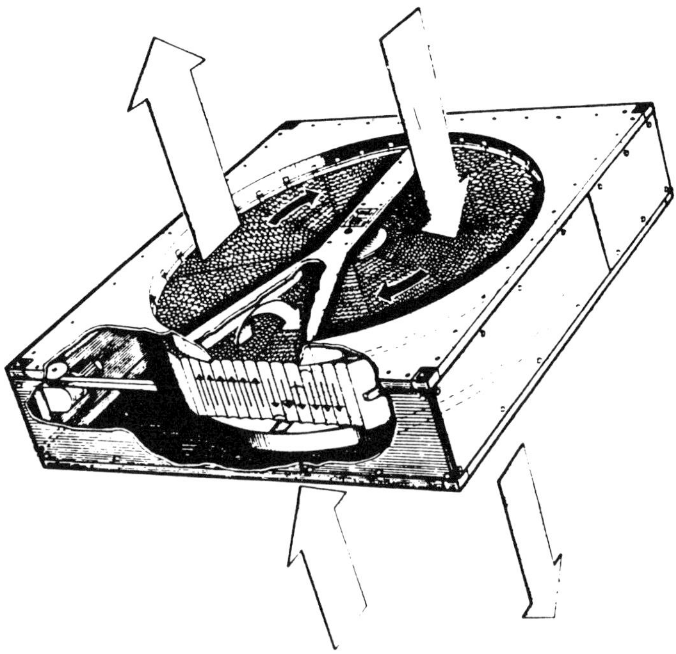

FIGURE 7. Schematic drawing of a heat exchanger.

is exchanged because the rotor segments are alternating between the warm and cold streams when the rotor rotates. The control problem is to adjust the angular velocity of the rotor in such a way that the room temperature is constant. The disturbances are due to sunshine, heat generated from people and other heat sources. In the particular case there are also sensor noise because the sensors are thermistors in the air streams. It is desirable to keep temperature fluctuations small but in contrast with the paper machine example there is no natural loss function. For a fixed operating condition it is easy to obtain good control simply by using integrating feedback from room temperature to rotor speed. The major difficulty is that the gain of the process changes drastically with operating conditions. It was not possible to find an auxiliary variable to schedule the controller gain since the gain depends on moisture temperature and rotor speed. A particular regulator structure was chosen. This structure was derived based on knowledge of the physics of the process.

The primary controlled variable is chosen as the thermal efficiency defined by

$$v = \frac{T_{cout} - T_{cin}}{T_{win} - T_{wout}} \tag{6.2}$$

where the subscript c stands for cold and w for warm air. The thermal efficiency v is first computed based on measurements of T_{cin}, T_{win}, T_{wout}, and knowledge of the desired T_{cout}. The advantage of choosing this variable as the controlled variable rather than the cold outlet temperature is that the effect on the control loop of some of the process nonlinearities are eliminated at the price of three extra thermistors. The thermal efficiency is then controlled by a feedback from the computed thermal efficiency to commanded rotor rotation speed.

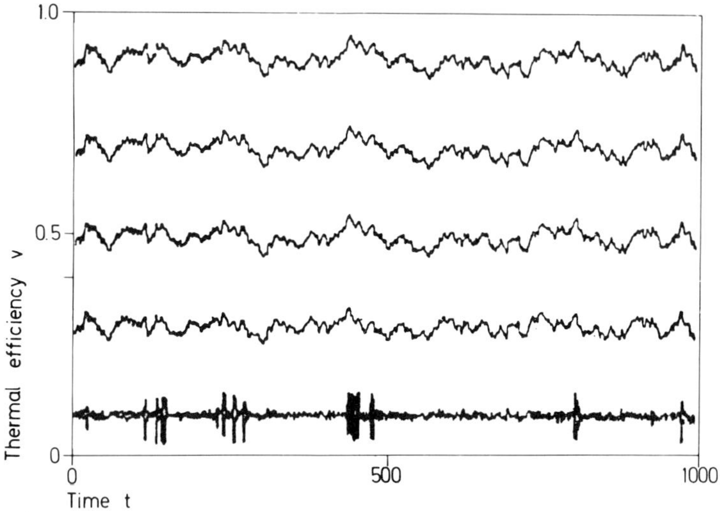

FIGURE 8. Thermal efficiency of a heat exchanger controlled by an integrating controller having fixed gain.

The dynamics relating thermal efficiency to rotor speed is characterized by a time delay and low order dynamics. The major difficulty in controlling the heat exchanger is that the static gain varies considerably. In steady state the relation between rotor speed u and thermal efficiency v is approximately given by

$$v = f(u) = \frac{au}{1 + au} . \tag{6.3}$$

If the units are chosen in such a way that $0 =< u =< 1$ then a may be 20 in a typical case. The static gain thus varies between 1 and 20.

The heat exchanger can be controlled by an integrating controller. Due to the gain variations there are, however, difficulties when a regulator with constant gain is used. This is illustrated by the simulation in Fig. 8. It is clearly seen from this figure that the loop gain is too high at low levels and too low at high levels of efficiency. If the relation (6.3) was accurate and did not change with time then Equation (6.3) could be used to schedule the gain of the controller as a function of the thermal efficiency. This is unfortunately not possible because the parameter a in (6.3) changes with time, temperature, and moisture content. A simple self-tuning regulator was therefore used to eliminate the gain variations. The self-tuning regulator is based on estimation of the parameter b in the model

$$v(t+1) - v(t) = b[u(t) - u(t-1)] = b\nabla u(t) \tag{6.4}$$

by least squares. In this case the estimate is particularly simple since only one parameter is estimated. The estimate is given by the following equations:

$$\hat{b}(t+1) = \hat{b}(t) + P(t+1)\nabla u(t)\varepsilon(t+1)$$

$$\varepsilon(t+1) = \nabla v(t) - \hat{b}(t)\nabla u(t)$$

$$P(t+1) = \frac{P(t)}{\{\lambda + P(t)[\nabla u(t)]^2\}}.$$

The variable v denotes the thermal efficiency and u denotes the angular velocity of the rotor. Having obtained the estimate $\hat{b}$ of b the following control law is then used:

$$u(t) = u(t-1) - (k_0/\hat{b})v(t) \tag{6.5}$$

where k_0 is an empirical constant. The 'cautious' control law

$$u(t) = u(t-1) - \frac{k_0\hat{b}}{\hat{b}^2 + P}v(t) \tag{6.6}$$

was also tried but there was little difference in performance compared to the certainty equivalence control (6.5).

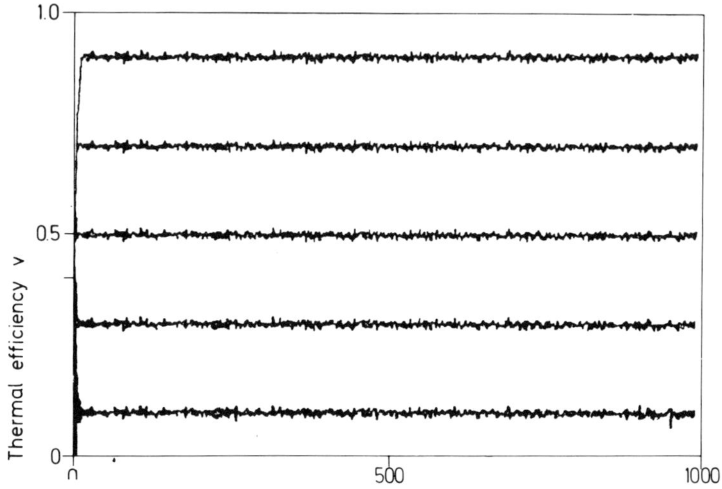

FIGURE 9. Thermal efficiency of the heat exchanger with a self-tuning integrating regulator.

The performance of the self-tuning regulator is illustrated by the simulation results shown in Fig. 9. The models and the disturbances were the same as when generating the results shown in Fig. 8. The Figures 8 and 9 are thus directly comparable. It is clear that the self-tuning regulator handles the gain variations very well. The behaviour of the self-tuning regulator on the actual plant is similar to that shown in the simulations. See [79].

Control of an Ore Crushing Plant

This example is described in detail in [80]. A schematic drawing of the process is shown in Fig. 10. The plant consists of an ore bin, a feeder, two screens, an ore crusher, and conveyor belts. The ore is transported from the bin to a screen, where the

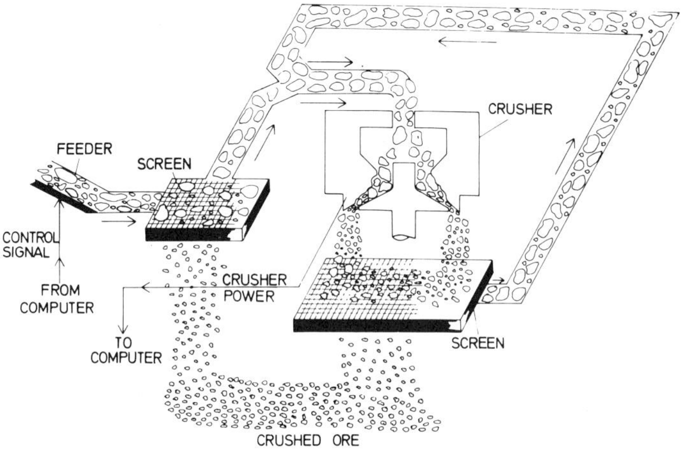

FIGURE 10. Schematic drawing of an ore crushing plant. From [80].

small ore lumps are separated from the large lumps. The larger ore lumps are transported to the crusher where the lumps are crushed. After the crusher there is another separation screen. Lumps with a diameter larger than 2.5 cm are recirculated to the crusher. The crusher is driven by an electric motor via a slip clutch which releases the motor and stops the line if the torque is too high. The control variable is the amount of ore fed into the line and the controlled variable is the power of the crusher motor. The goal of the control is to keep production as high as possible while avoiding overloading. This can be formulated approximately as to reduce the variance in the crusher power. By the usual argument, illustrated by Fig. 5, the set point of the crusher power can be moved closer to the target and the average production is increased as a consequence. The trade-off between high production and risk for overload is reflected in the choice of set point. The disturbances are due to variations in lumpsize, crushability, and variations of the crusher characteristics due to wear. The plant dynamics is characterized by a time delay of

70-80 s in the recycle loop and time constants of 10-20 s in the crusher itself.

When the experiments were started there was very little apriori knowledge of the characteristics of the process and its environment. Some step responses were therefore determined initially. An interesting aspect of the experiments was that they were performed using teleprocessing between a plant in Kiruna in northern Sweden and a computer in Lund in southern Sweden. The distance between the two places is about 1800 km.

It was decided to try the simple self-tuning regulator based on minimum variance control (Algorithm IMV). Based on the time delays of the process it was decided that a sampling period of 20 s was reasonable. Since the time delay of the process was 40-50 s the parameter k in the self-tuner should have the value 3 or 4.

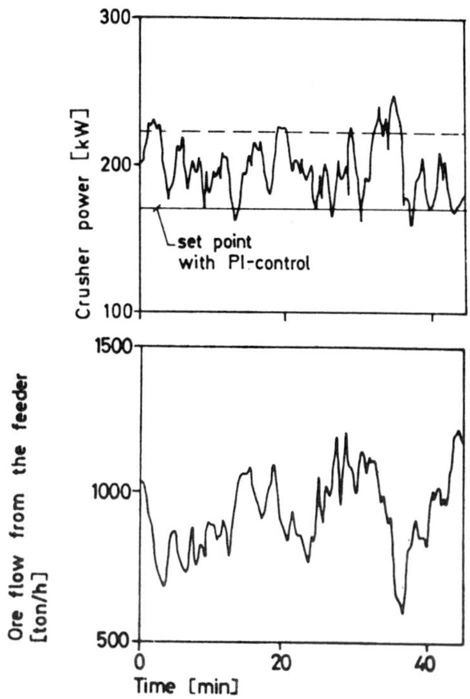

FIGURE 11. Results of an ore crusher experiment [79].

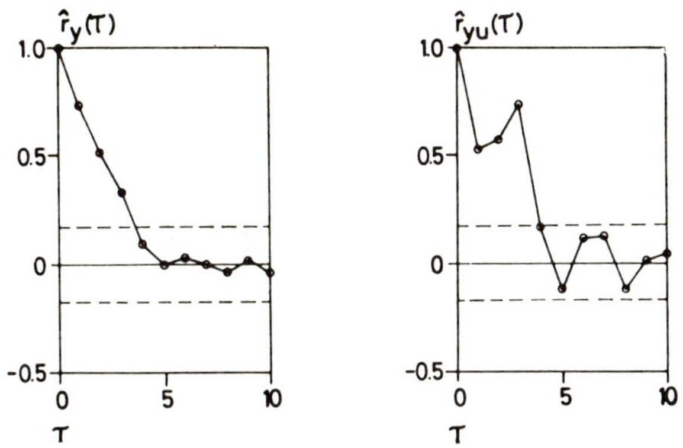

FIGURE 12. Covariance functions. From [80].

It was found experimentally that k = 4 worked better than k = 3. The complexity of the regulator was determined experimentally by controlling the plant with regulators having different complexity. The sample covariance function and the cross covariance between the output and the control variable were determined. The complexity of the regulator was increased until the conditions

$$\begin{aligned} r_y(\tau) &= 0, \quad \tau > k \\ r_{yu}(\tau) &= 0, \quad \tau > k \end{aligned} \tag{6.7}$$

which hold for the minimum variance controller, were fulfilled. See [56]. It was found that a simple self-tuning regulator with deg R = 3 and deg S = 4 was performing well. The forgetting factor λ was also determined empirically. The value λ = 0.99 was chosen after some experimentation. The value λ = 0.95 was found to be slightly better during start-up and during periods with a high variability in the ore properties. The results of one experiment are illustrated in Figs. 11, 12, and 13.

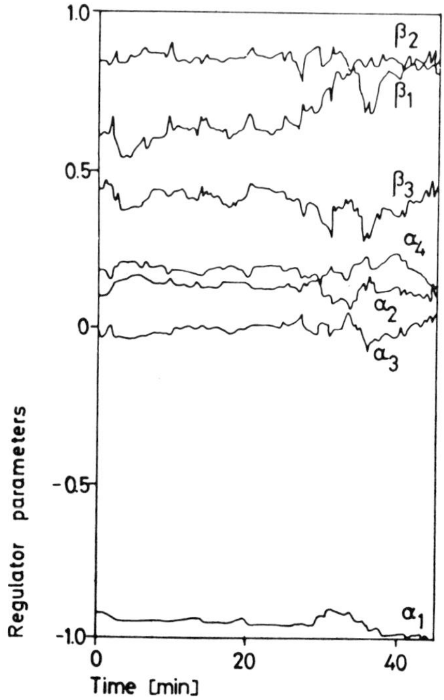

FIGURE 13. Parameter estimates. The parameter r_0 is fixed. From [79].

An Autopilot for Ship Steering

This application is described in detail in [12] and [81] where many additional references are given. An ordinary autopilot for a ship is based on feedback from measurements of heading (and possibly also the rate of change of the heading) to the rudder angle. A PID algorithm is commonly used. An autopilot has different functions. It should be able to maintain the ship at a constant course and it should be able to handle maneuvers. The dynamics of a ship will change with speed, trim, loading, and water depth. The characteristics of the disturbances will also change considerably with weather and wind. Although, it is in many cases possible to find constant settings of an ordinary autopilot which will guarantee stability over a wide range of operating conditions, there is a considerable advantage in having an adaptive autopilot. It is a common experience on tankers that ordinary autopilots do not work well in bad weather. The reason is partly that the PID algorithm is too simple to handle the requirements and partly that proper tuning to different weather conditions is required.

The design of an autopilot for straight course keeping can be formulated as a stochastic control problem. The ship dynamics can be described as a linear dynamical system and the disturbances can be characterized as random processes. Fortunately there is also a natural loss function which fits well into the stochastic control formulation. It can be shown by hydrodynamic theory that the average increase in drag due to yawing and rudder motion can be approximately described by

$$\frac{\Delta R}{R} = \mu(\overline{\psi}^2 + \zeta\overline{\delta}^2), \tag{6.8}$$

where R is the drag, ψ the heading deviation, δ the rudder angle, and $\overline{\psi}^2$ denotes the quadratic mean value. The values $\mu = 0.014$ deg^{-2} and $\zeta = 0.1$ are typical for a tanker. It is thus natural to use the criterion

$$V = \frac{1}{T}\int_0^T \{[\psi(t) - \psi_{ref}(t)]^2 + \zeta\delta^2(t)\}dt \tag{6.9}$$

as a basis for the design and evaluation of autopilots for steady state course keeping. One unit of the loss function would then correspond to an increase of 1.4% of the average drag or about the same increase in fuel consumption.

Since the design of an autopilot can be formulated as a linear quadratic stochastic control problem an adaptive autopilot can be designed using the corresponding self-tuning regulator. Several such designs have been made, simulated and field tested. The

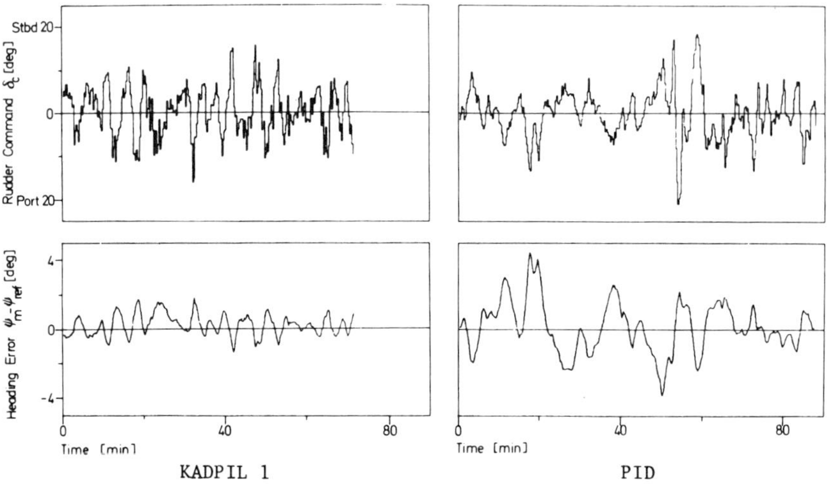

FIGURE 14. Results from an experiment with a 255,000 ton tanker comparing a well-tuned PID regulator with a self-tuning regulator. The values of the loss function (6.9) are 3.7 for the adaptive and 5.6 for the PID control. From [81].

experiments have been carried out on several large tankers. Available on board computers were used in the experiments. Due to memory constraints several simplifications were made. Since the LQG self-tuners require the solution of Riccati equations, which is space consuming, it was also attempted to use the simple self-tuner based on minimum variance control. This was reasonably successful provided that the prediction horizon was chosen appropriately. Extensive comparisons between different regulator structures including well-tuned PID-regulators were made. One comparison is illustrated in Fig. 14. It follows from (6.8) that the adaptive autopilot reduces the drag by 2.7% in comparison with an ordinary autopilot based on PID control. In these comparisons the ordinary autopilot was at all times retuned for best performance.

The magnitude of the improvements depends on the operating conditions. The largest improvements are found in bad weather when the ship is fully loaded. The self-tuning ship steering autopilot has been tried on several different tankers. On one ship it has been in continuous operation for several years.

7. CONCLUSIONS

A systematic way to design self-tuning regulators has been presented. The basic idea is very straightforward. A design procedure for a system with known parameters which fits the particular application is first chosen. When the parameters of the process are not known they are estimated recursively and the regulator parameters are recalculated at each step using the updated estimates. In some cases the model can be reparametrized so that the design calculations are avoided. Since there are many different design methods and many different parameter estimation methods, there are consequently a large variety of self-tuning regulators. So far only a small number of the available combinations have been explored. Some comments on the theory and practice of self-tuning regulators will now be given.

Theory

There are many interesting and important theoretical problems associated with self-tuning regulators. Stability, convergence, and performance are some of the key problems, but there are also many other questions like convergence rate and parametrization which are of considerable interest.

The stability problem is, of course, of major interest both theoretically and practically. There have recently been major advances in stability theory for self-tuners. Most of the results are, however, limited to the simple self-tuner based on least squares estimation and minimum variance control [59] or its direct multivariable generalization [82],[83]. The simple self-tuner will work well only for minimum-phase systems. Conditions for stability when there are no disturbances are given in [20]. The corresponding results for the case of bounded disturbances are found in [19]. Mean square stability for the stochastic case is investigated in [21]. Similar results for the continuous time problem are given in [22] and [23]. The theoretical results are also limited because of the assumptions made. The unpleasant assumption is that an upper bound of the order of the system must be known. This bound also determines the complexity of the self-tuner. In practice this is highly unrealistic because the process to be controlled will be of very high order (probably infinite dimensional) and the self-tuning regulator will be based on a simplified model. It is shown in [84] that regulators based on drastically simplified models can actually work very well.

The convergence problem has also been investigated for the simple self-tuners [58] based on least squares and minimum variance control. The surprising result that the correct regulator is the only possible equilibrium point, even if there is structural mismatch between the model and the process, is also given in [59]. Ordinary differential equations, which describe the mean paths of the parameter estimates, are derived in [17]. Necessary conditions for local stability around the equilibrium points are given in [85].

In summary there have recently been very interesting results in the theory of self-tuning regulators. It would be useful to eliminate some of the assumptions made. It would also be highly desirable to extend the analysis to other than the most simple self-tuning algorithms.

Practice

The basic algorithm for a PID regulator is very simple:

$$u = K[e + \frac{1}{T_I}\int^t e(s)\,ds + T_D\frac{de}{dt}].$$

An implementation of this algorithm in analog or digital hardware does, however, not necessarily give a good controller. In practice it is also necessary to consider operator interface, filtering of the signals, automatic/manual transfer, bumpless parameter changes, reset windup, nonlinear output, like gap, and saturation, etc. Notice that many of these functions involve nonlinearities which are not easily analyzed. Whether a PID regulator works well in an industrial environment depends very much upon how well the problems discussed above are handled. Similar things also apply to self-tuning regulators. The list is, however, longer because the basic algorithm is more complicated than the PID algorithm. For example, windup occurs in a PID regulator because the integrator in the algorithm could achieve large values if the control value saturates or if it is driven manually. In a self-tuner with a forgetting factor windup can also occur in the estimator. The self-tuning regulator can operate in many different modes like estimation only, tuning etc. The problem of operator interface is particularly important. A key problem is how the specifications are entered and how an operator should interact with the controller. There are many different possibilities ranging from the case where there are no buttons at all on the panel to fairly complicated operator interfaces. Certainly there are many interesting possibilities as is illustrated on self-tuning regulators which are already on the market or which are in the process of coming out.

Uses of Self-Tuners

Self-tuning regulators can be used in many different ways. Since the regulator becomes an ordinary constant gain feedback if the parameter estimates are kept constant, the self-tuner can be used as a tuner to adjust the parameters of a control loop. In such an application the self-tuner is connected to the process and run until satisfactory performance is obtained. The self-tuner is then disconnected and the system is left with the constant parameter regulator obtained. This mode of using the self-tuner is convenient to implement in a package for direct digital control (DDC-package). The DDC-package is simply provided with a tuning routine which can be connected to an arbitrary loop in the package.

The self-tuner can also be used to build up a gain schedule. In such a case the system is run at different operating points and the controller parameters obtained are stored. When the process has been run at a sufficient number of operating points a table for scheduling the controller parameters can be generated by interpolation and smoothing of the values obtained.

The self-tuner can, of course, also be used as a truly adaptive controller for systems with varying parameters. In cases where rapid adaptation over widely varying operating conditions are required combinations between gain-scheduling and self-tuning can also be considered.

Abuses of Self-Tuners

Compared with a three-term controller the self-tuner is a sophisticated controller. Such a controller can, of course, be misused. The self-tuner should certainly not be used if a simpler controller will do the job. Before considering a self-tuning regulator it is, therefore, useful to check if a constant parameter regulator or a regulator with gain scheduling will do the job. When designing a self-tuning regulator it is also useful to consider the particular application carefully and decide upon a design method which is suitable for the particular problem if a model for the process and its environment are known. A parameter estimation scheme which works wehl for the particular situation should also be chosen before the details of the design are considered.

Summary

The word self-tuning regulator may lead to the false conclusion that such regulators can be switched on and used blindly without any apriori considerations. This is definitely not true. The self-tuning regulator is a fairly complex control law. A proper design involves the choices of gross features like underlying design and estimation methods and decisions on details like initialization, selection of parameters, and safeguard methods. Proper choices require insight and knowledge. There are known cases where bad choices have been disastrous. All theory required is definitely not available. Based on experiences from a few applications I believe, however, that self-tuning regulators can and will be used profitably, even if some of their properties are not fully understood theoretically. It is thus my hope that this paper may inspire some of you to acquire the appropriate knowledge and try some schemes of your own. It is also my hope that some of you will tackle the important theoretical problems that remain.

ACKNOWLEDGMENTS

I have had the pleasure to work with adaptive and self-tuning regulators for a long time. Much insight and knowledge, which is reflected in this paper, was obtained when working with students who did their Ph.D. work in this area: B. Wittenmark, L. Ljung, T. Söderström, I. Gustavsson, U. Borisson, J. Sternby, L. Jensen, J. Holst, C. G. Källström, and B. Egardt. The benefit derived from this is gratefully acknowledged.

REFERENCES

1. Gregory, P. C. (editor), "Proceedings of the Self Adaptive Flight Control Systems Symposium," WADC Technical Report 59-49, Wright Air Development Center, Wright-Patterson Air Force Base, Ohio, March 1959.
2. Mishkin, E., and Braun, L., "Adaptive Control Systems," McGraw-Hill, 1961.
3. Unbehauen, H., and Schmid, Chr., "Status and Application of Adaptive Control Systems," Automatic Control Theory and Applications, 3 pp. 1-12, 1975.
4. Asher, R. B., Adrisani, D., and Dorato, P., "Bibliography on Adaptive Control Systems," Proc. IEEE, 64, pp. 1126-1240, 1976.
5. Clarke, D. W., and Gawthrop, P. J., "Implementation and Application of Microprocessor-Based Self-Tuners," Preprints 5th IFAC Symposium on Identification and System Parameter Estimation, Darmstadt, September 1979, pp. 197-208.
6. Neuman, C. P., and Morris, R. L., "Mini and Micro Computer Implementation of Adaptive Control Systems," Proc. First International Symposium on Mini- and Microcomputers in Control, San Diego, January 1979.

7. Neuman, C. P., and Baradello, C. S., "Microcomputer Implementation of Adaptive Controllers," Proc. First International Symposium on Mini- and Microcomputers in Control, San Diego, January 1979.

8. Åström, K. J., Borisson, U., Ljung, L., and Wittenmark, B., "Theory and Applications of Self-Tuning Regulators," Automatica, 13, pp. 457-476, 1977.

9. Dumont, G. A., and Bélanger, P. R., "Self-Tuning Control of a Titanium Dioxide Kiln," IEEE Trans., AC-23, pp. 532-537, 1978.

10. Kurz, H., Isermann, R., and Schumann, R., "Development, Comparison and Application of Various Parameter-Adaptive Digital Control Algorithms," Proc. 7th IFAC World Congress, Helsinki, 1978.

11. Keviczky, L., Hetthéssy, J., Hilger, M., and Kolostori, J., "Self-Tuning Adaptive Control of Cement Raw Material Blending," Automatica, 14, pp. 525-532, 1978.

12. Källström, C. G., Åström, K. J., Thorell, N. J., Eriksson, J., and Sten, L., "Adaptive Autopilots for Large Tankers," Proc. 7th IFAC World Congress, Helsinki, 1978.

13. Irving, E., "Improving Power Network Stability and Unit Stress with Adaptive Generator Control," Automatica, 15, pp. 31-46, 1979.

14. Buchholt, F., Clement, K., and Bar Jorgensen, S., "Self-Tuning Control of a Fixed Bed Chemical Reactor," Preprints 5th IFAC Symposium on Identification and System Parameter Estimation, Darmstadt, September 1979, pp. 1213-1218.

15. Haber, R., Hetthéssy, J., Keviczky, L., Vajk, I., Fehér, A., Czeiner, N., Császar, Z., and Turi, A., "Identification and Adaptive Control of a Glass Furnace by a Portable Process Computer Laboratory," Preprints 5th IFAC Symposium on Identification and System Parameter Estimation, Darmstadt, September 1979, pp. 209-220.

16. Zinober, A.S.I., "The Self-Adaptive Control of Overhead Crane Operations," Preprints 5th IFAC Symposium on Identification and System Parameter Estimation, Darmstadt, September 1979, pp. 1161-1168.

17. Ljung, L., "Analysis of Recursive Stochastic Algorithms," IEEE Trans. AC-22, pp. 551-575, 1977.

18. Ljung, L., "On Positive Real Transfer Functions and the Convergence of Some Recursive Schemes," IEEE Trans. AC-22, pp. 539-551, 1977.

19. Egardt, B., "Stability of Model Reference Adaptive and Self-Tuning Regulators," Thesis, CODEN: LUTFD2/(TFRT-1017)/1-163/, Department of Automatic Control, Lund Institute of Technology, Lund, Sweden, 1978.

20. Goodwin, G. C., Ramadge, P. J., and Caines, P. E., "Discrete Time Multivariable Adaptive Control," Report Div. of Applied Science, Harvard University, Cambridge, Mass., U.S.A., November 1978.

21. Goodwin, G. C., Ramadge, P. J., and Caines, P. E., "Discrete Time Stochastic Adaptive Control," Report Div. of Applied Science, Harvard University, Cambridge, Mass., U.S.A., December 1978.

22. Narendra, K. S., Lin, Y.-H., and Valavani, L. S., "Stable Adaptive Controller Design - Part II: Proof of Stability," S&IS Report 7904, Yale University, April 1979.

23. Morse, A. S., "Global Stability of Parameter-Adaptive Control Systems," S&IS Report 7902R, Yale University, March 1979.

24. Bellman, R., "Adaptive Processes - A Guided Tour," Princeton University Press, 1961.

25. Bohlin, T., "Optimal Dual Control of a Simple Process with Unknown Gain," Report TP 18.196, IBM Nordic Laboratory, Lidingö, Sweden, 1969.

26. Patchell, J. W., "On the Structure of Stochastic Control Laws," Dr. Phil. Dissertation, University of Oxford, May 1971.

27. Åström, K. J., Wittenmark, B., "Problems of Identification and Control," J. Mathematical Analysis and Applications, 34, pp. 90-113, 1971.

28. Sternby, J., "Topics in Dual Control," Thesis, CODEN: LUTFD2/(TFRT-1012)/1-135/, Department of Automatic Control, Lund Institute of Technology, Lund, Sweden, 1977.

29. Florentin, J. J., "Optimal Probing, Adaptive Control of a Simple Bayesian System," Int. J. on Control, 13, pp. 165-177.

30. Feldbaum, A. A., "Dual Control Theory I - IV," Automation & Remote Control, 21, pp. 874-880, 21, pp. 1033-1039, 22, pp. 1-12, 22, pp. 109-121, 1960.

31. Horowitz, I. M., "Synthesis of Feedback Systems," Academic Press, New York, 1963.

32. Åström, K. J., "Reglerteori (Control Theory)," Almqvist & Wiksell, Stockholm 1967, 2nd edition 1976 (in Swedish).

33. Åström, K. J., "Algebraic System Theory as a Tool for Regulator Design," in Halme et al. (eds.), "Topics in System Theory," Publication in honour of Professor Hans Blomberg on the occasion of his sixtieth birthday on December 18th, 1979. Acta Polytechnica Scandinavica, Ma 31, Helsinki 1979. Also available as report CODEN: LUTFD2/(TFRT-7164)/1-023/, Department of Automatic Control, Lund Institute of Technology, Lund, Sweden, 1979.

34. Söderström, T., Ljung, L., and Gustavsson, I., "A Comparative Study of Recursive Identification Methods," Report TFRT-3085, Department of Automatic Control, Lund Institute of Technology, Lund, Sweden, 1974.

35. Lawson, C. L., and Hanson, R. J., "Solving Least Squares Problems," Prentice-Hall, Englewood Cliffs, New Jersey, 1974.

36. Peterka, V., "A Square-Root Filter for Real-Time Multivariable Regression," Kybernetica, 11, pp. 53-67, 1975.

37. Bierman, G. J., "Factorization Methods for Discrete Sequential Estimation," Academic Press, New York, 1977.

38. Thornton, C. L., and Bierman, G. J., "Filtering and Error Analysis via the UDU Covariance Factorization," IEEE Trans., AC-23, pp. 901-907, 1978.

39. Levinson, N., "The Wiener RMS (Root Mean Square) Error in Filter Design and Prediction," J. Math. Phys., 28, pp. 261-278, 1947.

40. Kailath, T., "Some New Algorithms for Recursive Estimation in Constant Systems," IEEE Trans., IT-19, pp. 750-760, 1973.

41. Lindquist, A., "Optimal Filtering of Continuous-Time Stationary Processes by Means of the Backward Innovation Process," Siam J. Control, 12, pp. 747-754, 1974.

42. Fortescue, T. R., Kershenbaum, L. S., and Yostie, B. E., "Implementation of Self Tuning Regulators with Variable Forgetting Factors," Report, Dept. of Chem. Eng. and Chem. Technology, Imperial College, London, 1978.

43. Wittenmark, B., "A Self-Tuning Regulator," Thesis, TFRT-1003, Department of Automatic Control, Lund Institute of Technology, Lund, Sweden, 1973.

44. Morris, A. J., Fenton, T. P., and Nazer, Y., "Application of Self-Tuning Regulator to the Control of Chemical Processes," in H. R. van Nauta Lemke and H. B. Verbruggen (eds.), "Digital Computer Applications to Process Control," preprints 5th IFAC/IFIP Int. Conf., The Hague, Netherlands, 1977.

45. Narendra, K. S. and Valavani, L. S., "Direct and Indirect Adaptive Control," Automatica, to appear.

46. Åström, K. J., Westerberg, B., and Wittenmark, B., "Self-Tuning Controllers Based on Pole-Placement Design," Report, CODEN: LUTFD2/(TFRT-3148)/1-052/, Department of Automatic Control, Lund Institute of Technology, Lund, Sweden, 1978.

47. Åström, K. J., "New Implicit Adaptive Pole-Zero Placement Algorithms for Nonminimum-Phase Systems," Report, CODEN: LUTFD2/(TFRT-7172)/1-021/, Department of Automatic Control, Lund Institute of Technology, Lund, Sweden, 1979.

48. Åström, K. J., "Simple Self-Tuners I," Report CODEN: LUTFD2/(TFRT-7184)/1- /, Department of Automatic Control, Lund Institute of Technology, Lund, Sweden, 1979.

49. Clarke, D. W., and Gawthrop, B. A., "Self-Tuning Controller," Proc. IEE 122, pp. 929-934, 1975.

50. Åström, K. J., and Wittenmark, B., "Self-Tuning Controllers Based on Pole-Zero Placement," Report, CODEN: LUTFD2/ (TFRT-7180)/1-43/, Department of Automatic Control, Lund Institute of Technology, Lund, Sweden, 1979.

51. Clarke, D. W., and Gawthrop, P. J., "Self-Tuning Control," Proc. IEE 126, pp. 633-640, 1979.

52. Elliott, H., and Wolovich, W. A., "Parameter Adaptive Identification and Control," IEEE Trans., AC-24, pp. 592-599, 1979.

53. Wouters, W.R.E., "Adaptive Pole Placement for Linear Stochastic Systems with Unknown Parameters," Preprints IEEE Conference on Decision and Control," New Orleans, 1977.

54. Wellstead, P. E., Edmunds, J. M., Prager, D., and Zanker, P., "Self-Tuning Pole/Zero Assignment Regulators," Int. J. Control, 30, pp. 1-26, 1979.

55. Wellstead, P. E., Prager, D., and Zanker, P., "Pole Assignment Self-Tuning Regulator," Proc. IEE 126, pp. 781-787, 1979.

56. Åström, K. J., "Introduction to Stochastic Control Theory," Academic Press, New York, 1970.

57. Åström, K. J., "Stochastic Control Problems," in Coppel, W.A., (ed.), "Mathematical Control Theory. Lecture Notes in Mathematics," Springer-Verlag, Berlin, 1978.

58. Åström, K. J., and Wittenmark, B., "On the Control of Constant but Unknown Systems," Proc. 5th IFAC World Congress, Paris, 1972.

59. Åström, K. J., and Wittenmark, B., "On Self-Tuning Regulators," Automatica, 9, pp. 185-199, 1973.

60. Peterka, V., "Adaptive Digital Regulation of Noisy Systems," Proc. 2nd IFAC Symposium on Identification and Process Parameter Estimation, Prague, 1970.

61. Åström, K. J., "A Self-Tuning Regulator for Non-Minimum Phase Systems," Report TFRT-3113, Department of Automatic Control, Lund Institute of Technology, Lund, Sweden, 1974.

62. Åström, K. J., and Wittenmark, B., "Analysis of a Self-Tuning Regulator for Nonminimum Phase Systems," Preprints IFAC Symposium on Stochastic Control, Budapest, September 1974.

63. Whitaker, H. P., Yamron, J., and Kezer, A., "Design of Model-Reference Adaptive Control Systems for Aircraft," Report R-164, Instrumentation Laboratory, MIT, Cambridge, Mass., September 1958.

64. Parks, P. C., "Lyapunov Redesign of Model Reference Adaptive Control Systems," IEEE Trans., AC-11, pp. 362-367, 1966.

65. Monopoli, R. V., "The Kalman-Yakubovich Lemma in Adaptive Control System Design," IEEE Trans., AC-18, pp. 527-529, 1973.

66. Lindorff, D., and Carroll, R., "Survey of Adaptive Control Using Lyapunov Design," Int. J. Control, 18, pp. 897-914, 1973.

67. Popov, V. M., "Hyperstability of Control Systems," Springer-Verlag, 1973.

68. Landau, I. D., "A Survey of Model-Reference Adaptive Techniques: Theory and Applications," Automatica, 10, pp. 353-379, 1974.

69. Monopoli, R. V., "Model Reference Adaptive Control with an Augmented Error Signal," IEEE Trans., AC-19, pp. 474-484, 1974.

70. Ionescu, T., and Monopoli, R. V., "Discrete Model Reference Adaptive Control with an Augmented Error Signal," Automatica, 13, pp. 507-517, 1977.

71. Landau, Y. D., "Adaptive Control - The Model Reference Approach," Marcel Dekker, New York, 1979.

72. Narendra, K. S., and Lin, Y.-H., "Design of Stable Model Reference Adaptive Controllers," Proc. of the Workshop on Applications of Adaptive Control, Yale University, Aug. 1979.

73. Egardt, B., "Unification of Some Continuous-Time Adaptive Control Schemes," Proc. IEEE AC-24, pp. 588-592, 1979.

74. Egardt, B., "Unification of Some Adaptive Control Schemes - Part II, Discrete Time," Proc. IEEE, AC, to appear.

75. Egardt, B., "Stability of Adaptive Controllers," Lecture Notes in Control and Information Sciences, Vol. 20, Springer-Verlag, Berlin, 1979.

76. Åström, K. J., "Computer Control of a Paper Machine - An Application of Linear Stochastic Control Theory," IBM J. Res. & Dev., 11, pp. 389-405, 1967.

77. Åström, K. J., "Applications to Paper Making," in Curriculum and Methods in Control and Systems Engineering, Vol. II, IBM Corp., 1970.

78. Borisson, U., and Wittenmark, B., "An Industrial Application of a Self-Tuning Regulator," Proc. IFAC Symp. on Digital Computer Applications to Process Control, Zürich, 1974.

79. Jensen, L., and Hänsel, R., "Computer Control of an Enthalpy Exchanger," Report TFRT-3081, Department of Automatic Control, Lund Institute of Technology, Lund, Sweden, 1974.

80. Borisson, U., and Syding, R., "Self-Tuning Control of an Ore Crusher," Automatica, 12, pp. 1-7, 1976.

81. Källström, C. G., Åström, K. J., Thorell, N. E., Eriksson, J., and Sten, L., "Adaptive Autopilots for Tankers," Automatica, 15, pp. 241-254, 1979.

82. Peterka, V., and Åström, K. J., "Control of Multivariable Systems with Unknown but Constant Parameters," Proc. 3rd IFAC Symposium on Identification and System Parameter Estimation, The Hague, June 1973.

83. Borisson, U., "Self-Tuning Regulators for a Class of Multivariable Systems," Automatica, 15, pp. 209-215, 1979.

84. Åström, K. J., "Self-Tuning Control of a Fixed Bed Chemical Reactor System," Report, CODEN: LUTFD2/(TFRT-3151)/1-066/, Department of Automatic Control, Lund Institute of Technology, Lund, Sweden, 1978.

85. Holst, J., "Local Convergence of Some Recursive Stochastic Algorithms," Preprints 5th IFAC Symposium on Identification and System Parameter Estimation, Darmstadt, September, 1979.

86. Landau, I. D., and Lozano, R., "Design and Evaluation of Discrete Time Explicit MRAC for Tracking and Regulation," Report LAG 79.14, Laboratoire d'Automatique de Grenoble, June, 1979.

DESIGN OF STABLE MODEL REFERENCE ADAPTIVE CONTROLLERS

Kumpati S. Narendra
Yuan-Hao Lin

Department of Engineering and Applied Science
Yale University
New Haven, Connecticut

The paper deals with the principles developed in recent years for the design of stable model reference adaptive observers and controllers. The emphasis throughout the paper is on concepts rather than on mathematical formalism. A detailed description of error models is first presented since they provide the basis for much of the design. Recent important advances in adaptive control, particularly those related to the stability problem, are described qualitatively and open problems in the area are indicated towards the end of the paper.

ISBN 0-12-514060-6

I. INTRODUCTION

Conventional control theory, for the most part, deals with the control of dynamical systems whose mathematical representations are completely known. In contrast to this, adaptive control refers to the control of partially known systems. For many years there has been an increasing interest in adaptive control which can be attributed to the fact that there is invariably some uncertainty in the dynamic characteristics of most practical systems. The tools of conventional control theory, even when used efficiently in the design of controllers for such systems, are inadequate to achieve satisfactory performance in the entire range over which the characteristics of the system may vary. Hence some type of monitoring of the system's behavior followed by the adjustment of the control input i.e., feedback, is needed and is referred to as adaptive control. Obviously it is possible to monitor different system characteristics and take different control actions and hence there is a large class of nonlinear feedback systems which can be referred to as adaptive control systems.

In the thirties when the concept of feedback was being refined, it was realized that while feedback in general improved the performance of a system it could also make the system unstable. Hence, although feedback systems were being designed before this period, it was only after the stability theory of linear feedback systems developed by Nyquist, Bode and Black was established and well understood that such systems were designed extensively on a systematic basis. Since adaptive control systems are nonlinear feedback systems, there is the distinct possibility that such systems can become unstable. Even though there has been interest in this area for over twenty years, due to the lack of a well developed stability theory of such systems until recently, the application of adaptive control to practical control systems has not been attempted on a large scale.

In this paper, we describe one particular form of adaptive control called Model Reference Adaptive Control (MRAC). Based on the conceptual framework of stability theory it has in the past decade developed into a systematic procedure for the design of stable adaptive systems. Recently it was shown [1], [2] that for a class of such systems with a single input and single output stability in the large can be demonstrated so that the controller parameters will converge to stable values from any arbitrary initial conditions. This marks an important stage in the evolution of the field. While we are still very far from a comprehensive theory of adaptive control of multivariable and stochastic systems, this new result provides a firm analytic foundation for further exploration in this area.

We attempt in the following sections to present some of the major ideas in model reference adaptive control using a stability approach without the use of too much mathematical formalism. After describing some of the important theoretical results of recent years we consider questions which are relevant for the theory to be applied to practical problems. In section II a broad classification of the problems in this area is presented. The stability approach to the design of adaptive systems is outlined in section III and a detailed analysis of the stability of error models is undertaken in section IV. The contents of this latter section may be considered to form the mathematical basis for all the principal results known in this area at the present time. The applications of these results to adaptive identifiers and observers and adaptive controller design are treated in section V and VI respectively. In section VII the present state of adaptive control theory is indicated by describing several questions that still remain unanswered. The resolution of these questions is essential if the theory is to be effectively applied to the design of practical controllers.

II. CLASSIFICATION OF PROBLEMS

Ever since the fifties many attempts have been made to define an adaptive system but even at the present time there is no universally accepted definition. This is due to the profusion of meanings that can be attached to such a term depending on the prior information that is assumed about the system to be controlled. For the purposes of this paper, by the problem of adaptive control we shall mean the specific problem of control of a linear time-invariant system with unknown parameters.

a) The Control Problem

The input and output of a linear time-invariant plant with unknown parameters are $u(\cdot)$ and $y_p(\cdot)$ respectively (Figure 1). A linear time-invariant model and a reference input $r(t)$ are specified which result in a model output $y_M(t)$. From all available on-line data it is desired to determine the control input $u(\cdot)$ such that the error $e_1(t)$ between $y_p(t)$ and $y_M(t)$ tends to zero asymptotically.

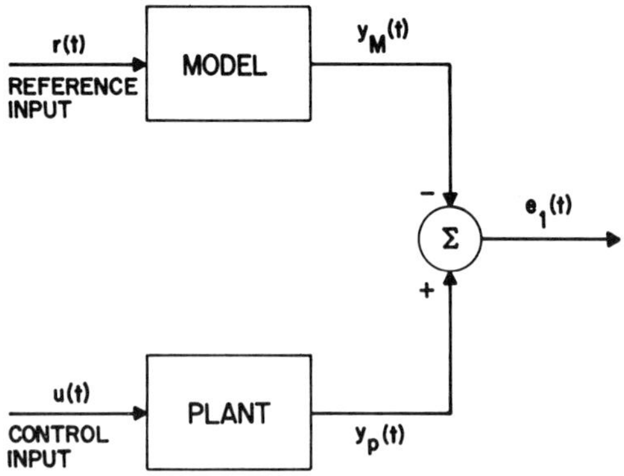

FIGURE 1. The adaptive control problem.

Our interest then is in determining the prior information needed to solve the problem and generating a method for realizing the controller. The parametrization of the plant, the structure of the controller and the manner in which the controller parameters have to be adjusted to achieve stable control are all found to be important aspects as shown in section VI. The first two constitute the algebraic part and the third the analytic part of the adaptive control problem.

b) Indirect and Direct Control

Two philosophically different approaches exist for the solution of the above problem. In the first approach called "Indirect Control" in this paper, the plant parameters are estimated and the control parameters are adjusted based on these estimates so that the overall plant transfer function matches that of the reference model. This has also been referred to as explicit control in the literature [3]. In "Direct Control" no effort is made to identify the plant parameters but the control parameters are directly adjusted to minimize the error between plant and model outputs. This has also been referred to as implicit control. The relation between the two approaches is obviously of interest and is treated in section VI.

A somewhat different approach to the field of adaptive control from that described in this paper is via Self Tuning Regulator (STR) theory. In this approach the basic idea is to select any known procedure for the design of a controller for the system by which the control input to the plant can be computed if the plant parameters are known. In the absence of such knowledge regarding the plant parameters they are recursively estimated and the control input is generated at every instant using these estimates. It is obvious that indirect control as described earlier is a form of

STR in which the objective is to reduce the error between plant and model outputs. The relation between the two methods is discussed in section VII.

c) Discrete and Continuous Systems

Since the plant to be controlled can be either a discrete or continuous time system there is interest in both cases. Continuous systems can give rise to analytic problems which do not have discrete counterparts and are theoretically more interesting. However, with improved digital computational facilities and the availability of inexpensive microprocessors discrete control is becoming increasingly attractive. We shall discuss both types of systems in a unified manner in this paper and attempt to indicate the similarities as well as the differences between them.

d) Single Variable and Multivariable Systems

Most of the principal theoretical results currently known in MRAC apply to single variable systems. From both practical and theoretical considerations it is important to extend these results to the multivariable case. In section VII some of the difficulties encountered in this context are elucidated.

e) Deterministic and Stochastic Systems

The error models described in section IV as well as the stability results given in sections V and VI deal with cases in which noise is absent. Very few precise results are available in stochastic systems where input or observation noise is present. In view of their obvious importance there is considerable interest in them at the present time. In particular the nature of the noise signals, their effect on the parameter estimates and hence on the convergence of the control algorithms are currently under investigation. In section VII some of these aspects are discussed briefly.

From the above comments it is clear that the emphasis in this paper is on the control of deterministic single input - single output discrete and continuous systems for which precise theoretical results are available. Corresponding advances in multivariable and stochastic systems are essential before the methods can be applied to practical problems.

III. THE STABILITY APPROACH

A linear time-invariant system is known to be asymptotically stable if the poles of its transfer function lie in the open left-half of the complex plane. Hence in linear system design the control parameters are constrained so that this condition is satisfied and adjusted within these constraints to optimize a performance criterion. In the adaptive literature of the last two decades several techniques have been developed including

(i) Parameter Perturbation

(ii) Sensitivity Methods

(iii) Nonlinear Estimation and Control

(iv) Self-tuning Regulators

and (v) Stability (Lyapunov and Hyperstability) Methods

In (i) and (ii) laws are first derived for adjusting the parameters to optimize a performance index and later attempts are made to show that the resulting nonlinear system is locally stable. In (iii) and (iv) the parameters of the unknown plant are estimated and the control parameters are adjusted using these estimates to optimize a performance index. The choice of the estimation and control procedures is determined by many considerations which are not necessarily related to the stability of the overall system. The fifth approach using either Lyapunov's Direct Method or Popov's Hyperstability Theory [4] is the only one in which the global stability of the entire adaptive system is the principal consideration. The regions in the parameter space where such stability is guaranteed are first determined and the parameters are then adjusted to optimize system performance. In this sense it is seen that the stability approach for the design of adaptive systems runs along the same conceptual lines as those used in the design of linear time-invariant systems.

a) Error Equations

Let the state of the plant at any time t be $x_p(t)$ and that of the model $x_M(t)$. Let the adaptive controller have a fixed structure with a vector $\theta(t)$ of adjustable parameters. Further let us assume that for some constant value θ^* of $\theta(t)$, the transfer function of the plant together with the controller is identical to that of the model. In the adaptive control problem the state error $e(t) \triangleq x_p(t) - x_M(t)$ and the parameter error $\phi(t) \triangleq \theta(t) - \theta^*$ are of prime interest and the objective is to determine adaptive laws for adjusting the controller parameters $\theta(t)$ in such a manner that $e(t) \to 0$ as $t \to \infty$.

The adjustment of $\theta(t)$ has to be accomplished using all the available signals from the model and plant, but without the use of differentiators, so that all the control signals and parameters are uniformly bounded. Since $\phi(t)$ is not known it is obvious that the law for updating $\theta(t)$ should not explicitly depend on the value of this vector.

Mathematically the problem may be stated as follows:

If the evolution of the state vector e(t) is described by the differential equation

$$\dot{e}(t) = f_1[e(t),\phi(t),t] \tag{1}$$

determine the control law

$$\dot{\theta}(t) = \dot{\phi}(t) = f_2[e(t),t] \tag{2}$$

such that the system described by (1) and (2) is globally stable. For discrete systems the corresponding equations have the form

$$e(k+1) = g_1[e(k),\phi(k),k] \tag{3}$$

and

$$\Delta\phi(k) = \phi(k+1) - \phi(k) = g_2[e(k),k] \tag{4}$$

and once again the objective is to determine equation (4) so that the overall system is globally stable.

The above procedure reduces the adaptive control problem to the stability problem of a set of nonlinear error differential or difference equations. The next section which deals with the stability of error models in detail consequently provides the mathematical basis for all the design procedures outlined in the following sections.

The error differential or difference equations (1-4) are time-varying due to the presence of the reference input. The stability theory of Lyapunov and hyperstability theory which are the principal tools available for stability analysis when applied to such systems yield merely stability (hyperstability) rather than asymptotic stability (asymptotic hyperstability). The difficulties encountered in resolving the principal stability question of adaptive control described in section VI may be attributed to this fact.

IV. THE ERROR MODELS

As mentioned in the previous section, the study of adaptive systems invariably reduces to a study of error equations between plant and model. In this section we describe three error models [5], which are prototypical of those that arise in adaptive control theory. Since we are interested in discrete and continuous time systems both types of error models are discussed.

a) Prototypes

PROTOTYPE I: (CONTINUOUS TIME). The first error model consists of a vector input $u(t)$, a vector of parameter errors $\phi(t)$ and a scalar output error $e_1(t)$ related by the algebraic equation

$$\phi^T(t)u(t) = e_1(t) \tag{5}$$

PROTOTYPE II. In the second error model (Figure 2b) $\phi^T(t)u(t)$ is the input to a stable dynamical system whose state vector $e(t)$ can be measured.

PROTOTYPE III. The third error model is shown in Figure 2c and consists of a dynamical system with a strictly positive real transfer function $W(s)$. The input vector $u(t)$ and the output $e_1(t)$ are related by

$$W(s)\phi^T(t)u(t) = e_1(t) \tag{6}$$

The three prototypes can be readily extended to the multivariable case in which $\phi(t)$ and $e_1(t)$ are matrices and vectors of appropriate dimensions. The analysis carried out when $e_1(t)$ is a scalar can be directly extended to such systems.

In all three prototypes we shall assume that u(t) and $e_1(t)$ (or e(t)) can be measured, that the parameter error vector $\phi(t)$ is unknown but that $\dot{\phi}(t)$ can be adjusted using any signal that can be either measured or generated. The rule for adjusting $\dot{\phi}(t)$ will be referred to as the adaptive law and the objective will be to derive stable adaptive laws so that $\lim_{t\to\infty} e_1(t) = 0$ (or $\lim_{t\to\infty} e(t) = 0$). Discrete analogs of error models of prototypes I and II can also be derived and have the same form shown in Figures 2a and 2b. A search for a discrete analog for prototype III eventually led to a new model suggested by the authors [6]. This model contains an additional feedback term and possesses the necessary properties for proving the stability of the adaptive control problem in the discrete case (section VI). The additional feedback term can also be used with the other two prototypes or with all three prototypes discussed earlier for the continuous case. In the following sections for simplicity, we shall refer to them as error models I, II or III whether they are discrete or continuous and whether or not they contain the additional feedback term; from the context the reader can easily decide which of the models is being referred to.

Table I shows three such error models I-III for both discrete and continuous systems.

As is clear from the error equations in Table I, the only independent variable is the input vector u(·) and the only parameters that can be adjusted are the adaptive gains - the elements of the matrix Γ - in the equations defining $\dot{\phi}(t)$ or $\Delta\phi(k)$. Hence the nature of u(·) and Γ have to be specified before a detailed study of the error models is undertaken. The following represent the special cases of interest:

Input u(·) (i) Uniformly bounded.

(ii) "Sufficiently rich" and uniformly bounded.

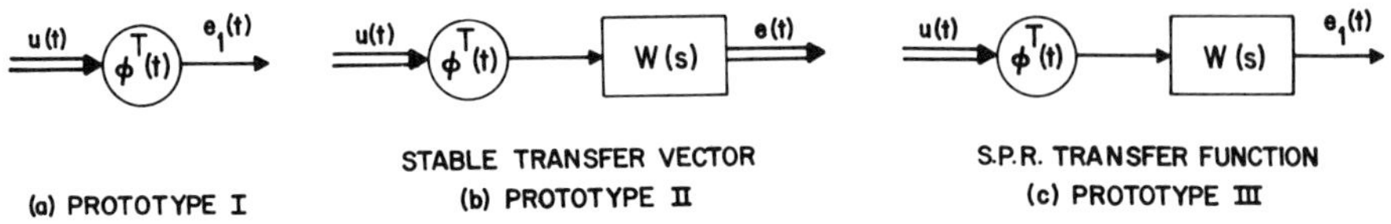

FIGURE 2. Error models.

(iii) Unbounded but $u \in L_e^{\infty}$.[1]

(iv) A stationary random process or sequence.

Adaptive Gain

(i) A positive scalar constant i.e. $\Gamma = I$.

(ii) A positive definite constant matrix $\Gamma = \Gamma^T > 0$.

(iii) A time-varying matrix.

An exhaustive study of all the error models described in Table I with different choices of $u(\cdot)$ and Γ would be both tedious and unproductive. Instead we shall discuss in detail the discrete versions of the first and a continuous version of the third error models. Qualitatively the same ideas also carry over to all the other models.

---

[1] u(t) is defined for all time in the interval $[0,\infty)$ but is not uniformly bounded.

TABLE I

ERROR MODEL	ADAPTIVE LAWS: CONTINUOUS	ADAPTIVE LAWS: DISCRETE	STRUCTURE
I	$\dot{\phi}(t) = \dfrac{-\alpha\Gamma e_1(t)u(t)}{1+u^T(t)\Gamma u(t)} \quad \alpha > 0$	$\Delta\phi(k) = \dfrac{-\alpha\Gamma e_1(k)u(k)}{1+u^T(k)\Gamma u(k)}$ $0 < \alpha < 2$	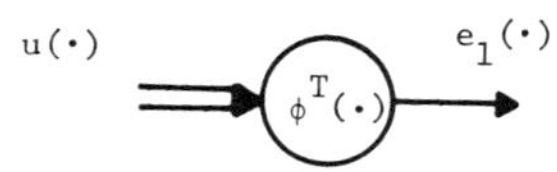
II	$\dot{\phi}(t) = \dfrac{-\alpha e^T(t)Pbu(t)}{1+u^T(t)u(t)} \quad \alpha > 0$	$\Delta\phi(k) = \dfrac{-\alpha e^T(k)Pbu(k)}{1+u^T(k)u(k)}$ $0 < \alpha < 2$	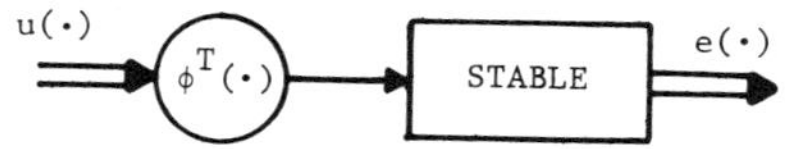
III	$\dot{\phi}(t) = -\alpha\Gamma e_1(t)u(t)$ $\alpha > 0$	$\Delta\phi(k) = -\alpha\Gamma e_1(k)u(k)$ $0 < \alpha < 2$	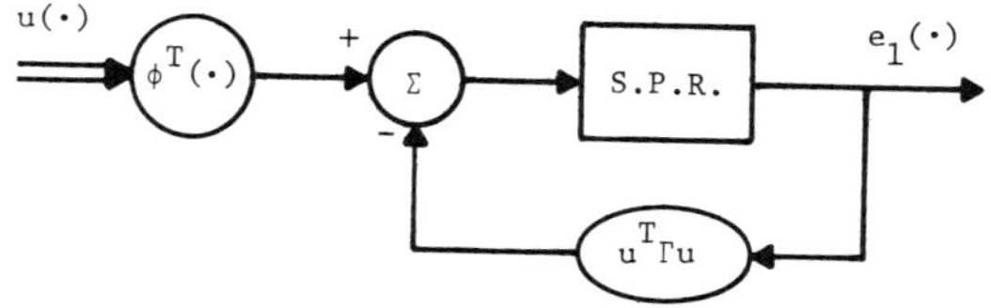

b) Error Model I - Discrete Case

The relation between the input u(k) and output $e_1(k)$ in this model is described by

$$\phi^T(k)u(k) = e_1(k) \tag{6}$$

The adaptive law is given by

$$\Delta\phi(k) \triangleq \phi(k+1) - \phi(k) = \frac{-\alpha\Gamma e_1(k)u(k)}{1 + u^T(k)\Gamma u(k)} \tag{7}$$

$$\Gamma = \Gamma^T > 0 , \qquad 0 < \alpha < 2$$

Choosing a Lyapunov function candidate

$V(k) = \frac{1}{\alpha}\phi^T(k)\Gamma^{-1}\phi(k)$, we obtain

$$\Delta V(k) \leqslant \frac{-e_1^2(k)}{1 + u^T(k)\Gamma u(k)}(2-\alpha)$$

$$\leqslant 0 \qquad \text{if } 0 < \alpha < 2 \tag{8}$$

From (8) it follows that $\phi(k)$ is bounded for any bounded initial vector $\phi(0)$. Further since V(k) is a monotonically non-increasing function which is bounded below, it converges to a limit as $k \to \infty$. Hence

$$\lim_{k\to\infty} \phi^T(k)\Gamma^{-1}\phi(k) = V_\infty < \infty \tag{9}$$

Thus we can conclude that the norm of the parameter error vector $\phi(k)$ will converge to a limit but very little else can be concluded about the convergence of $\phi(k)$.

(i) BOUNDED $\{u(k)\}$. If the input vector sequence $\{u(k)\}$ is uniformly bounded, since

$$\lim_{k\to\infty} \sum_{i=0}^{k} |\Delta V(i)| = V(0) - V_\infty < \infty , \tag{10}$$

it follows from (8) that

$$\sum_{k=0}^{\infty} \frac{e_1^2(k)}{1 + u^T(k)\Gamma u(k)} < \infty \tag{11}$$

When $u(k)$ is uniformly bounded, we obtain

$$e_1(k) \to 0 \quad \text{as} \quad k \to \infty$$

From equation (7) it also follows

$$\Delta\phi(k) \to 0 \text{ as } k \to \infty$$

In summary, when the input $u(k)$ is uniformly bounded $e_1(k)$, $\Delta\phi(k)$ and $\phi^T(k)u(k) \to 0$ as $k \to \infty$. $\Delta\phi(k) \to 0$ implies that the parameter error vector changes slowly for large values of k but does not necessarily assure that $\phi(k) \to \phi^*$ a constant vector; $\phi^T(k)u(k) \to 0$ implies that the two vectors are asymptotically orthogonal.

(ii) "SUFFICIENTLY RICH" $\{u(k)\}$. It has been known for a long time that when the input is sufficiently rich in frequencies the parameter error vector will tend to zero. In the identification problem (section V) this corresponds to perfect identication. The conditions for "sufficiently rich" input have been quantified by Yuan and Wonham [7], Morgan and Narendra [8], Sondhi and Mitra [9] for continuous time systems and by Weiss and Mitra [10] for discrete time systems.

Definition. A bounded input sequence $\{u(k)\}$ is said to be sufficiently rich if there exist numbers T and $\beta > 0$ such that for any constant non-zero n-vector d and any time k,

$$\frac{1}{T}\sum_{j=0}^{T-1}\{d^T u(k+j)\}^2 \geq \beta\|d\|^2. \tag{12}$$

Condition (12) is referred to as a mixing condition in [9]. It has been shown in [10] that when u(k) is "sufficiently rich" according to (12) the error model (6),(7) is uniformly asymptotically stable in the large and $\lim_{k\to\infty} \phi(k) = 0$. Similar results were derived earlier in [7],[8] for continuous systems. Condition (12) implies qualitatively that the norm of the vector u(k) does not become arbitrarily small and that its component along any fixed direction is periodically large resulting in $\phi(k)$ decreasing uniformly in every direction.

The richness condition (12) is both necessary and sufficient for the exponential convergence of $\phi(k)$ to zero. Assuming that the input is sufficiently rich, an important question, particularly from a practical viewpoint, is the rate of convergence of $\phi(k)$. This has been treated in detail in [9] and [10] where the following extremely interesting though counterintuitive result is given. By equations (6) and (7)

$$\phi(k+1) = \left[\, I - \frac{\alpha\Gamma u(k)u(k)^T}{1 + u^T(k)\Gamma u(k)}\,\right]\phi(k) \qquad 0 < \alpha < 2 \tag{13}$$

and it would appear that the speed of convergence would increase as α tends to 1. However it is shown in [10] that for $\alpha \approx 0$ the speed of convergence is proportional to α and for $\alpha \approx 1$ the speed is inversely proportional to α. Hence an optimal value of α exists in the interval [0,1] corresponding to the maximum speed of response of the system. As pointed out in [10] this is generally small and is inversely proportional to the interval T in equation (12). Since it takes at least n vectors to span an n-dimensional space, $T \geq n$. Hence for higher dimensional systems the optimal gain tends to be smaller.

In the algorithm described so far a normalization of the input signal involving a term $\frac{1}{1 + u^T(k)\Gamma u(k)}$ is used. The quadratic form requires the summation of $\frac{n(n+1)}{2}$ terms for its realization which may be practically infeasible for large values of n. In such cases a modified algorithm is desirable which may trade speed of response for simplicity even while assuring global stability. Such an algorithm was originally proposed by Widrow [11] for discrete systems and Monopoli and Subba Rao [17] for continuous systems of type III. Widrow'a algorithm is given by:

$$\begin{aligned} e_1(k) &= \phi^T(k)u(k) \\ \Delta\phi(k) &= -\mu e_1(k)u(k) \end{aligned} \tag{14}$$

where $\mu \lesssim \frac{1}{M^2}$ and $\|u(k)\| \lesssim M$

(iii) UNBOUNDED $\{u(k)\}$. So far we have assumed that the input sequence $\{u(k)\}$ is uniformly bounded. Such an assumption is reasonable when u(k) is an independent input as in the case of adaptive identifiers and observers treated in section V. However, when u(k) is a signal generated by the system whose stability is under consideration it is no longer possible to make such an assumption. In such cases it is necessary to know precisely what can be expected when u(k) is defined for all $k \in N$ but is not uniformly bounded i.e. $u(k) \in \ell_e^\infty$ and $u(k) \notin \ell^\infty$.

A recent report [12] deals with the problem of unbounded inputs in continuous time error models. Very little work corresponding to this has been reported in the adaptive control literature for unbounded discrete inputs. However for the model under consideration some precise statements can be made about the output $e_1(k)$ as well as $\Delta\phi(k)$.

From equation (8) it follows that even if u(k) is not uniformly bounded

$$\frac{e_1(k)}{[1 + u^T(k)\Gamma u(k)]^{1/2}} \to 0 \qquad \text{as } k \to \infty$$

or we can write

$$e_1(k) = o(\|u(k)\|) \tag{15}$$

By (15) if u(k) tends to infinity, and the adaptive law (7) is used, $e_1(k)$ can at worst grow at a slower rate. Further since $\Delta\phi(k) \to 0$ the corrections to the parameter error vector tend to zero. These features of the error model are found to be useful in proving the stability of the discrete adaptive control problem.

(iv) STOCHASTIC {u(k)}. The analysis of error models with stochastic instead of deterministic inputs u(k) has recently been reported by Bitmead and Anderson [13] and Bitmead [14]. Two different models - the Least Mean Square (LMS) (equation 14) and the normalized LMS (equation 13) - are treated in [14]. The input can be either bounded or unbounded (e.g. Gaussian input). For the normalized LMS model which has been discussed thus far in this section it is shown that if the input is ergodic and for some finite integer n the condition

$$E\{\lambda_{min}(\sum_{i=1}^{n-1} \frac{u(i)u(i)^T}{u(i)^T u(i)})\} > 0 \tag{16}$$

is satisfied then $\phi(k)$ will converge to zero exponentially with probability one. It is further shown that if the covariance of {u(k)} is of full rank and the fourth moment exists then condition (16) is satisfied. Hence condition (16) may be considered to be the stochastic counterpart of the richness conditions derived in [7-10] for deterministic inputs.

(v) OUTPUT DISTURBANCES. So far we have analyzed the effect of different types of inputs in the error model described by (6) and (7). This corresponds to the analysis of the homogeneous error equation

$$\phi(k+1) = [I - \frac{\alpha\Gamma u(k)u^T(k)}{1 + u^T(k)\Gamma u(k)}]\phi(k).$$

In many practical situations due to the presence of measurement noise, parameter variations, truncation errors etc. an additional input n(k) is present in the error model and equation (6) is modified to

$$\phi^T(k)u(k) + n(k) = e_1(k) \qquad (17)$$

Using the same adaptive law (7) as before yields the non-homogeneous equation

$$\phi(k+1) = [I - \frac{\alpha\Gamma u(k)u^T(k)}{1 + u^T(k)\Gamma u(k)}]\phi(k) + \frac{\alpha\Gamma u(k)n(k)}{1 + u^T(k)\Gamma u(k)} \qquad (18)$$

The importance of having a sufficiently rich input u(k) becomes apparent when the behavior of (18) is considered. Since the homogeneous equation is uniformly asymptotically stable with a sufficiently rich {u(k)} a bounded disturbance n(k) will produce a bounded parameter error $\phi(k)$. However, if the richness condition is not satisfied and u(k) and n(k) are correlated the analysis becomes considerably more difficult. Questions of this type are treated in [10].

Adaptive observers and controllers treated in sections V and VI lead to error models of the form (17) when output disturbances are present and hence this model merits further investigation.

(vi) ADAPTIVE GAINS. When the matrix Γ in equation (7) is the identity matrix we have a constant scalar gain α which is easiest to implement. It is found from simulations that for uniform convergence in all directions of the parameter space a matrix of gains is preferable. However no efficient method for the choice of Γ is known at present. For stochastic environments a time-varying gain $\Gamma(k)$ has been used extensively where

$$\Gamma(k+1) = \Gamma(k) - \frac{\Gamma(k)u(k)u^T(k)\Gamma(k)}{1 + u^T(k)\Gamma(k)u(k)}$$

or equivalently

$$\Gamma(k+1)^{-1} = \Gamma(k)^{-1} + u(k)u^T(k) \tag{19}$$

This corresponds to the recursive least squares method and it is clear that $\Gamma(k) \to 0$ as $k \to \infty$. While such an approach is satisfactory for noise smoothing it suffers from the obvious disadvantages of re-initialization when used in the context of time-varying systems. As a compromise between the two desired characteristics the following modification of (19) has been suggested in [3] and [15]:

$$\Gamma^{-1}(k+1) = \lambda_1\Gamma(k)^{-1} + \lambda_2 u(k)u^T(k)$$

$$0 < \lambda_1 < 1 \; ; \; 0 < \lambda_2 < 2\lambda_1$$

c) Error Model III - Continuous Case

We describe briefly in this section a typical error model of the third prototype for continuous systems. The model contains an additional error feedback term and has the structure indicated in Table I. If $e(t)$ represents the state error vector of dimension 'n', the error equations are described by

$$\dot{e}(t) = Ae(t) + bv(t)$$

$$v(t) = \phi^T(t)u(t) - u^T(t)\Gamma u(t)e_1(t) \qquad \Gamma = \Gamma^T > 0$$

$$e_1(t) = c^T e(t)$$

or equivalently, if $c^T(sI-A)^{-1}b \triangleq W(s)$

$$W(s)v(t) = e_1(t) \tag{20}$$

The transfer function W(s) is assumed to be strictly positive real.

The adaptive law

$$\dot{\phi}(t) = -\alpha\Gamma e_1(t)u(t) \qquad \alpha > 0 \tag{21}$$

is chosen so that a Lyapunov function candidate $V(e,\phi) = e^T(t)Pe(t) + \frac{1}{\alpha}\phi^T(t)\Gamma^{-1}\phi(t)$ yields a time derivative $\dot{V}(e,\phi)$ where

$$\begin{aligned}\dot{V}(t) &= -e^T(t)L_1 e(t) - 2\alpha[e_1(t)u(t)]^T\Gamma[e_1(t)u(t)] \\ &\leqslant 0 \qquad L_1 = L_1^T > 0.\end{aligned} \tag{22}$$

From (22) it follows that the state and parameter error vectors e(t) and $\phi(t)$ are uniformly bounded and furthermore e(t) and $\dot{\phi}(t)$ belong to L^2.

When the input u(t) and its time derivative are uniformly bounded so that $\ddot{V}(t)$ exists it follows that $e(t) \to 0$ as $t \to \infty$ and $\dot{\phi}(t) \to 0$. Again the latter does not imply that $\phi(t)$ will converge to a constant vector ϕ^*.

When u(t) is sufficiently rich as defined in [7-9] the system described by (20) and (21) is uniformly asymptotically stable and $\phi(t) \to 0$ as $t \to \infty$.

This error model arises both in adaptive observers and controllers. While in the former it can be assumed that u(t) is uniformly bounded such an assumption cannot be made in the latter case. Hence in the control problem we are only assured of the boundedness of $e(t)$ and $\phi(t)$ and that $e(t)$, $\dot{\phi}(t) \in L^2$. The fact that $\dot{\phi}(t) \in L^2$ plays an important role in the proof of stability of the adaptive control problem of continuous systems.

d) A Special Error Model

In the adaptive control of a general linear system an error model arises frequently which cannot be cast into any of the forms considered so far. In view of its importance in the proof of stability a typical case is briefly described here. The extension of the model to the discrete time case is trivial and is indicated through a simple example.

Let $\phi(t)$ be the parameter error and $W(s)$ a stable transfer function. Let $u(t)$ and $e_1(t)$ be an input-output pair related by

$$W(s)\phi(t)u(t) = e_1(t) \tag{23}$$

The objective again is to adjust $\dot{\phi}(t)$ such that $\lim_{t\to\infty} e_1(t) = 0$. Since $W(s)$ is not a strictly positive real transfer function the model (23) is not of the third type. Hence adaptive laws cannot be easily generated. However if it is <u>assumed</u> that an operator $\phi(t)W(s) - W(s)\phi(t)$ can be constructed then defining

$$[\phi(t)W(s) - W(s)\phi(t)]u(t) \overset{\Delta}{=} e_2(t)$$

we have

$$\begin{aligned} \phi(t)W(s)u(t) &= e_1(t) + e_2(t) = \varepsilon_1(t) \\ \text{or} \quad \phi(t)\,\zeta(t) &= \varepsilon_1(t) \end{aligned} \tag{24}$$

in Figure 3.

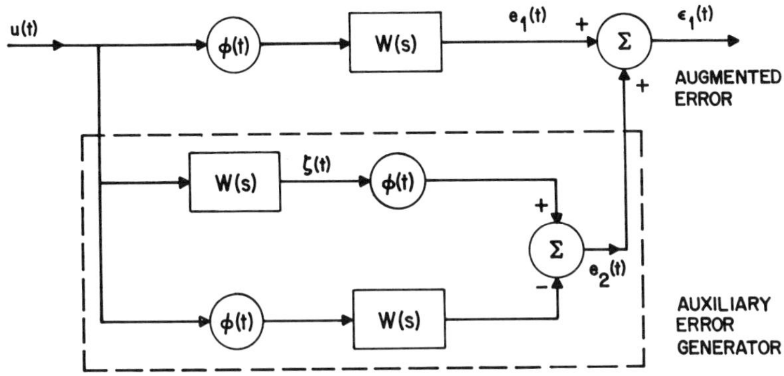

FIGURE 3. A special error model.

$e_2(t)$ is referred to as the "auxiliary error signal". By the addition of $e_2(t)$ to $e_1(t)$ an augmented error is generated and an error model I is realized so that the adaptive law

$$\dot{\phi}(t) = -\alpha\varepsilon_1(t)\zeta(t) / (1+\zeta^2(t)) \qquad \alpha > 0$$

can be written by inspection. As described earlier if $\zeta(t)$ is uniformly bounded it can be shown that $\varepsilon_1(t)$, $e_2(t)$ and hence $e_1(t)$ tend to zero as $t \to \infty$ and $\phi(t)$ will be bounded.

To develop the stable adaptive law it was assumed that an operator $\phi(t)W(s) - W(s)\phi(t)$ can be constructed. But this poses a problem since the vector $\phi(t)$ is unknown. If $\theta(t)$ is a control parameter which can be adjusted and θ^* its unknown desired value and

$$\phi(t) = \theta(t) - \theta^*$$

then $\phi(t)W(s) - W(s)\phi(t) \equiv \theta(t)W(s) - W(s)\theta(t)$. (25)

and the r.h.s. of (25) can be physically realized. While only the simple case where $\phi(t)$ is a scalar has been described the results also carry over to the general cases where $\phi(t)$ is a vector or a matrix.

The approach outlined here is used in [1] and [16] to derive the adaptive control laws for the general control problem in the continuous case. The same arguments carry over to the discrete model as well.

EXAMPLE-(DISCRETE CASE). Let W(z) in the discrete version of (23) be z^{-d} (a d-step pure delay). The input and output u(k), $\varepsilon_1(k)$ are then related by

$$\varepsilon_1(k) = \phi(k-d)u(k-d) \tag{26}$$

The procedure described in this section yields the augmented error equation of type I

$$e_1(k) = \phi(k)u(k-d) \tag{27}$$

The adaptive law for this is

$$\phi(k+1) - \phi(k) = \Delta\phi(k) = -\frac{\alpha e_1(k)u(k-d)}{1 + u^2(k-d)}, \quad 0 < \alpha < 2 \tag{28}$$

The importance of this model is described in section VI.

V. ADAPTIVE IDENTIFIERS AND OBSERVERS

The error models described in the previous section were first developed in the context of adaptive observers. Given the input and output of a linear time-invariant system with a known transfer function an observer is a device which generates asymptotically the state of the system. Parameter identification on the other hand involves the determination of the parameters of a specific representation of a system from its input-output data. When both parameters and states of a system are unknown, an adaptive observer is used which estimates them simultaneously.

A vast literature currently exists on both discrete and continuous time adaptive observers and recently a description of such observers was given in [5]. Since the emphasis in this paper is on stable adaptive control, an exhaustive survey of adaptive observers is not attempted here, but instead, attention is confined to two specific structures [22], [20] which have proved useful in the control context. An attempt is also made to indicate how the error models described in section IV form the basis of the design of such systems.

Before discussing the two adaptive observers mentioned earlier we shall consider a simple structure which has been used for over two decades in system identification.

a) Identification

Let W(s) be the transfer function of an unknown linear time-invariant plant and let

$$W(s) = \sum_{i=1}^{N} c_i^* W_i(s) \quad [2] \tag{29}$$

---

[2] Throughout the paper both differential equations and transfer functions are used. Depending on the context 's' will be used as a differential operator or the Laplace transform variable. Further, for simplicity, it is assumed in the following sections that all initial conditions are identically zero when they do not affect the arguments.

where $W_i(s)$ are known stable transfer functions but the parameters c_i^* are unknown. If u(t) and y(t) are respectively the input and output of the plant, $W(s)u(t) = y(t)$. To identify the parameters c_i^* a model of the plant is constructed as shown in Figure 4 with adjustable parameters $c_i(t)$.

If $W_i(s)u(t) = \zeta_i(t)$, $c^T(t) = [c_1(t), c_2(t), \ldots, c_N(t)]$ and $\zeta^T(t) = [\zeta_1(t), \ldots \zeta_N(t)]$ the plant and model outputs are given by $y(t) = c^{*T}\zeta(t)$ and $\hat{y}(t) = c^T(t)\zeta(t)$ respectively. Hence the error equation is given by

$$e_1(t) = \phi^T(t)\zeta(t) \tag{30}$$

where $\hat{y}(t) - y(t) \triangleq e_1(t)$ and $c(t) - c^* = \phi(t)$. Equation (30) is of type I discussed in section IV and the adaptive law may be expressed as

$$\dot{c}(t) = \dot{\phi}(t) = -\Gamma e_1(t)\zeta(t) \qquad \Gamma = \Gamma^T > 0 \tag{31}$$

From section IV it then follows that if u(t) is uniformly bounded $\lim_{t\to\infty} e_1(t) = 0$. Further if it is sufficiently rich $\lim_{t\to\infty} \phi(t) = 0$ or $\lim_{t\to\infty} c(t) = c^*$.

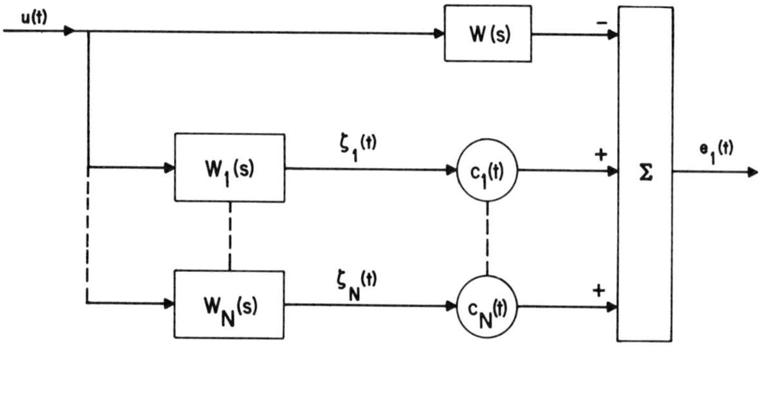

FIGURE 4. An identifier.

As mentioned earlier, this model is ubiquitous in the area of system identification and we give three typical examples below:

(i) If the state equations of the plant are $\dot{\zeta} = A\zeta + bu$; $y = c^{*T}\zeta$ the above procedure can be used to identify the zeros of the plant transfer function when its poles are known.

(ii) If $h_i(t)$ is the impulse response of $W_i(s)$ and $h_i(t)$ $(i = 1,2,\ldots N)$ are orthogonal over the interval $[0,\infty)$ the method described yields an orthogonal expansion of the impulse response of the plant.

(iii) For discrete time plants if $W_i(z) = z^{-i}$ the model represents a tapped delay line and the coefficients c_i^* are the amplitudes of the impulse response of the plant at the instant i $(i = 1,2\ldots)$.

In all the above simple cases the model output can match the plant output exactly and $e_1(t) \to 0$ as $t \to \infty$. However when the output of the plant is corrupted with noise n(t) and the adaptive law (31) is used, the parameters do not converge to zero as discussed in section IV. This is also the case when truncation error is present due to the use of a finite number of terms $(W_i(s))$ in an orthogonal expansion. As indicated in section IV time-varying adaptive gains converging to zero have to be used when noise is present to make the parameters converge to their true values.

The above approach is effective when the poles of the plant transfer function W(s) are known. When both poles and zeros of W(s) are unknown the models described in the next subsection can be used.

b) Adaptive Observers

Adaptive observers were suggested in 1974 by Carroll and Lindorff [21] and Luders and Narendra [19] and later generalized in [23]. The extensions to multivariable systems by Anderson [24], the analysis of related stability questions by Yuan and Wonham, Morgan and Narendra and the study of the speed of response of the observer by Kim [25] succeeded in establishing the area on a firm footing by 1975. In recent years Kreisselmeier [20] and Landau [26],[27] have added significantly to our knowledge of adaptive observer theory.

The basic idea of the early adaptive observers was to use a Luenberger observer whose parameters can be updated even while it is in operation. The updating is carried out in such a manner that the state observation error asymptotically goes to zero with time. In some of the early schemes certain auxiliary signals had to be fed into the observer to achieve global asymptotic convergence. Such auxiliary signals were later eliminated by Luders and Narendra [22] by the use of a new canonical form for linear systems. In view of its simplicity this observer has found application in the design of stable adaptive controllers and is discussed in this section as Adaptive Observer I.

A somewhat different approach to adaptive observer design was suggested by Kreisselmeier [20] who proposed an equivalent but structurally different representation of the Luenberger observer. This observer also appears to have attractive features for use in adaptive control and is discussed briefly here as Adaptive Observer II.

(i) ADAPTIVE OBSERVER I (LUDERS-NARENDRA). Any transfer function W(s) with n poles and m zeros ($\leq$n-1) can be represented in the form

$$W(s) = \frac{Q(s)}{R(s)}\left[\frac{1/(s+\lambda_0)}{1+\frac{P(s)}{R(s)(s+\lambda_0)}}\right] \tag{32a}$$

where R(s) is a known Hurwitz polynomial of degree (n-1) and Q(s) and P(s) are (n-1) degree polynomials. An alternate representation of W(s) is

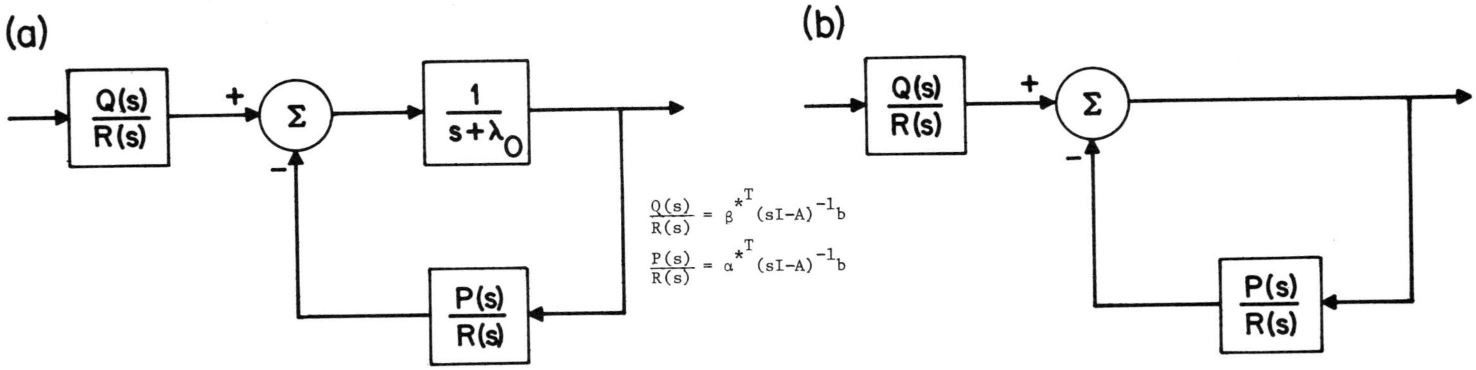

$$\frac{Q(s)}{R(s)} = \beta^{*T}(sI-A)^{-1}b$$

$$\frac{P(s)}{R(s)} = \alpha^{*T}(sI-A)^{-1}b$$

(a) (b)

FIGURE 5. Representation of an unknown plant.

$$W(s) = \frac{Q(s)}{R(s)} \left[\frac{1}{1 + \frac{P(s)}{R(s)}} \right] \tag{32b}$$

where R(s) is an nth degree Hurwitz polynomial and P(s) and Q(s) are polynomials of degree n. Both representations can be used for the identification of the unknown parameters of a linear system. The feedback models corresponding to (32a) and (32b) are shown in Figure 5.

To identify the unknown plant the coefficients of the polynomials P(s) and Q(s) have to be determined. Denoting these by the components of the n-vectors α^* and β^*, the identification procedure would consist in estimating these vectors. A model of the plant having the same structure as those shown in Figure 5a or 5b can be constructed with adjustable parameter vectors $\alpha(t)$ and $\beta(t)$ respectively to identify the plant parameters. However simple adaptive laws for stable identification cannot be easily determined in this case.

An important contribution of [22] was the realization that the plant output rather than the model output should be used in the model feedback path to simplify the generation of the adaptive laws. This led to the configuration shown in Figure 6 for the

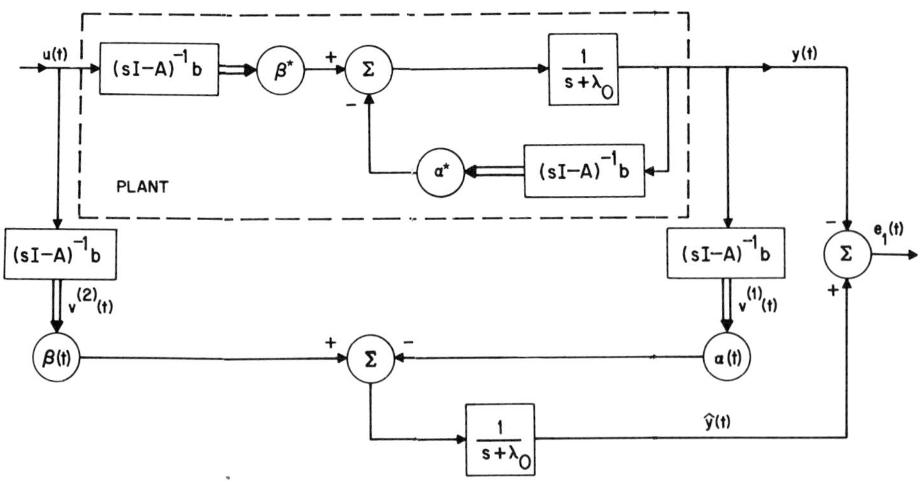

FIGURE 6. Adaptive observer I.

plant representation shown in Figure 5a. If $v^{(1)}(t)$ and $v^{(2)}(t)$ are the state vectors of the feedback and feedforward compensators as shown in Figure 6 the plant and model output can be represented by

$$y(t) = \frac{1}{s+\lambda_0} \theta^{*T} w(t) \text{ and}$$

$$\hat{y}(t) = \frac{1}{s+\lambda_0} \theta^T(t)w(t) \text{ respectively where } \theta^{*T} = [-\alpha^{*T}, \beta^{*T}],$$

$\theta^T(t) = [-\alpha^T(t), \beta^T(t)]$ and $w^T(t) = [v^{(1)^T}(t), v^{(2)^T}(t)]$. The output error equation is then given in terms of the parameter error vector $\phi(t) \triangleq \theta(t) - \theta^*$ as

$$e_1(t) \triangleq \hat{y}(t) - y(t) = \frac{1}{s+\lambda_0} \phi^T(t)w(t) \tag{33}$$

which is in the form of the third error model since $\frac{1}{s+\lambda_0}$ is a strictly positive real transfer function. Similarly if the plant representation shown in Figure 5b is used, the error equation has the form

$$e_1(t) = \phi^T(t)w(t)$$

and is of the first prototype. In both cases the adaptive control laws can be written by inspection as

$$\dot{\theta}(t) = -\Gamma e_1(t)w(t)$$

For any bounded input to the plant, as shown in section IV, the error $e_1(t)$ will tend to zero. However the input has to be sufficiently rich for the parameters $\alpha(t)$ and $\beta(t)$ to converge to α^* and β^* respectively.

Comment. $\frac{1}{s + \lambda_0}$ and unity are strictly positive real transfer functions which are used in the two representations shown in Figure 5. Replacing $\frac{1}{s + \lambda_0}$ by any general strictly positive real transfer function other representations can be obtained which, in turn, will yield an error model of the third prototype. Such a model was first used in [28] to show the equivalence of direct and indirect control [refer to section VI].

(ii) THE DISCRETE CASE. The discrete version of the representation (32b) has been used extensively for the identification of discrete systems for a long time. If the plant is described by the ARMA model

$$y(k) = \sum_{i=1}^{n-1} a_i^* \, y(k-i) + \sum_{i=1}^{m-1} b_i^* \, u(k-i) \tag{34}$$

and the estimates of a_i and b_i at time k are $a_i(k)$ and $b_i(k)$, the model output is given by

$$\hat{y}(k) = \sum_{i=1}^{n-1} a_i(k) y(k-i) + \sum_{i=1}^{m-1} b_i(k) u(k-i) \tag{35}$$

If $\hat{y}(k) - y(k) \triangleq e_1(k)$, from (34) and (35) we obtain the error model

$$e_1(k) = \phi^T(k) w(k) \tag{36}$$

where

$$\phi^T(k) = [a_1(k) - a_1^*, \ldots, a_{n-1}(k) - a_{n-1}^*, b_1(k) - b_1^*, \ldots \\ \ldots b_{m-1}(k) - b_{m-1}^*]$$

$$w^T(k) = [y(k-1), \ldots y(k-n+1), u(k-1), \ldots u(k-m+1)]$$

(36) is an error model I and the adaptive laws can be easily determined.

(iii) PARAMETRIZED OBSERVER (KREISSELMEIER). Let a single-input single-output system be represented in the form

$$\dot{x} = Fx + g^* x_1 + h^* u \tag{37}$$

$$y = c^T x = x_1$$

where F is a known stable matrix, c^T and F are in observable canonical form and g^* and h^* are two unknown constant parameter vectors. The aim of the observer is then to estimate g^* and h^* as well as the state x(t) of the system asymptotically. If $e_1, e_2, \ldots e_n$ represent n unit vectors and the 2n n-dimensional vector functions $\zeta_1(t), \zeta_2(t), \ldots, \zeta_{2n}(t)$ are generated using

$$\begin{aligned} \dot{\zeta}_i(t) &= F\zeta_i(t) + e_i y(t) \\ \dot{\zeta}_{n+i}(t) &= F\zeta_{n+i} + e_i u(t) \qquad i = 1,2,\ldots n \end{aligned} \tag{38}$$

by linearity the state x(t) in (37) can be obtained as a linear combination of the vectors $\zeta_i(t)$ (i = 1,2,...,2n). If the nx2n matrix Z(t) is defined as

$$Z(t) = [\zeta_1(t), \zeta_2(t), \ldots, \zeta_{2n}(t)] \tag{39}$$

then a constant vector p^* exists such that

$$Z(t)p^* = x(t) \tag{40}$$

$$\text{or} \quad c^T Z(t)p^* = x_1(t) = y(t) \text{ where } p^* = \begin{bmatrix} g^* \\ h^* \end{bmatrix}. \tag{41}$$

If $c^T Z(t) \triangleq z^T(t)$ then

$$z^T(t)p^* = y(t) \ . \tag{42}$$

Since p^* is the vector to be identified, a model can be set up with a time-varying vector $p(t)$ to yield an estimate $\hat{y}(t)$ such that

$$z^T(t)p(t) = \hat{y}(t). \tag{43}$$

From (42) and (43)

$$z^T(t)[p(t) - p^*] = \hat{y}(t) - y(t) = e_1(t)$$

and represents an error equation which is of type I. The adaptive equations for updating the estimates are then given by

$$\dot{p}(t) = -\Gamma e_1(t)z(t) \tag{44}$$

and the estimate $\hat{x}(t)$ of the state of the system may be computed using equation (40) as

$$Z(t)p(t) = \hat{x}(t) \tag{45}$$

From the brief description given above it is seen that the matrix $Z(t)$ and the vector $z(t) = c^T Z(t)$ can be generated using fixed filters while the estimate $\hat{x}(t)$ and $p(t)$ of the states and the parameters can be derived independently from them. These features have been used by Kreisselmeier [29],[30] to design adaptive controllers for unknown linear systems.

In contrast to Adaptive Observer I where the state estimate $\hat{x}(t)$ can be obtained only as a functional of the parameters, it is seen that the estimate of the state $\hat{x}(t)$ is a linear function of p(t) here. However 2n vectors $\zeta_i(t)$ have to be generated in this case in the place of the two vectors $v^{(1)}$ and $v^{(2)}$ used in Adaptive Observer I.

c) Adaptive Observers with Output Noise

In the two observers described earlier if the plant output can be measured only in the presence of noise, then both the output error as well as the signals $v^{(1)}(t)$ (in observer I) and Z(t) (in observer II) contain noise components which are correlated. This results in biased estimates of $\theta_i(t)$ and p(t) respectively. To avoid this, Landau [26],[27] has suggested a "parallel model" observer in which the model output rather than the plant output is used to generate the signals (corresponding to $v^{(1)}(t)$), used in the adaptive law. Furthermore the error between plant and model is filtered and used in the place of $e_1(t)$ to adjust the observer parameters. While this observer reduces bias effects due to noise its global stability has not been established so far.

VI. ADAPTIVE CONTROLLER DESIGN

The problem of Model Reference Adaptive Control as stated in section II is to determine the control input to a linear time-invariant plant with unknown parameters so that the output evolves asymptotically to the output of a reference model. When the theory of adaptive observers was well understood, it was felt that identification of the plant parameters using an adaptive observer followed by an adaptive controller could solve the problem in a relatively straightforward fashion. If the observer parameters converged to the desired values, it was argued, so would the control parameters, making the overall system asymptotically stable. However, as pointed out in [31], it soon became apparent that in view of the feedback that exists, it is no longer possible to assume that the input to the plant is bounded, since it is precisely the stability of the feedback system that has to be established. In view of this, the observer parameters cannot be shown to converge to their true values and hence the stability question remains unresolved.

The above stability problem was until recently perhaps the single most important problem in the area of deterministic MRAC during the period 1976-1979. Over the last two decades many novel adaptive schemes were suggested but none of them could be shown to be globally stable. Recently it was proven independently by Morse [2] and Narendra, Lin and Valavani [1] that a modified version of an adaptive procedure originally suggested by Monopoli [16] for the control of a linear time-invariant plant possesses such globally stable properties. At about the same time it was also shown by Narendra and Lin [32] and Goodwin, Ramadge and Caines [33] that globally stable discrete adaptive controllers could also be realized. It was also demonstrated by Narendra and Valavani [28] that using a specific parametrization of the plant Indirect and Direct Control lead to identical error equations and hence are equivalent. These major developments in the field of MRAC now enable one to consider more complex and practically more interesting problems of multivariable and stochastic control from a firmer analytic foundation. In view of their importance we shall first discuss the principal ideas which led to these developments.

a) The Problem

A plant P is completely represented by the input-output pair $\{u(t), y_p(t)\}$ and can be modeled by a transfer function $W_p(s) = \frac{k_p Z_p(s)}{R_p(s)}$ where Z_p and R_p are monic polynomials of degrees

$m(\leq n-1)$ and n respectively. A stable reference model is represented by the input-output pair $\{r(t), y_M(t)\}$ and has a transfer function $W_M(s) = k_M \frac{Z_M(s)}{R_M(s)}$. The error between plant and model outputs is defined as

$$e_1(t) \triangleq y_p(t) - y_M(t)$$

The problem in the continuous case is to determine the control input u(t), so that $\lim_{t\to\infty} e_1(t) = 0$.

The problem for the discrete case can be stated in a similar manner replacing 't' by the stage number 'k' and the Laplace transform variable 's' by 'z'.

As mentioned in section 2, this problem can be conveniently divided into an algebraic part which is concerned with establishing conditions under which a control function u(t) would exist and an analytic part for generating such an input function.

b) Minimal Information

If the plant and model transfer functions $W_p(s)$ and $W_M(s)$ have n_1 poles and m_1 zeros and n_2 poles and m_2 zeros respectively and $n_1^* \triangleq n_1 - m_1$, $n_2^* \triangleq n_2 - m_2$, the minimal information required to be able to construct an adaptive controller having the structure described later may be expressed in terms of $W_p(s)$, $W_M(s)$, n_1^* and n_2^*. n_1^* and n_2^* will be referred to as the relative degrees of the plant and the model.

(i) n_1^* should be known exactly.

(ii) $n_2^* \geq n_1^*$.

(iii) An upper bound on the order n_1 of the plant should be known.

(iv) The sign of the gain k_p of the plant must be known.

(v) The zeros of the plant must lie in the open left half of the complex plane.

For the discrete case n_1^* and n_2^* represent the effective delays in the plant and model respectively and the same conditions (i)-(v) have to be satisfied. In addition in condition (iv) an upper bound on the unknown gain k_p must be known.

It is clear that condition (i) is quite restrictive. Since the relative degree of the plant cannot be decreased by dynamic feedforward or feedback compensation we obtain condition (ii), condition (v) is required to make the overall feedback system structurally stable since pole-zero cancellation is used to match plant and model transfer functions asymptotically.

c) Structure of the Controller (Direct Control)

The plant transfer function has a maximum of $2n_1$ unknown parameters which are the coefficients of $k_p Z_p(s)$ and $R_p(s)$. Since these are assumed to be unknown the controller structure must have ade-

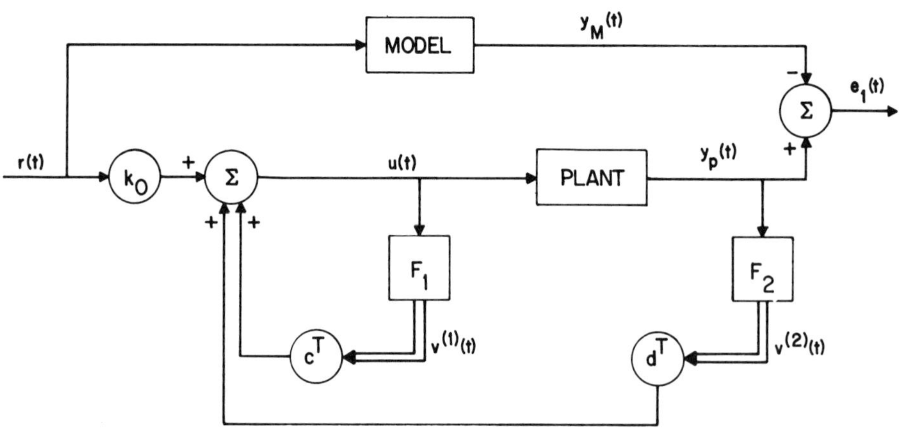

FIGURE 7. Direct control.

quate freedom so that by the adjustment of the control parameters the transfer function of plant together with the controller can match that of any specified model.

For direct control, the configuration shown in Figure 7 has evolved as the basic one for the controller. The input u(t) and the output $y_p(t)$ of the plant are made the inputs to two identical filters (as in the adaptive observer described in section V) whose state vectors $v^{(1)}(t)$ and $v^{(2)}(t)$ are of dimension (n_1-1). Together with r(t) and the output $y_p(t)$ they constitute the $2n_1$ signals whose linear combination yields the desired input u(t). If $\theta(t)$ is a control parameter vector with $2n_1$ values

$$u(t) = \theta^T(t)w(t) \tag{46}$$

where

$$\theta^T(t) = [\theta_1(t),\theta_2(t),\ldots,\theta_{2n_1}(t)],$$

$$w^T(t) = [r(t),v^{(1)^T}(t),y_p(t),v^{(2)^T}(t)]$$

Under the conditions specified in the previous section it can be shown algebraically that a constant vector θ^* of dimension $2n_1$ exists such that when $\theta(t) \equiv \theta^*$, the transfer function of the plant will match that of the model. Hence it only remains to show how $\theta(t)$ is to be adjusted so that $\lim_{t\to\infty} \theta(t) = \theta^*$. This is treated later in this section.

d) Structure of the Controller (Indirect Control)

Indirect control involves identification of the plant parameters using an adaptive observer followed by an adaptive controller. The control parameters at any instant are then adjusted using the

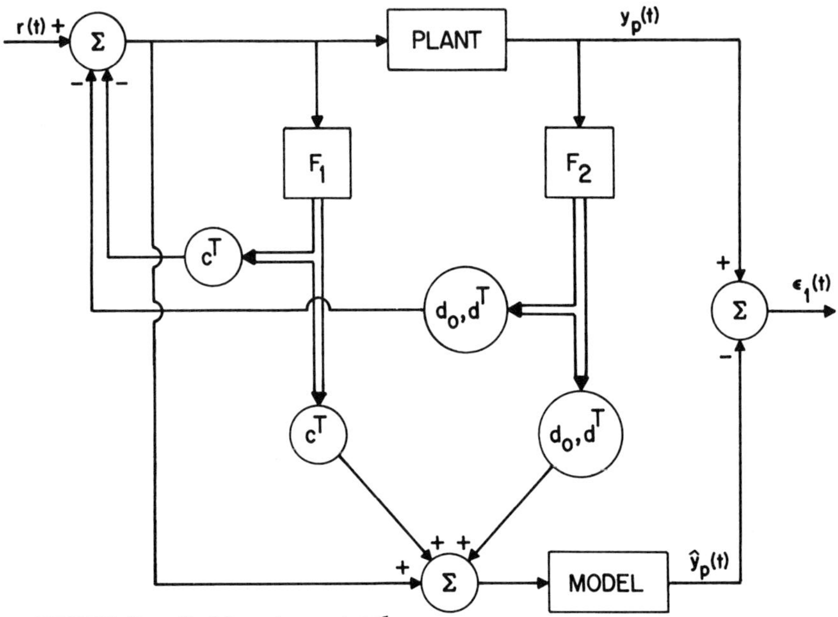

FIGURE 8. Indirect control.

estimates of the plant parameters as given by the observer. Unfortunately, the parametrization of the plant required to simplify the control problem is not the same as the parametrization required to simplify the observer structure. Hence, in general, the control parameters are nonlinearly related to the observer parameters and the error differential equations of the controller become nonlinear and intractable. For proper choice of observer and controller structures where the controller and observer parameters are linearly related this problem can be circumvented. It was shown in [28] by Narendra and Valavani that if the plant is represented as shown in Figure 8 with the reference model embedded in it, this condition would be satisfied. In such a case, it is found that indirect control is equivalent to direct control. Hence the stability arguments described in the following section apply to both direct and indirect control systems.

e) Adaptive Laws

Once the structure of the controller is specified it only remains to determine the manner in which the controller parameters are to be updated so that the overall system is globally stable. It was shown in section VIc that $2n_1$ parameters of the controller have to be adjusted simultaneously. For ease of explanation we shall consider the case of a single adjustable parameter in the controller; the same ideas also carry over to the general case.

In Figure 9 the reference model has a transfer function $W_M(s)$. The plant has an input u(t) and two outputs $y_p(t)$ and v(t) and the transfer function from u(t) to $y_p(t)$ is $W_p(s)$. $\theta(t)$ is an adjustable parameter in the plant feedback path as shown in Figure 9b and it is assumed that a constant θ^* exists such that when $\theta(t) = \theta^*$ the plant transfer function matches the model.

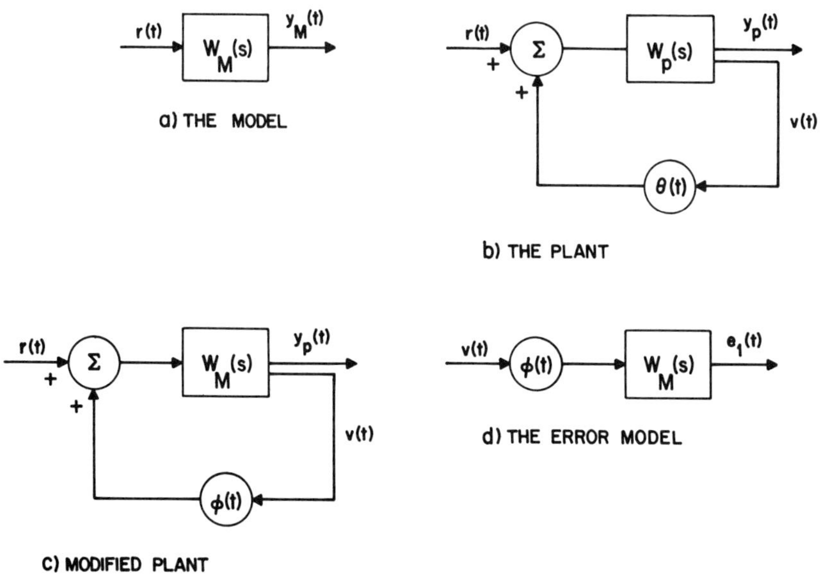

FIGURE 9. Controller with a single parameter.

If $\theta(t) - \theta^* = \phi(t)$, the plant together with the controller can be represented as shown in Figure 9c; the error model relating output error $e_1(t) = y_p(t) - y_M(t)$ to parameter error $\phi(t)$ is given in Figure 9d.

(i) MODEL TRANSFER FUNCTION $W_M(s)$ SPR. If $W_M(s)$ is strictly positive real the error model in Figure 9d is of type III and the adaptive law

$$\dot{\theta}(t) = \dot{\phi}(t) = -\alpha\, e_1(t)v(t) \tag{46}$$

can be used. From section IV it follows that $e_1(t)$ will be uniformly bounded. Since $y_p(t) = y_M(t) + e_1(t)$ the plant output and hence $v(t)$ are also uniformly bounded and hence $\lim_{t\to\infty} e_1(t) = 0$.

(ii) MODIFIED ERROR MODEL. If in addition to the feedback signal $\theta(t)v(t)$ in Figure 9b the signal $-e_1(t)v^2(t)$ is also fed back to the plant we have the modified error model described in section IV which can be represented as shown in Figure 10. Using the same procedure as before it can be shown that $\lim_{t\to\infty} e_1(t) = 0$. In addition, we can also conclude that $\dot{\phi}(t) \in L^2$.

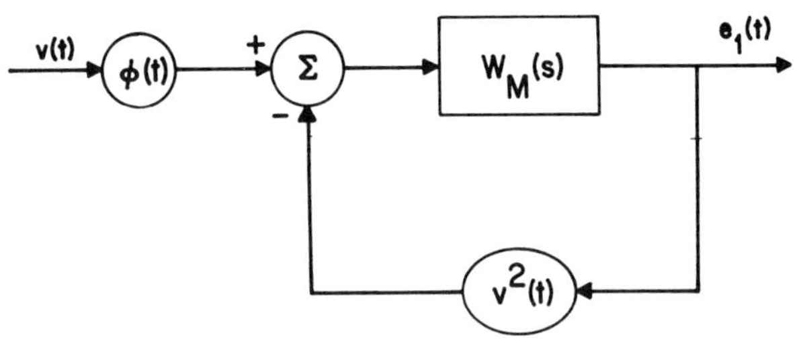

FIGURE 10. Modified error model.

(iii) $W_M(s)$ NOT SPR. When $W_M(s)$ is SPR the error equations were seen to be of type III. When $W_M(s)$ is not SPR it is no longer possible to determine the adaptive laws by inspection and an auxiliary signal has to be used. We propose here two methods (based on section IV) which lead to error models of the first and third prototypes respectively enabling adaptive laws to be generated in a simple fashion.

Error Model III. Let $L(s)$ be an operator in 's' such that $W_M(s)L(s)$ is SPR. Define the signal $e_2(t)$ as

$$e_2(t) \triangleq W_M(s)L(s)[\theta(t)L^{-1}(s) - L^{-1}(s)\theta(t)]v(t) \tag{47}$$

and let

$$e_1(t) + e_2(t) = \varepsilon_1(t) \tag{48}$$

$\varepsilon_1(t)$ is called the "augmented error" and satisfies the relation

$$W_M(s)L(s)\phi(t)L^{-1}(s)v(t) = \varepsilon_1(t)$$

or $$W_M(s)L(s)\phi(t)\zeta(t) = \varepsilon_1(t) \tag{49}$$

where $L^{-1}(s)v(t) = \zeta(t)$. Equation (49) is of type III and the adaptive laws are given by

$$\dot{\phi}(t) = -\alpha\varepsilon_1(t)\zeta(t) \qquad \alpha > 0 \tag{50}$$

Hence the adaptive procedure consists in generating the auxiliary error signal given in (47) adding it to the true error $e_1(t)$ to obtain the augmented error $\varepsilon_1(t)$ and using the latter to derive the adaptive laws.

This crucial idea of adaptive control using an augmented error signal was first suggested by Monopoli [16] and later cast in the form described here by Narendra and Valavani [34]. Figure 11 shows the complete structure for the adaptive control problem with a single adjustable parameter $\theta(t)$.

Error Model I. If $L^{-1}(s)$ in the above discussion is chosen to be $W_M(s)$ a simplified representation of the controller as well as error model is obtained. In such a case we have the equations:

$$e_2(t) = [\theta(t)W_M(s) - W_M(s)\theta(t)]v(t)$$

$$c_1(t) = \phi(t)W_M(s)v(t) \tag{51}$$

and $\dot{\theta}(t) = -\alpha\varepsilon_1(t)\zeta(t) \qquad \alpha > 0$

While this model is obviously better for purposes of analysis it involves the use of more integrators than Error Model III. The latter consideration becomes significant when many parameters have to be adjusted and n_1^* is only slightly greater than 1 and $n_1 >> 1$.

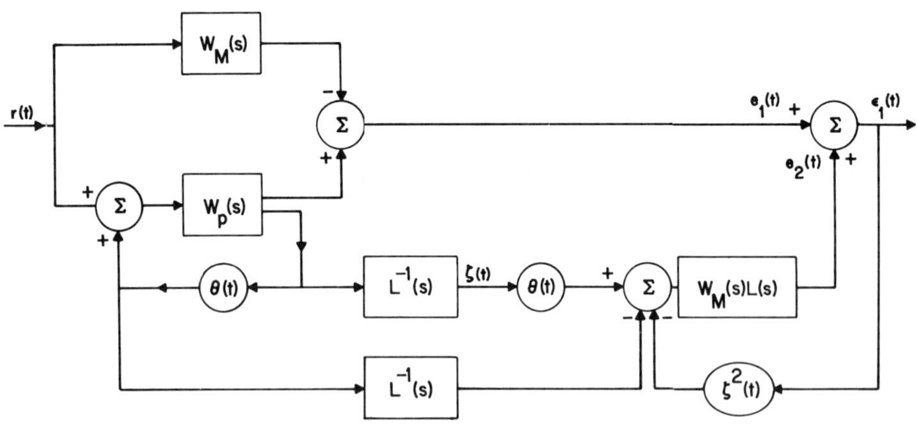

FIGURE 11. Stable adaptive controller (single parameter).

Comments.

(i) For the general adaptive control problem of a single-input single-output system it was shown earlier that $2n_1$ control parameters are needed of which $(2n_1-1)$ are in the feedback path. The procedure outlined in this section for the adjustment of a single parameter carries over directly to all $(2n_1-1)$ feedback parameters.

(ii) The adjustment of the gain $\theta_1(t)$ however poses problems which are not encountered in the other cases. Hence the control problem is analytically considerably simpler, when k_p is known. When k_p is not known an additional gain parameter has to be used in the generation of the auxiliary signal $e_2(t)$ to obtain the error equation in the desired form. In direct control this implies the simultaneous adjustment of $(2n_1+1)$ parameters. If indirect control is used k_p has to be assumed to be known; the generated plant input is proportional to $1/k_p$ and if an estimate of $k_p (= \hat{k}_p)$ is used, theoretical questions of convergence arise when $\hat{k}_p \approx 0$. The equivalence of the direct and indirect approaches to the control problem shown in [28] consequently holds only for the case where k_p is known.

(iii) In section VIe an additional feedback signal $-\varepsilon_1(t)\zeta^2(t)$ as shown in Figure 11 could have been included without affecting the discussion. For the vector case (where $\zeta(t)$ is a vector) this signal is of the form $-\varepsilon_1(t)\zeta^T(t)\Gamma\zeta(t)$. Such a feedback signal is found to be essential to prove the stability of the adaptive loop as indicated in the next section.

THE STABILITY PROBLEM. In section VIe when $W_M(s)$ is SPR the boundedness of the output of the plant followed directly from the boundedness of $y_M(t)$ and $e_1(t)$. However, when $W_M(s)$ is not SPR we can only conclude that $y_M(t)$ and the augmented error $\varepsilon_1(t)$ are uniformly bounded. Hence theoretically the true error $e_1(t)$ and the auxiliary error $e_2(t)$ can grow in an unbounded fashion even while their sum is bounded. That this cannot occur was first proved for the continuous time problem in [1] and [2] by using the additional feedback signal $-\varepsilon_1(t)\zeta^T(t)\Gamma\zeta(t)$. This major step establishes the adaptive controller described so far as the first efficient globally stable one in the adaptive literature. In view of its theoretical importance, we shall attempt to describe qualitatively some of the ideas used in proving it.

PROOF OF STABILITY. Given that the augmented error $\varepsilon_1(t)$ is uniformly bounded and that the adaptive law (50) is used our objective is to show that the plant outputs $y_p(t)$ and $v(t)$ as well as the vector $\zeta(t)$ are also uniformly bounded. This is achieved by assuming $\zeta(t)$ to be unbounded and demonstrating that this results in a contradiction.

Figure 12 shows the relation between the plant feedback loop and the auxiliary error model. From section IV it can be concluded that $\varepsilon_1(t)$ and $\phi(t)$ will be bounded and that $\dot{\phi}(t) \in L^2$. This implies that the signal $\nu(t)$ in Figure 12 is such that $W_M(s)L(s)\nu(t) = \varepsilon_1(t)$ is uniformly bounded. The output of the plant $y_p(t)$ can be shown to be of the form

$$y_p(t) = W_M(s)r(t) + W_M(s)L(s)[L^{-1}(s)\phi(t)L(s)]\zeta(t) \tag{52}$$

which can be expressed as $W_M(s)r(t) + W_M(s)L(s)v(t) + v_1(t)$ where $v_1(t)$ is due to feedback signals involving a gain $\dot{\phi}(t)$. Since the

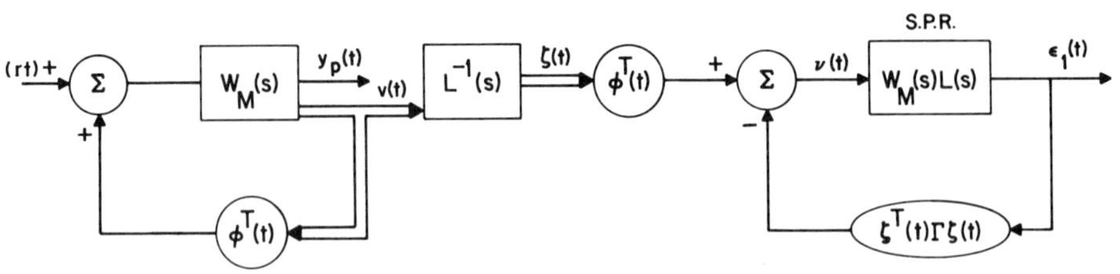

FIGURE 12. Plant feedback loop and error model.

first two terms are known to be uniformly bounded, the boundedness of $y_p(t)$ depends on the nature of $v_1(t)$. However it can be shown that a feedback system with an exponentially stable linear system in the forward path and a gain which belong to the space L^2 in the feedback path is asymptotically stable. Since $\dot{\phi}(t) \in L^2$ we conclude that $v_1(t)$ is uniformly bounded and hence $y_p(t)$ is also uniformly bounded. The fact that $\dot{\phi}(t) \in L^2$ is seen to play a central role in the proof of stability and accounts for the importance of the feedback term used in the error model.

DISCRETE SYSTEMS. The results derived so far for continuous time systems can also be derived for discrete time systems using the corresponding error models described in section IV. In [35] Ionescu and Monopoli presented a discrete analog of the adaptive controller in [16]; the discrete model however contained the additional feedback term. Recently Narendra and Lin [32] studied the behavior of this system and established its global stability. This analysis also led to the proof of stability of the continuous time problem in [1].

Proof of Stability (Discrete Case).

(i) Narendra and Lin [32]. In Figure 11, if continuous time functions are replaced by discrete time functions, and z-transforms are used in place of Laplace transforms the discrete solution to the adaptive control problem is obtained. While the proof of stability outlined for the continuous case applies to the discrete case also the following, which was first suggested in [32], is somewhat simpler.

In the third error model in Figure 12, $\varepsilon_1(k)$ and $\nu(k) \to 0$ as $k \to \infty$ and

$$\phi^T(k)\zeta(k) - \varepsilon_1(k)\zeta^T(k)\Gamma\zeta(k) = \nu(k) \to 0 \tag{53}$$

Since $\Delta\phi(k) = -\alpha\Gamma\varepsilon_1(k)\zeta(k) \to 0$, equation (53) implies

$$\phi(k)^T(k)\zeta(k) = o[\|\zeta(k)\|] \tag{54}$$

$\Delta\phi(k) \to 0$ as $k \to \infty$ implies that $\phi(k)$ is almost a constant for large k. This fact together with (54) assures that $\phi^T(k)v(k) = o[\sup_{i\le k} \|v(i)\|]$ in the plant feedback loop and hence $y_p(k)$ is uniformly bounded.

(ii) Goodwin, Ramadge and Caines [33]. If $\{u(k)\}$ and $\{y(k)\}$ denote the input-output sequences of a plant and

$$A(z^{-1})y(k) = z^{-d}B(z^{-1})u(k)$$

where A and B are polynomials in z^{-1}, d a specified time delay and $\{y^*(k)\}$ a given reference sequence the objective in [33] is to design a feedback control law which will stabilize the system and cause the output y(k) to track $y^*(k)$ asymptotically so that

$$\lim_{k\to\infty} y(k) - y^*(k) = 0. \tag{55}$$

The problem is first cast in the form

$$y(k+d) = \alpha(z^{-1})y(k) + \beta(z^{-1})u(k) \tag{56}$$

where the coefficients of the polynomials α and β are related to those of A and B. This is referred to as the d-step ahead linear prediction method. The objective now is to determine at stage k the estimates of the coefficients of α and β such that the error between the predicted value of y(k+d) and the desired value $y^*(k+d)$ is minimized.

It was mentioned in section VI that in [28] it was shown that for a specific parametrization of the plant indirect and direct control lead to identical error equations and are hence equivalent. The structure of such an indirect controller is reproduced from [28] in Figure 13 and can be used for both discrete and continuous time control. For application to a discrete plant F_1 and F_2 can be chosen to be tapped delay lines and L(z) can be chosen so that LM = 1 where M is the model transfer function. It is further assumed that the gain of the plant k_p is known as discussed in section VI. To apply the above model to the problem considered in [33] we make the following substitutions: $M(z) = z^{-d}$; $L(z) = z^{d}$; $r(k) = y^*(k+d)$ is the reference input, F_1 and F_2 are tapped delay lines $c^T = \beta^T, (d_0, d^T) = \alpha^T$ and obtain from Figure 13 directly the control input as

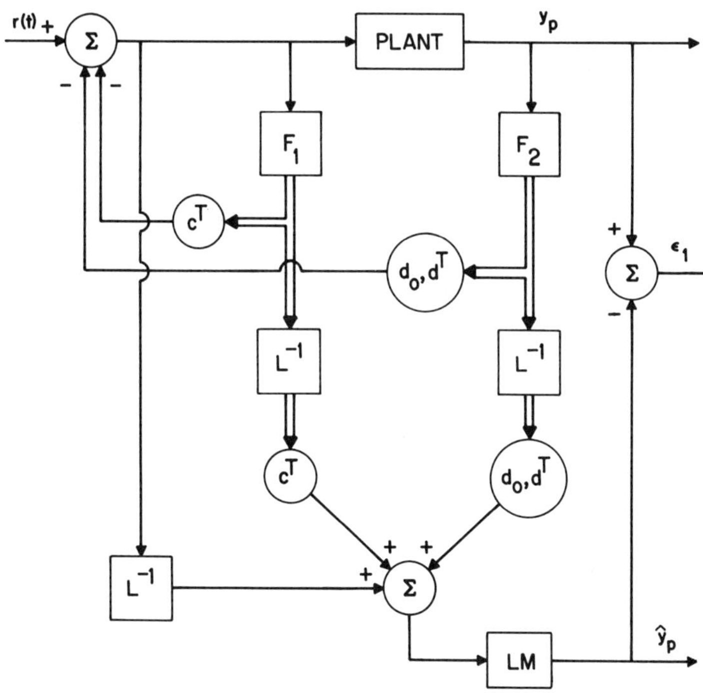

FIGURE 13. Indirect control.

$$u(k) = y^*(k+d) - [\alpha(k),\beta(k)]^T w(k)$$

where $w(k) = [y(k),..,y(k-n+1),u(k),...,u(k-m-d+1)]$. The adaptive laws for updating α and β are

$$\Delta\theta(k) = \begin{bmatrix} \Delta\alpha(k) \\ \Delta\beta(k) \end{bmatrix} = + \frac{\gamma\varepsilon_1(k)w(k-d)}{1 + w(k-d)^T w(k-d)} \qquad 0 < \gamma < 2$$

which is identical to that obtained in [33] using the first method.

The second method in [33] can be shown in a similar fashion to be identical to the direct control approach used in [32] with reference model transfer function z^{-d} and reference input $y^*(k)$. [However, the authors of [33] use a d-step interlacing recursive algorithm, which can be further simplified to yield the same results that would be given by [32].] This is not surprising since the two methods in [33] are equivalent and the results in [28] and [32] are equivalent as indicated in [28].

From the above it is clear that the MRAC approach and the STR approach to stable adaptive control of discrete systems are equivalent.

VII. COMMENTS AND CONCLUSIONS

In the previous sections some recent developments in the design of stable adaptive controllers for single-input single-output Model Reference Adaptive Systems were considered in detail. While the theoretical results presented provide a sound basis for further exploration, many questions have to be resolved before adaptive control theory emerges as a truly powerful tool. In this section we comment briefly on some of the principal features of MRAC and relevant questions which require further investigation.

a) MRAC and STR

Model Reference Adaptive Control and Self Tuning Regulators represent two broad areas of adaptive control and for a long time advances in each area were made independently. In MRAC the first step is the choice of a reference model; this is followed by the selection of a suitable controller structure. In STR a design procedure is first chosen which can be used when the plant parameters are known and this is applied to the unknown plant using recursively estimated values of these parameters. MRAC has tended to proceed from considerations of stability of the overall system to optimization of control parameters while interest in STR has to a large extent been in the opposite direction. During the last year the two approaches have tended to come together and researchers in both areas are more aware of the interrelation between them. Efforts have also been made to show that the two schemes are in some sense "equivalent". In [36], [39] and [42] the basic algorithms for MRAC and STR are shown to be the same. In [28] it was shown that direct and indirect control of an unknown plant lead to identical error equations. More recently, after the stability results described in section VI were established, it has also been recognized that the assumptions made, the structure of the controller used, the error equations which result and the arguments for global stability are similar in the two approaches.

The entire literature on STR deals with discrete systems while research interest in MRAC has always been in both discrete and continuous cases. However, STR is applicable to a wider class of problems though the stability of such schemes may be harder to prove.

b) Reference Model and Reference Input

As mentioned in section (a), in the design of MRAC it is assumed that a reference model and a reference input are specified. In many practical applications it is well-known that the specification of a suitable model is a difficult task. Hence MRAC can be applied efficiently only in those cases where such a model is already available and satisfies all the conditions specified in section VI.

In many cases the designer may have little control over the choice of the reference input and hence on the "richness condition" stated in section IV. If the reference input is identically zero we have a regulator problem and the choice of the model is not critical provided its transfer function has a relative degree greater than or equal to that of the plant.

c) Separation Principle

In linear feedback systems the use of state estimation followed by state feedback control has proved invaluable. In many adaptive control schemes an attempt is also made to use a similar separation principle either explicitly or implicitly. Since estimates of parameters are invariably involved, all of them result in nonlinear control systems. The problem of stable adaptive control is then to determine which of these different schemes would be globally stable. So far, only those schemes in which the control and estimation parameter errors are linearly related have been shown to be globally stable. Other schemes in which the control parameter and plant parameter errors are nonlinearly related have given rise to intractable stability problems. The importance of globally stable adaptive schemes is that they are more likely to perform satisfactorily under real conditions involving noise, nonlinearities, time-varying plant parameters and perhaps most important of all due to truncation errors resulting from mismatching between plant and model.

Attempts have been made in the past [37],[30] to identify the plant parameters and use linear feedback to control it. As mentioned earlier, it is argued that if the input to the plant is sufficiently rich the estimates of the plant parameters will converge to their true values and hence the control parameters will, in turn, converge to the desired values. However, all of them suffer from the same shortcomings discussed in section V. The most promising effort in this direction is made by Kreisselmeier who, using a stable state and parameter observer, computes a feedback matrix. This matrix together with the observed state is shown to be asymptotically stable in the sense of Lyapunov provided that the input to the plant is sufficiently excited by means of an external command signal. But in [30] only how such an excitation can be assured locally is given and global excitation is left as an open question.

The only indirect (continuous time) control system known at the present time to be globally stable is the one discussed in [28] provided the additional feedback term described in section IV is used. This nonlinear system is globally stable even when the reference input is identically zero. It is the authors' opinion that the requirement of persistent excitation may be a serious limitation since in most practical cases the input is rarely sufficiently rich.

d) Assumptions for Stable Adaptive Control

In section VIb the minimum information needed for the design of a stable adaptive controller for an unknown single-input single-output plant was given. Condition (iii) implies that in its present form the theory cannot be extended to distributed parameter systems. Condition (i) is perhaps the most restrictive of all the assumptions made and seriously limits the applicability of the theory to practical problems. Since the cancellation of the zeros of the plant transfer function is used to match the system zeros with those of the model the plant zeros must be located in the left half plane as required by (v). If, however, the plant zeros are known exactly and the model transfer function is chosen to have the same zeros, adaptive pole placement may be theoretically possible. However in practice the resulting system is usually not globally stable. Suggestions have been made to continuously estimate the zeros of the plant and adjust the zeros of the model accordingly. However, no proof of global stability for this scheme exists at present.

e) Multivariable Systems

The extension of the results presented in this paper to the multivariable control problem is probably the next most important question that has to be resolved in this area. As mentioned in section II every adaptive control problem consists of two parts - the algebraic part concerned with the realizability of a suitable controller structure and the analytic part which deals with the generation of stable adaptive laws. While both aspects are complex in the multivariable case the analytical part appears to be relatively simpler. If the model transfer matrix $W_M(s)$ and its inverse have poles in the left half plane the analytic part developed for the single variable case appears to carry over to the multivariable case. However, a precise solution to the algebraic part of the multivariable problem is not currently available. Perhaps the first step in this direction would be to derive conditions under which a controller can be determined so that the transfer matrix of a known plant together with the controller would match that of a known model. Monopoli and Hsing [38], Goodwin, Ramadge and Caines [33] have made the first attempts to extend methods developed for single variable systems to the multivariable case. [38] treats the continuous time case while [33] deals with discrete time systems and both appear to apply to restricted classes of multivariable systems.

f) Adaptive Control with Noise

The MRAC systems described so far are both nonlinear and time-varying and the analysis of even the deterministic control problem is quite complex. Needless to say, a complete analysis of the stability problem in the presence of noise is significantly more difficult, though such efforts are currently underway. In the work that has been in progress on self-tuning regulators there has been an interest in the control of noisy systems from the very beginning and the work of Ljung in this area is well-known. In view of the growing interaction between MRAC and STR we can naturally expect attempts in the near future to obtain similar results for MRAC as well.

The presence of noise in dynamical systems may be due to many sources such as input noise, measurement noise, discretization error, truncation error etc. Since in the control problem the output of the plant is processed to generate the control input, the presence of output noise results in biases in the estimates of the plant parameters. Hence if the same control laws as in the noise free case are used the feedback system can become unstable [41]. To reduce or eliminate the effect of noise the controller structure as well as the adaptive laws may have to be modified. Prefiltering the signals used in the adaptive laws and reducing the adaptive gains are both being currently investigated. For example, Landau [36] has suggested various structures for adaptive observers and controllers to reduce the effect of noise; Egardt [39] has shown a method for using time-varying adaptive gains which do not go to zero to assure that the plant input and output will be uniformly bounded provided the control parameters can be assumed to be uniformly bounded. Ljung [40], [41] has provided a general tool to analyze the asymptotic behavior of recursive stochastic algorithms through the analysis of related ordinary differential equations.

In view of the insight that was gained about adaptive observers and controllers through the use of error models it is the authors' opinion that similar error models should be set up and analyzed for the noisy case as well. A first attempt in this direction has been made by Bitmead and Weiss and Mitra as indicated earlier in section IV.

g) Imperfect Model Matching, Global Stability, Local Stability

The emphasis throughout this paper has been on perfect model matching and the global stability of the adaptive control system. For perfect model matching the adaptive control systems described result in $y_p(t) - y_M(t) \overset{\Delta}{=} e_1(t) \to 0$ as $t \to \infty$ for all reference inputs $r(t)$. The restrictive nature of the assumptions in section (c) may be attributed to this stringent requirement on $e_1(t)$. It is the authors' opinion that emphasis in the future should be on the design of adaptive controllers which merely result in bounded error $e_1(t)$. Such controllers, it is felt would be applicable to a much wider range of problems than those described in earlier sections. However even a precise statement of the imperfect model matching problem is not known at the present time.

Global stability implies that for any arbitrary parameter error vector at time t_0 the state as well as the parameter error are uniformly bounded for all $t \geq t_0$. In practice a priori information is also usually available regarding the region R in which the initial parameter error and state error vector may lie. In such cases stability in the region R rather than global stability may be all that is required. The methods for realizing such practically stable schemes are also currently not available.

h) Other Aspects of Adaptive Control

To design adaptive controllers for practical systems, it is essential that a number of related topics such as effect of nonlinearity, transient response, rate of convergence, etc. be well understood. While the transient behavior and rate of convergence of adaptive observers have been studied to some extent [7-12] very little is known about such characteristics of adaptive control systems. Further adaptive control is mainly required in those cases where the plant parameters vary with time. However the controllers available at the present time were developed for unknown time invariant plants. How effective these schemes will prove for the control of time-varying plants has yet to be investigated.

REFERENCES

1. Narendra, K. S., Lin, Y-H., Valavani, L. S., "Stable Adaptive Controller Design - Part II: Proof of Stability," S&IS Report No. 7904, Yale Univ., April 1979, to appear in IEEE Trans. Auto. Contr. , June 1980.

2. Morse, A. S., "Global Stability of Parameter-Adaptive Control Systems," S&IS Report No. 7902R, Yale University, March, 1979, to appear in IEEE Trans. Auto. Contr., June 1980.

3. Astrom, K. J., "Self-Tuning Regulators-Design Principles and Applications," Proceedings of the Workshop on Applications of Adaptive Control, Yale University, Aug., 1979.

4. Popov, V. M., "Hyperstability of Control Systems," Springer-Verlag, New York, 1973.

5. Narendra, K. S., "Stable Identification Schemes," "System Identification: Advances and Case Studies," Academic Press, New York, 1976.

6. Lin, Y-H., and Narendra, K. S., "A New Error Model for Discrete Systems and its Application to Adaptive Identification and Control," S&IS Report No. 7802, Yale University, Oct., 1978, to appear in IEEE Trans. Auto. Contr., June 1980.

7. Yuan, J.S-C., and Wonham, W. M., "Probing Signals for Model Reference Identification," IEEE Trans. Auto. Control, Vol. AC-22, No. 4, pp. 530-538, Aug., 1977.

8. Morgan, A. P., and Narendra, K. S., "On the Uniform Asymptotic Stability of Certain Linear Nonautonomous Differential Equations," SIAM J. Contr. and Opt., Vol. 15, No. 1, pp. 5-24, Jan., 1977.

9. Sondi, M. M., and Mitra, D., "New Results on the Performance of a Well-Known Class of Adaptive Filters," Proc. IEEE, Vol. 64, No. 11, pp. 1583-1597, 1976.

10. Weiss, A., and Mitra, D., "Digital Adaptive Filters: Conditions for Convergence, Rates of Convergence, Effects of Noise and Errors Arising from the Implementation," to be published.

11. Widrow, B., McCool, J. M., Larimore, M. G., and Johnson, C. R., Jr., "Stationary and Nonstationary Learning Characteristics of the LMS Adaptive Filters," Proc. IEEE, Vol. 64, No. 8, pp. 1151-1162, Aug., 1976.

12. Morgan, A. P., and Narendra, K. S., "On the Uniform Asymptotic Stability of Certain Linear Time-Varying Differential Equations with Unbounded Coefficients," S&IS Report No. 7807, Yale Univ., Nov., 1978.

13. Bitmead, R. R., and Anderson, B.D.O., "Exponentially Convergent Behaviour of Simple Stochastic Estimation Algorithms," Proc. 17th IEEE Conf. on Decision and Control, San Diego, 1979.

14. Bitmead, R. R., "Convergence Properties of Discrete-Time Stochastic Adaptive Estimation Algorithms," Ph.D. Thesis, the University of Newcastle, Australia, April, 1979.

15. Landau, I. D., and Silveira, H. M., "A Stability Theorem with Application to Adaptive Control," IEEE Trans. Auto. Contr., Vol. AC-24, No. 2, pp. 305-311, April, 1979.

16. Monopoli, R. V., "Model Reference Adaptive Control with an Augmented Error Signal," IEEE Trans. Auto. Contr., Vol. AC-19, pp. 474-484, Oct. 1974.

17. Monopoli, R. V., and Subbarao, V. N., "A Simplified Algorithm for Model Reference Adaptive Control," ECE-SY-79-2, University of Massachusetts, Amherst, Ma., June, 1979.

18. Morgan, A. P., and Narendra, K. S., "On the Stability of Nonautonomous Differential Equations $\dot{x} = [A+B(t)]x$ with Skew-Symmetric Matrix B(t)," SIAM J. of Contr. and Opt., Vol. 15, No. 1, pp. 163-176, Jan. 1979.

19. Luders, G., and Narendra, K. S., "An Adaptive Observer and Identifier for a Linear System," IEEE Trans. Auto. Contr., Vol. AC-18, pp. 496-499, Oct. 1973.

20. Kreisselmeier, G., "Adaptive Observers with Exponential Rate of Convergence," IEEE Trans. Auto. Contr., Vol. AC-22, No. 1, pp. 2-8, Feb., 1977.

21. Carroll, R. L., and Lindorff, D. P., "An Adaptive Observer for Single-Input Single-Output Linear Systems," IEEE Trans. Auto. Contr., Vol. AC-18, pp. 428-435, Oct., 1973.

22. Luders, G., and Narendra, K. S., "A New Canonical Form for an Adaptive Observer," IEEE Trans. Auto. Contr., Vol. AC-19, pp. 117-119, April, 1974.

23. Narendra, K. S., and Kudva, P., "Stable Adaptive Schemes for System Identification and Control - Part I, Part II," IEEE Trans. on System, Man and Cybernetics, Vol. SMC-4, pp. 541-560, Nov., 1974.

24. Anderson, B.D.O., "An Approach to Multivariable System Identification," Automatica, Vol. 13, pp. 401-408, 1977.

25. Kim, C., "Convergence Studies for an Improved Adaptive Observer," Ph.D. Thesis, Univ. of Connecticut, Storrs, Ct., 1975.

26. Landau, I. D., "Unbiased Recursive Identification Using Model Reference Adaptive Techniques," IEEE Trans. Auto. Contr., Vol. AC-21, No. 2, April, 1976.

27. Dugard, L., Landau, I. D., and Silveira, H. M., "Adaptive Estimation Using MRAS Techniques, Convergence Analysis and Evaluation," Note LAG N° 79-06, Universite de Grenoble, France, Mars, 1979.

28. Narendra, K. S., and Valavani, L. S., "Direct and Indirect Model Reference Adaptive Control," Automatica, Vol. 15, pp. 653-664, 1979.

29. Kreisselmeier, "Algebraic Separation in Realizing a Linear State Feedback Control Law by Means of an Adaptive Observer," Technical Report, Institut fur Dynamik der Flugsysteme, Oberpfaffenhofen, F. R. Germany, 1979, to appear in IEEE Trans. Auto. Control.

30. Kreisselmeier, G., "Adaptive Control via Adaptive Observation and Asymptotic Feedback Matrix Synthesis," Technical Report, Institut fur Dynamik der Flugsysteme, Oberpfaffenhofen, F. R. Germany, 1979.

31. Narendra, K. S., and Valavani, L. S., "Stable Adaptive Observers and Controllers," Proc. IEEE, Vol. 64, pp. 1198-1208, Aug., 1976.

32. Narendra, K. S., and Lin, Y-H., "Stable Discrete Adaptive Control," S&IS Report No. 7901, Yale University, March, 1979, to appear in IEEE Trans. Auto. Contr., June 1980.

33. Goodwin, G. C., Ramadge, P. J., and Caines, P. E., "Discrete Time Multi-Variable Adaptive Control," Technical Report, Harvard University, Nov., 1978, to appear in IEEE Trans. Auto. Contr., June 1980.

34. Narendra, K. S., and Valavani, L. S., "Stable Adaptive Controller Design - Direct Control," IEEE Trans. Auto. Contr., Vol. AC-23, No. 4, pp. 570-583, Aug., 1978.

35. Ionescu, T., and Monopoli, R. V., "Discrete Model Reference Adaptive Control with an Augmented Error Signal," Automatica, Vol. 13, No. 5, pp. 507-518, Sept. 1977.

36. Landau, I. D., "Model Reference Adaptive Control and Stochastic Self Tuning Regulators - Towards Cross Fertilization," Note interne LAG 79.13, Universite de Grenoble, France, Juin, 1979.

37. Elliot, H., and Wolovich, W. A., "Parameter Adaptive Identification and Control," IEEE Trans. Auto. Contr., Vol. AC-24, No. 4, Aug., 1978.

38. Monopoli, R. V., and Hsing, C. C., "Parameter Adaptive Control of Multivariable Systems," Int. J. Control, Vol. 22, No. 3, pp. 313-327, 1975.

39. Egardt, B. O., "Stability of Model Reference Adaptive and Self-Tuning Regulators," Technical Report, Department of Automatic Control, Lund Institute of Technology, Dec., 1978.

40. Ljung, L., "Analysis of Recursive Stochastic Algorithms," IEEE Trans. Auto. Contr., Vol. AC-22, No. 4, pp. 551-575, Aug., 1977.

41. Ljung, L., "On Positive Real Transfer Functions and the Convergence of Some Recursive Schemes," IEEE Trans. Auto. Contr., Vol. AC-22, No. 4, pp. 539-551, Aug., 1977.

42. Johnson, C. R., Jr., "Input Matching, Error Augmentation, Self-Tuning, and Output Error Identification: Algorithmic Similarities in Discrete Adaptive Model Following," to appear in IEEE Trans. Auto. Contr.

AN OVERVIEW OF OPTIMUM-ADAPTIVE CONTROL IN SONAR ARRAY PROCESSING

N. L. Owsley

Naval Underwater Systems Center
New London, Connecticut

As an important example of adaptive signal processing in sonar, the subject of optimum (minimum variance) and adaptive arrays for passive acoustic reception is discussed. The issues of basic performance, multiple constraint minimization, minimum complexity configuration and adaptive realization alternatives are addressed. The passive sonar adaptive array processing problem is presented as a case study of adaptive system design wherein achievable performance and potential system complexity suggest that a suboptimum configuration is indeed the most practical. As a second example, the subject of constrained maximization as a control criteria is considered in the context of some fundamental sonar signal processing objectives.

ISBN 0-12-514060-6

INTRODUCTION

The theory of optimum, minimum variance processing for sonar receiving arrays first appeared in the literature in the early 1960's [1,2]. Techniques for the adaptive realization of the optimum, discrete element sonar array processor in an environment of time-varying and directional noise received considerable attention in the mid to late 1960's [3-5]. In recent years, the interest in adaptive antenna arrays within the sonar community has focused on the practicalities of implementation. The prospect of fabricating multibeam sonars that utilize 100 or more hydrophones has forced a trend to suboptimum adaptive-array processor design that is typified by hardware efficiency while maintaining near optimum performance [6]. For most directional acoustic noise environments, it has been recognized that array processors need not be fully adaptive, in the sense that the outputs of each hydrophone in an array be adaptively filtered. Rather, partially adaptive systems that are properly constrained with respect to preserving the signal in the main lobe seem to offer near-optimum processor performance compatible with minimal hardware [7]. In fact, these suboptimum adaptive array processors have a firm basis in the optimal, minimum variance (MV) processor but bear little resemblance to the configuration implicit in the so-called fully adaptive array. Such systems have been termed either null-steering or noise-cancelling devices depending on the specific type of coherent noise process to be discriminated against. In this paper, the fully optimum adaptive MV array processor is derived and its performance for both a plane wave, point source interference and noise of limited spatial coherence length is investigated. The basis for a null-steering system is established and performance in a typical sonar situation is considered.

Several alternatives exist in both the configuration and control of optimum-adaptive MV array processors in sonar. Certainly it is appropriate to present the issues involved to the experience of adaptive control specialists in the hope of obtaining some guidance in these areas. The first issue involves the configuration of a minimum complexity system, in terms of the number of adaptive array filter elements, required to suboptimally process an acoustic environment of a given intrinsic dimensionality. The second issue presented is that of closed-loop adaptive control as typified by the host of gradient descent schemes available [9] versus open-loop control using either matrix inversion [10] or the matrix inversion lemma [11]. Finally, selection of either the time domain or the frequency domain for implementation of the adaptive control process is discussed. Certain computational efficiencies are accrued in the frequency domain [12,13]. However, steady state misadjustment, response time and energy spreading effects must also be considered [14]. Of course, other issues exist in such areas as the application of constraints to the control process. However, the three presented above are considered important enough to warrant attention.

With respect to new applications of adaptive control techniques to underwater acoustics, the use of adaptive principle component analysis will be mentioned. It is proposed to use these techniques for general signal field decomposition [15] array focusing [16] coherent multipath recombination [17], high resolution frequency (or wave-number) analysis [18,19] and signal extraction from a time-frequency data history by techniques such as dynamic programming.

THE OPTIMUM MINIMUM VARIANCE (MV) ARRAY PROCESSOR

In this section the optimum processor, in a minimum variance (MV) sense, for a discrete element receiving array is derived. This MV array processor, referred to as the MV beamformer, is derived for a rather general signal spatial covariance matrix. The performance of the MV beamformer is considered for the case of a single plane wave interference in additive uncorrelated noise and for the case of partially coherent noise with finite angular extent. A suboptimum configuration suggested by the MV beamformer which uses only the outputs of a conventional delay and sum (DS) beamformer is also considered. Examples of practical situations in sonar are presented.

Since the remainder of this paper relies heavily on matrix notation, certain matrix definitions and operations must be introduced. All upper-case letters are matrices, and the underlined upper-case letters are column matrices (vectors). Lower-case letters are scalars. Upper and lower case indices and parameters are specifically defined. The superscripts "*", "T", "'" and "-1" indicate, respectively, the matrix operations of conjugation, transposition, conjugate transposition, and inversion.

Let the output from an N-element discrete array at frequency f be given by the vector

$$\underline{X}(f)^{T} = [x_1(f)x_2(f)\ldots x_N(f)]. \tag{1}$$

The frequency domain output from the k-th sensor is hereafter referred to as

$$x_k(f) = \int_{-\infty}^{\infty} dt\ x_k(t)\ \exp[-j2\pi ft]. \tag{2}$$

In (2), $x_k(t)$ is the stationary, zero-mean temporal output from the k-th sensor. The cross-spectral density matrix for the array is herein defined as

$$R(f) = E[\underline{X}(f)\underline{X}(f)'] \tag{3}$$

$$= \sigma_s^2(f)P(f) + \sigma_n^2(f)Q(f) \tag{4}$$

where E[·] indicates the statistical expectation operation. In addition, P(f) and Q(f) are the signal and noise cross-spectral density matrices, normalized so that Trace [P] = Trace [Q] = N. The input signal and noise spectral densities are $\sigma_s^2(f)$ and $\sigma_n^2(f)$ respectively.

An N-element vector filter $\underline{H}(f)$ to be applied to the array data at frequency f is now defined as

$$\underline{H}(f)^T = [h_1(f)h_2(f)\ldots h_N(f)] \tag{5}$$

where

$$h_k(f) = \int_{-\infty}^{\infty} dt\ h_k(t)\ \exp[-j2\pi ft] \tag{6}$$

is the transfer function for the k-th channel and $h_k(t)$ is the corresponding impulse response. The beamformer output for this vector filter is

$$y(f) = \underline{H}'(f)\underline{X}(f). \tag{7}$$

In the following, the frequency f is deleted in all expressions unless it is needed explicitly. Let the signal cross-spectral density matrix P be a rank L matrix expressed in diagonalized form as

$$P = DAD' \tag{8}$$

where D is an N-by-L matrix with L<N orthogonal vectors $\underline{D}_m$, m = 1,2,...L as columns which span the L-dimensional signal subspace Ω_s. A is a real diagonal matrix with mth element a_m (Trace [A] = 1) giving the relative level of signal power associated with the mth minimum basis vector $\underline{D}_m$ ($|\underline{D}_m|^2 = N$) in the signal subspace. It is now desired to form a minimum variance beam which is defined to minimize the beamformer output power

$$E[|y|^2] = \underline{H}'R\underline{H} \tag{9}$$

subject to the constraint

$$D'\underline{H} = A^{1/2}\underline{1} \tag{10}$$

where $\underline{1}$ is a L-by-1 column vector of ones. Since equation (10) maintains the beamformer response at the proper level for each of the signal components, then minimization of (9) subject to (10) results in a minimization of only the noise contribution to (9). Accordingly, the real functional

$$z = \underline{H}'R\underline{H} + \underline{\lambda}_R^T(D'\underline{H}+\underline{D}'\underline{H}^*-2A^{1/2}\underline{1}) + J\underline{\lambda}_I^T(D'H-D'H^*) \tag{11}$$

can be minimized with respect to the real and imaginary parts of the filter vector $\underline{H}$ where $\underline{\lambda} = \underline{\lambda}_R + j\underline{\lambda}_I$ is a L-element complex Lagrange multiplier vector. The minimum of (11) is found by solving the set of N simultaneous complex equations given by the gradient

$$\frac{\partial z}{\partial \underline{H}} = 2R\underline{H}_o - 2D\underline{\lambda}^* \tag{12}$$

$$= \underline{o} \tag{13}$$

to yield

$$\underline{H}_o = R^{-1}D\underline{\lambda}^*. \tag{14}$$

Application of the constraint (10) to (14) gives

$$\underline{\lambda}^* = [D'R^{-1}D]^{-1}A^{1/2}\underline{1}. \tag{15}$$

Thus the optimum beamformer filter vector for the signal covariance matrix P = DAD' is

$$\underline{H}_o = R^{-1}D[D'R^{-1}D]^{-1}A^{1/2}\underline{1}. \tag{16}$$

The SNR gain at the beamformer output relative to a single sensor for the optimum MV beamformer can be expressed as

$$G_o = \frac{\underline{H}_o'P\underline{H}_o}{\underline{H}_o'Q\underline{H}_o} \tag{17}$$

which, in turn, can be evaluated in general by using (16) in (17). However, some special cases are of particular interest. In particular, the case of a single perfectly coherent signal wavefront is considered wherein $P = \underline{D}(\theta)\underline{D}'(\theta)$. Two cases of additive noise are examined in the following.

Plane Wave Interference with Spatially Uncorrelated Noise [20,21]

For this case

$$P = \underline{D}(\theta)\underline{D}'(\theta) \tag{18}$$

and

$$Q = \alpha \underline{D}(\phi)\underline{D}'(\phi) + (1-\alpha)I \tag{19}$$

where $(0 \leq \alpha < 1)$ specifies the relative amounts of noise at a sensor contained in either the plane wave interfering noise arriving from angle ϕ or the noise component which is of uniform level from sensor to sensor. Invoking the matrix inversion identity

$$[USV'+B]^{-1} = B^{-1} - B^{-1}UTV'B^{-1} \tag{20}$$

with

$$T = (V'B^{-1}U+S^{-1})^{-1}, \tag{21}$$

the MV beamformer filter for this interference and noise configuration can be shown to be

$$\underline{H}_o = g_1\underline{D}(\theta) - g_2\underline{D}(\phi). \tag{22}$$

In (22), g_1 and g_2 are real and complex scalars respectively given by

$$g_1 = \left(N \left[1 - \frac{|r|^2}{1 + (1/i)} \right] \right)^{-1}$$

$$g_2 = g_1 \frac{r^*}{1 + (1/i)}$$

such that

$$r = \underline{D}'(\phi)\underline{D}(\theta)/N \quad (|r| < 1)$$

$$i = N\alpha/(1-\alpha) . \tag{23}$$

The quantity r is the response of a conventional delay and sum (DS) beamformer steered at the signal angle θ to an interference wavefront of amplitude 1/N arriving from angle ϕ. Also, i is the ratio of interfering noise power to the uncorrelated noise power in the output of a DS beamformer steered directly at the interference angle ϕ. The physical interpretation of the process dictated by (22) is quite simple. Namely, (22) indicates that a conventional DS beam is formed at a desired angle θ to receive the signal. A second DS beam is then formed at an angle ϕ which is the arrival angle of the interference. The two beam outputs are then scaled and the interference directed beam output is time delayed before subtracting it from the output of the beam steered at the signal. The result is the phase coherent subtraction of a portion of the interference waveform which appears in the signal DS beam output as a result of the finite level of the signal DS beam response r in the direction of the interference wavefront arrival. The improvement in the ratio of signal power to total background noise power due to plane wave interfering noise and uncorrelated noise for the MV beamformer relative to a DS beamformer is called the gain improvement ratio

$$G = G_o/G_c \tag{24}$$

$$= \frac{\underline{H}_o'P\underline{H}_o/\underline{H}_o'Q\underline{H}_o}{\underline{D}'(\theta)P\underline{D}(\theta)/\underline{D}'(\theta)Q\underline{D}(\theta)} \tag{25}$$

$$= 1 + \frac{i^2}{1+i}\,|r|^2(1-|r|^2). \tag{26}$$

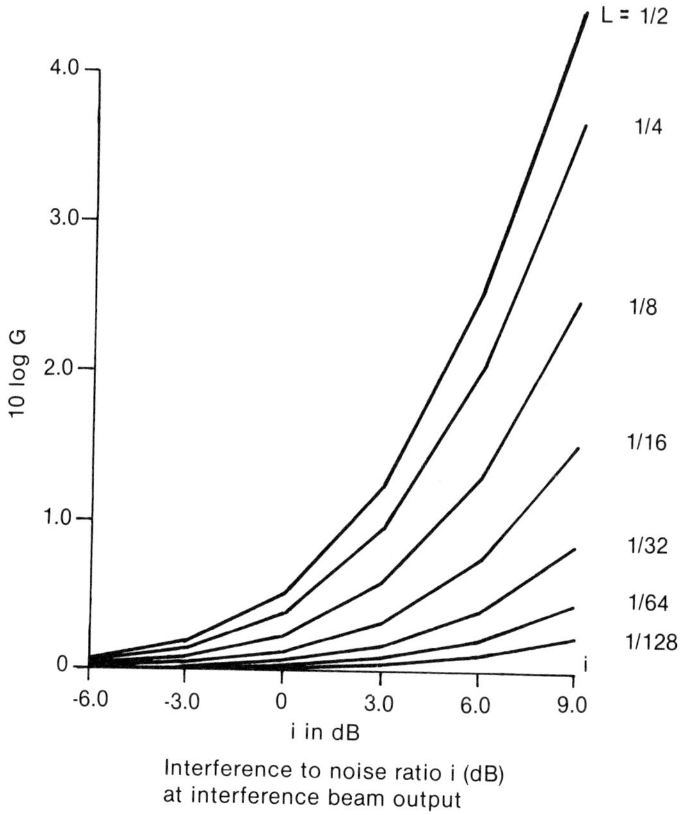

FIGURE 1. Array gain improvement ratio (G) of a MV relative to a DS beamformer as a function of sidelobe level $L = |r|^2$ and interference-to-noise ratio (i) measured at the output of a DS beam steered at the interference.

The array gain improvement ratio G of a MV beamformer relative to a DS beamformer is plotted in Figure 1 for some representative values of $L = |r|^2$ and i. It is noted that unless the interference level is at least 0 dB relative to the uncorrelated noise in the output of the DS beam steered at the interference ($i \geq 1$), then little gain results. G is a maximum with respect to $|r|^2$ for

$|r|^2 = 1/2$, i.e., when the interference arrival angle is at the 3 dB point of a DS beam steered at the signal.

Partially Coherent Noise [22,23]

In many cases of interest in sonar, multipath interference and random medium effects limit the spatial extent to which a propagating wavefront exhibits coherence. Such partially coherent wavefronts represent a type of sonar array received noise. This noise can be envisioned at the array as an angularly extended source. A particularly simple one dimensional model for this type of spatially stationary, partially coherent noise is characterized by the coherence function

$$q(x) = \exp[-\beta|x|] \tag{27}$$

where x is the separation of two points on a line array of sensors. The spatial frequency (ν) power spectrum for this noise is

$$b(\nu) = \int_{-\infty}^{\infty} q(x)\exp[-j\nu|x|]dx \tag{28}$$

$$= \frac{2\beta}{\nu^2 + \beta^2} \tag{29}$$

which can be recognized as the spectrum for a first order spatial Markov process with decorrelation distance $1/\beta$ and a half-power extent of 2β. For a line array of N sensors equally spaced at a distance d, the noise coherence between the ℓ-th and m-th sensor is

$$[Q]_{\ell,m} = \exp[-\beta d|\ell-m|] \tag{30}$$

where now ρ is defined as $\rho = \exp[-\beta d] < 1$. In (30) $[Q]_{\ell,m}$ is the ℓ,m-th element in the noise covariance matrix Q of (4). As in the previous case, the plane wave signal with covariance matrix given by (18) is assumed. For this case the m-th element of the signal wavefront direction vector $\underline{D}(\theta)$ is $\exp[-j2\pi fd(m-1)\sin\theta/c]$ where c is the speed of propagation of sound in water. The array gain for the MV and DS beamformers can be shown to be

$$G_o(\nu_s,\rho,N) = \frac{2[1-\rho\cos(\nu_s d)] + (N-2)[(1+\rho^2)-2\rho\cos(\nu_s d)]}{(1-\rho^2)} \tag{31}$$

and

$$\begin{aligned} G_c(\nu_s,\rho,N) &= N^2[N + (1-2\rho\cos\nu_s d+\rho^2)^{-1}\{2\rho(N-1)(\cos\nu_s d-\rho)\} \\ &+ 2(1-2\rho\cos\nu_s d+\rho^2)^{-2}\rho^2\{\rho^{N-1}[\cos([N+1]\nu_s d) \\ &- 2\rho\cos[N\nu_s d] + \rho^2\cos([N-1]\nu_s d) \\ &- [\cos(2\nu_s d) - 2\rho\cos(\nu_s d) + \rho^2]\}]^{-1} \end{aligned} \tag{32}$$

Evaluation of the gain improvement ratio $G = G_o/G_c$ at $\nu_s = 0$ and $\nu_o = \omega/c$ for $d = \pi/\nu_o$ gives an indication of performance bounds. Specifically, for large N>>1, G = 1 at $\nu_s = 0$ and $\nu_s = \nu_o$. These results are indicative of the fact that performance gains are modest when using optimum processing for large arrays with correspondingly narrow beams and low side lobes. Evaluation of the gain improvement ratio for angularly extended noise for a range of values for ρ and ν_s is given in Figure 2.

From the above considerations it is indicated that plane wave sources of interfering noise and not angularly extended noise justify the utilization of adaptive processing techniques. Furthermore, the optimum processor for plane wave interference reduces to

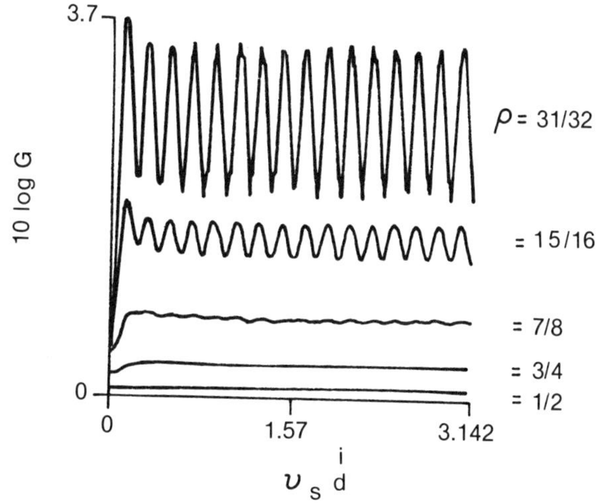

FIGURE 2. Array gain improvement ratio (G) of a MV relative to a DS beamformer for the case of angularly extended noise. The ratio is evaluated as a function of $\nu_s d$ for 32 (=N) sensors and various values of coherence (ρ).

a simple cancellation process at the outputs of conventional DS beams. This is contrasted to the complex multichannel adaptive filter as proposed in the beginning of this section. This theme of reduced complexity for optimal adaptive system implementation is pursued further in the following section.

MINIMUM COMPLEXITY IMPLEMENTATION OF THE MV BEAMFORMER

As pointed out above, with few exceptions, the expected sonar array gain improvement ratios will be unspectacular for a MV relative to DS beamformer operating in either a far-field interference or an angularly extended noise environment which is typical of distant shipping. The exceptions are either a point source of interference with levels which could be typical of jamming devices or sonar towing ship interference. For a point source of interfering noise, it was observed that the complex MV beamformer with N filters per beam reduced to two simple DS beamformers with a single filter and difference operation for beam output noise cancellation. This type of noise cancellation scheme is referred to commonly as post-beamformer interference cancellation (PIC) or simply null steering. A generalization of the beam output, single point source noise cancellation scheme is illustrated in Figure 3. Three basic functions are identifiable:

1. Conventional spatial signal estimation implemented by time-delay-and-sum (DS) beamforming;
2. Adaptive spatial noise estimation implemented by adaptive filtering of a linear combination of hydrophone outputs and any auxiliary noise only sensing device outputs; and
3. Coherent noise subtraction at the output of the DS beamformer.

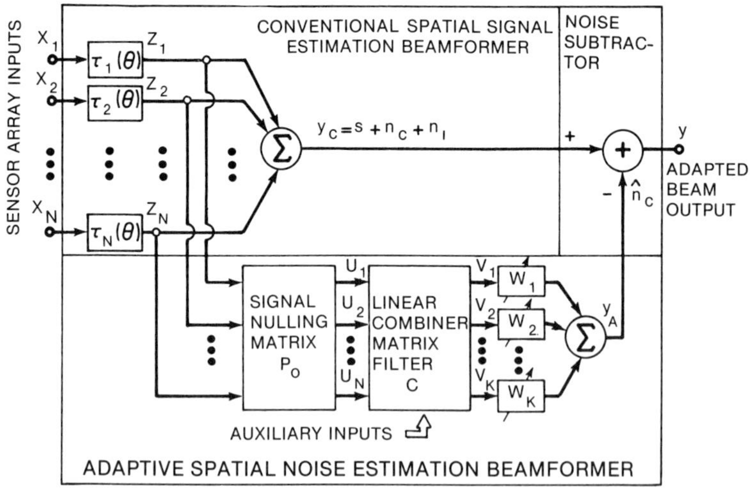

FIGURE 3. Generalized noise cancellation adaptive processor for sonar sensor arrays.

The DS beamformer output, y_C, consists of signal, s, coherent noise, n_C, and incoherent noise, n_I. The adaptive noise-estimation filter-beamformer provides an estimate of the coherent noise component, $\hat{n}_C$, and subtracts it from the conventional beamformer output. To preclude subtraction of any portion of the signal when n_C is removed, the adaptive beamformer must remove the signal components from the steered hydrophone data, z_k. This signal removal is accomplished by the signal-nulling matrix filter, P_o. The linear-combiner matrix filter, C, is applied to the nulled data, u_k, to reduce the number of adaptive filter channels required from N to K. Making $K(\leq N)$ as small as possible, commensurate with effective coherent noise subtraction, reduces implementation complexity.

The final adaptive noise estimation beamformer operation is that of multiplying the linear-combiner outputs, v_k, by the corresponding filter weights, w_k, and summing to give the adaptive beamformer output, $y_A = \hat{n}_C$. The quantity $\hat{n}_C$ is an LMS estimator of the coherent noise, n_C. This is because the filter weights, w_k, are selected so as to minimize the average power of the adapted beam output, y. Since y is formed as a difference between y_C, which contains signal s, and y_A, which contains no signal, the process of minimizing the power in y with respect to the weights, w_k, can only reduce n_C, which is coherent between the conventional beamformer output, y_C, and the linear-combiner outputs, v_k. The linear combiner matrix filter output and the adaptive noise estimation beamformer filter weight vector are given by

$$\underline{V}^T = [v_1 v_2 \cdots v_K] , \tag{33}$$

and

$$\underline{W}^T = [w_1 w_2 \cdots w_K] . \tag{34}$$

A presteering matrix filter for a plane wave nominal signal angle of arrival θ is defined as

$$D_o(\theta) = \begin{bmatrix} e^{-j2\pi f\tau_1(\theta)} & & & 0 \\ & e^{-j2\pi f\tau_2(\theta)} & & \\ & & \ddots & \\ 0 & & & e^{-j2\pi f\tau_N(\theta)} \end{bmatrix}, \tag{35}$$

where $\tau_N(\theta)$ is the delay applied to the nth hydrophone for steering a DS beam at angle θ. The signal-nulling and linear-combiner matrix filters are given, respectively, by P_o and C. As a result of the above, the DS beamformer output can be written as

$$y_C = s + n_C + n_I \tag{36}$$

$$= \underbrace{\underline{1}'\underbrace{D_o(\theta)^{-1}\underline{X}}_{\text{steering}}}_{\text{beamforming}} \tag{37}$$

and the adaptive noise estimation beamformer output as

$$y_A = \hat{n}_C \tag{38}$$

$$= \underbrace{\underline{W}'\underbrace{C'\underbrace{P_o'\underbrace{D_o(\theta)^{-1}\underline{X}}_{\text{steering}}}_{\text{signal nulling}}}_{\text{grouping}}}_{\text{beamforming}}. \tag{39}$$

Let the output of the presteering matrix filter have the cross-spectral density matrix R given by (3) with the signal and noise cross-spectral density matrices P and Q still given by (8). Removal of all signal from the noise estimation adaptive beamformer output requires a signal nulling matrix filter given by

$$P_o = I - \frac{1}{N} D D' , \tag{40}$$

where D is defined in (8) as opposed to the $D_o(\theta)$ of (35). This can be seen by computing the spectral density matrix for the linear combiner matrix filter output $\underline{V}$. Specifically,

$$R_{VV} = C'P_o'RP_oC \tag{41}$$

$$= \sigma_s^2 C'P_o'DAD'P_oC + \sigma_n^2 C'P_o'QP_oC \tag{42}$$

where it is noted that $P_o'D$ is a N-by-K null matrix of zeros which eliminates the contribution due to signal in the first term of (42). The adaptive filter weight vector that minimizes

$$E[|y|^2] = E[|y_C - y_A|^2] \tag{43}$$

is given by

$$\underline{W}_o = R_{vv}^{-1} \underline{P}_{vy_C} , \tag{44}$$

where R_{vv} is given above and $\underline{P}_{vy_C}$ is the cross-correlation between the linear combiner matrix filter output $\underline{V}$ and the DS beam output. With these relationships

$$\underline{W}_o = [C'P_o'QP_oC]^{-1} CP_oQ\underline{1} \tag{45}$$

which is the K-by-1 vector of MV optimum filter weights to be applied to the linear combiner output data.

Before discussing the alternatives available for the adaptive realization of (45), the question of the minimum dimensionality, K, of the linear combiner matrix filter output is considered. This question is significant in that the number of adaptive filters which must be implemented with finite cost is exactly K. The combined matrix filter $P_o C$ is a fixed structure which does not require the complexity of any type of adaptive control. For reasons of economy, therefore, it is most desirable to make K as small as possible without drastically affecting the system performance in terms of cancelling the significant coherent noise components represented in Q. With this in mind, let the noise cross-spectral density matrix Q be represented as a sum of two components

$$Q = \alpha Q_C + (1-\alpha)I \ . \tag{46}$$

The portion of Q which exhibits significant spatial coherence is described in terms of a rank M cross-spectral density matrix Q_C. Likewise, the spatially uncorrelated noise portion of Q is reflected in the identity matrix, I. The relative weighting of coherent and uncorrelated components is specified by $0 \leq \alpha \leq 1$. Let Q_C be represented in diagonal form by

$$Q_C = EBE' \tag{47}$$

where the M columns of E constitute an M-dimensional orthogonal eigenvector basis spanning a coherent noise subspace, Ω_N, and the diagonal elements of B are the corresponding eigenvalues. Notice that the example in the previous section of a plane wave interfering noise source is just a special one dimensional case of the above where $E = \underline{D}(\phi)$ and $B = 1$.

A necessary condition for $\underline{W}_o$ to process optimally against the coherent noise component Q_C of the cross-spectral density Q is that the rank, $r(Q_C)$, of the matrix Q_C be preserved under the transformation $C'P_oQ_CP_oC$. A choice of P_oC which clearly insures $r(C'QC) \geq r(Q_C)$ is $P_oC = EG$ wherein the columns of P_oC are a linear combination of the eigenvectors of Q_C. The simplest such combination is just $P_oC = E$ where the columns of P_oC are either the eigenvectors of the coherent portion of the noise only cross-spectral density matrix or a reasonable facsimile thereof.

An interesting example of the above is for M resolvable plane wave interfering noise sources. In this case,

$$C = [\underline{D}(\phi_1)\underline{D}(\phi_2)\ldots\underline{D}(\phi_M)] \tag{48}$$

represents M, DS beamformers steered at the interference angles with weights

$$\underline{W}_o = \frac{1}{N} E \, \underline{1}. \tag{49}$$

ADAPTIVE REALIZATIONS OF THE MV BEAMFORMER

Assuming stationary but statistically unknown data, an adaptive realization of the minimum complexity adaptive filter $\underline{W}_o$ given by (45) requires either a direct or indirect computation of the unknown correlation matrices.

$$R_{vv} = \begin{bmatrix} \text{a K-by-K matrix of cross-} \\ \text{correlations between the} \\ \text{linear combiner outputs} \end{bmatrix} \tag{50}$$

$$= E[\underline{V}\,\underline{V}'] \,, \tag{51}$$

and

$$\underline{P}_{vy_C} = \begin{bmatrix} \text{a K-by-1 vector of correlations between} \\ \text{the linear combiner (outputs) and conventional} \\ \text{beam output} \end{bmatrix} \tag{52}$$

$$= E[\underline{V}y_C^*] \ . \tag{53}$$

Three principal methods can be employed to realize $\underline{W}_o$ adaptively:

1. Estimate-Invert-Plug

The first method, referred to as an open-loop approach, is based on the "estimate-and-plug" philosophy. In particular, this approach computes estimators at time i by a continuous exponential averaging process expressed by

$$\hat{R}_{vv}(i) = (1-\nu)\hat{R}_{vv}(i-1) + \nu\, \underline{V}(i)\underline{V}(i)' \tag{54}$$

and

$$\hat{\underline{P}}_{vy_C}(i) = (1-\nu)\hat{\underline{P}}_{vy_C}(i-1) + \nu\underline{V}(i)y_C^*(i). \tag{55}$$

At intervals which would best correspond to a time constant of the integrator, the estimator $\hat{R}_{vv}(i)$ is inverted and plugged into the expression

$$\underline{W}_o(i) = \hat{R}_{vv}^{-1}(i)\hat{\underline{P}}_{vy_C}(i) \ . \tag{56}$$

Instead of exponential averaging as in (54) and (55) above, some versions of this estimate-and-plug approach would use uniformly weighted data (block averaging) over an interval of time i of length M, where M is sufficiently greater than K to insure stability [24]. Numerous methods exist for inverting the estimator $\hat{R}_{vv}(i)$ [25]. Particular care must be taken in selecting a procedure which is robust to the limitations of finite precision arithmetic and near singularity. In particular, it is necessary in cases of ill-conditioning in $\hat{R}_{vv}(i)$ to inject a minimal level of synthetic white noise into the linear combiner output vector to insure full rank and the resultant invertability. The Cholesky extension of the Crout decomposition method to Hermitian matrices is recommended from the standpoints of both computational efficiency and numerical accuracy [26].

2. Direct Inverse Estimate-Plug

The second method for the solution of (45) is by an application of the matrix inversion identity of (20) and (21) in the form

$$\hat{R}_{vv}^{-1}(i) = [(1-\nu)\hat{R}_{vv}(i-1) + \nu\, \underline{V}(i)\underline{V}(i)']^{-1} \tag{57}$$

$$= \frac{1}{1-\nu}\left[\hat{R}_{vv}^{-1}(i-1) - \frac{R_{vv}^{-1}(i-1)\underline{V}(i)V(i)'R_{vv}^{-1}(i-1)}{\frac{1-\nu}{\nu} + \underline{V}(i)'R_{vv}^{-1}(i-1)\underline{V}(i)}\right]. \tag{58}$$

The correlation vector $\underline{P}_{vy_C}$ could be estimated as in (55). This approach is probably the least well developed of the three adaptive realization schemes to be presented. It has been compared by Lunde [11] in terms of computational load, convergence rate and steady state variance to gradient descent realizations and found to perform better. It is reported by Lunde that the stability of (58) is heavily dependent on the linear combiner output word size (1) which, in turn, is related to the conditional number, $C = \lambda_{max}/\lambda_{min}$, which is the ratio of the maximum to minimum eigenvalues of $R_{vv}(i)$. For a given word size, the ratio C can be controlled by the addition of white noise to the linear combiner matrix filter output as in the estimate-invert-plug technique discussed above. Finally, it is observed that this approach is also an open-loop, feed-forward technique requiring no feedback of beam output data in the adaptation process.

3. Steepest Gradient Descent

The third method considered involves an unconstrained adjustment of the estimate $\underline{W}_o(i)$ at time i by a scaled instantaneous negative gradient of $|y(i)|^2$ (44) with respect to $\underline{W}(i)$ [7]. Specifically, there is defined the gradient vector

$$\underline{J}(i) = \frac{\partial |y(i)|^2}{\partial \underline{W}(i)} \tag{59}$$

$$= 2\underline{V}(i)y(i)^* \tag{60}$$

with the update relation

$$\underline{W}(i+1) = \underline{W}(i) + \mu\underline{J}(i) \tag{61}$$

giving the steepest gradient descent adaptive control algorithm. For stationary data and a properly selected step-size constant, μ, convergence of the mean of (61) to (45) as $i \to \infty$ is assured. For a complete discussion of convergence and stability, see Widrow [27] for the case of real input data. In the following, the results from [27] are used in a discussion of convergence. Finally, the steepest descent control algorithm is regarded as a feedback control process where the gradient $\underline{J}(i)$ serves as a vector error signal. If the control process is stable, then the mean value of the gradient (60) will tend to zero.

The issue of convergence rate deserves some consideration here in that it is basic to the differences between the two feedforward and one feedback approaches presented above. The estimate-invert-plug approach is considered here as representative of the two feed-forward (FF) schemes. It is desired to compare the rate of convergence of the mean for a FF approach to the gradient descent, feedback (FB) approach. In particular, let the mean value of the estimated cross-spectral density matrix $R_{vv}(i)$ after J iterations of (54) be G(J). By iteration of this mean value expression there results

$$G(J) = (1-\nu)^J G(o) + (1-(1-\nu)^J) R_{vv} \tag{62}$$

The first term in (62) is the transient term wherein an exponential envelope $\exp[-J/\tau]$ can be fit to the geometric expression $(1-\nu)^J$. For large τ, the convergence time in iteration cycles, $\exp(-1/\tau) = 1-1/\tau$ and $\tau \simeq 1/\nu$. Thus, for ν selected small enough to give acceptable variance in the estimator $R_{vv}(i)$ it is clear that convergence rate for the FF is only a function of ν and not any property of the data. On the other hand, for the gradient descent FB approach the convergence occurs independently along each principal axis defined by the eigenvectors of the cross-spectral density matrix Q in (46). The time constant associated with the p-th axis convergence process is

$$\tau_p = \begin{cases} 1/2\mu\sigma_n^2(\alpha(b_p-1) + 1), & p \leq M \\ 1/2\mu\sigma_n^2(1-\alpha), & p > M \end{cases} \tag{63}$$

where b_p is the p-th eigenvalue associated with the coherent noise cross-spectral density matrix Q_C. Thus, it is seen that the convergence rate for the FB approach is dependent on the noise level σ_n^2 such that the individual modal FB time constants are an inverse function of σ_n^2 and the relative amount of coherent noise power, b_p, in a mode. The τ_p are referred to simply as the modal closed loop time constants. For the FB system to converge in the mean

$$\frac{1}{\sigma_n^2(\alpha(b_1-1) + 1)} < \mu < 0 \; . \tag{64}$$

where the ordering $b_1 \geq b_2 \geq \ldots \geq b_M$ of the eigenvalues of Q_C is assumed. Therefore, convergence for FB control can either be very rapid and correspondingly noisy for high coherent noise levels or slow and stable for low coherent noise levels, whereas in FF control, the convergence will be of predetermined rate and noisiness. When the coherent noise is dominated by a strong mode, rapid convergence may be desirable if this mode represents an interfering source with a high angular velocity for example. On the other hand, if the increased adaptation noise due to rapid convergence succeeds in masking low level signals of interest, then rapid convergence in the mean is of little benefit. A resolution of this issue for a particular sonar system would in all likelihood depend on the mission to be performed. A number of configuration issues are currently of some interest to designers of adaptive signal processing systems in sonar. Several of these issues are presented in the following section and some further discussion is given.

IMPLEMENTATION ISSUES

The following three issues are currently of particular interest to designers of MV controllers for adaptive array processors:

1. Adaptation on either sensor or beamformer outputs (system complexity),
2. Method of filter realization by either matrix inversion (FF) or steepest gradient descent (FB), and
3. Adaptation in the time or frequency domain.

Issues 1 and 2 are discussed above. Vural [28] addressed issues 1 and 3 and also included the related issue of multiple component signal preservation constraints. In summary, the minimum complexity MV beamformer should perform the noise cancellation at DS beam

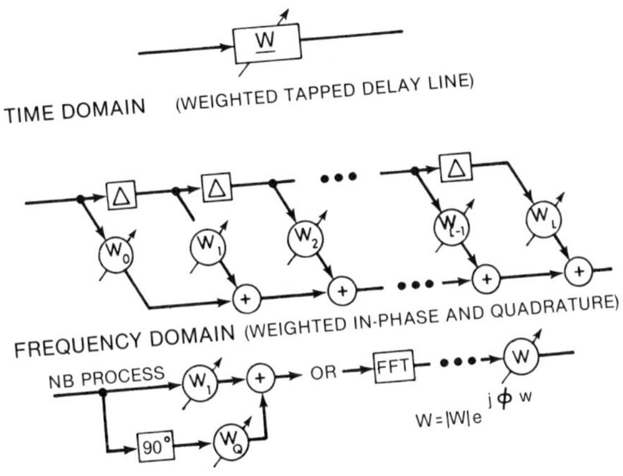

FIGURE 4. Broadband and narrowband realizations of the adaptive filter element.

outputs by means of an estimate-and-subtract scheme. The noise estimation "beams" are formed by "steering" the array with the eigenvectors of the noise only cross-spectral density matrix. Filter realization requires a study of the desirability of rapid convergence of the mean for high level coherent noise and the attendant increase in the noise of adaptation. By such a study, either a direct matrix inversion or gradient descent approach can be selected.

With respect to issue 3, the two basic realizations of an adaptive filter element which have been used in sonar are summarized in Figure 4. The time domain element consists of a weighted tapped delay line for the time-domain adaptive transversal filtering of the linear-combiner-output sampled data.

In contrast, the frequency domain approach utilizes either analog narrowband filter inputs with weighted in-phase and quadrature linear combining or digital narrowband filtering (e.g., fast Fourier transform, FFT) followed by complex weight multiplication for each resolution cell. In terms of sonar adaptive hardware implemented to date, no distinct trend has evolved that, in general, favors one over the other. Relatively small-aperture arrays have exploited time domain techniques. Historically, small-aperture arrays have operated on the basis of broadband radiated target energy and, therefore, no requirement for frequency partitioning has existed. However, because of linearity, complex spectrum analysis at the beam output and individual sensor level are theoretically interchangeable. Furthermore, for a given number of adaptive degrees-of-freedom, complex spectrum analysis at the sensor level followed by frequency domain adaptive control can offer certain advantages over broadband processing, if spectrum analyzed beam-output data are required. Specifically, frequency domain systems allow independent spatial adaptation, with a fixed number of adaptive weights assigned to each spectral bin. This feature is a form of spectral self-normalization. It assures that no portion of the band with high-level coherent noise will "capture" an inordinate share of the available adaptive degrees-of-freedom at the expense of spatial gain improvement in a local spectral region that might be strategically important. However, for a given number of adaptive weights, the time domain approach will distribute these weights over the processed band so that the performance averaged over the entire processed band is optimized. In addition, for a frequency domain implementation, a strong interference mode which is limited in bandwidth will not introduce adaptation noise outside of the frequency band to which the interference is confined.

Finally, perhaps the primary reason for a consideration of frequency domain instead of time domain adaptive filter realization is the possibility of significant hardware savings [13]. A time domain convolution is realized by the weighted tapped delay line, finite impulse response filter in Figure 4. It is well-known that convolution can be efficiently implemented using transform techniques. Savings in the number of multiplications per second are on the order of $L/\log_2 L$ where L is both the tapped delay line length and the FFT length. It should be noted, however, that little comparative work has been done with respect to time versus frequency domain techniques. Work is needed to establish relative levels of adaptation noise for equal response times over a range of signal and interference levels. In fact, this type of information is somewhat lacking in the context of all three of the issues raised here.

SOME OTHER APPLICATIONS

Although only the important example of processing a discrete array of sensors has been considered here, numerous other applications exist for the basic minimum variance adaptive filter. In addition to the basic noise cancellation application discussed above, applications in underwater sound exist in statistical linear prediction [29] and channel equalization in communication channels [30]. A capability for the adaptive prewhitening and gain adjustment of spectrally tapered and nonstationary noise in the presence of a narrowband signal is of current interest as an application of the adaptive MV filter. This application of the adaptive filter is called the "adaptive line enhancer" (ALE) and has been related to the computation of maximum entropy power spectra [31,32,33]. Extensions of the basic transversal adaptive filter structure to recursive configurations have also been reported [34]. These applications will not be treated here because they are all variations of the minimum variance criterion for adaptive filter design. Instead an application of adaptive filtering using constrained maximization will be summarized [35].

Consider the ratio of two quadratic forms

$$\text{SNR} = \frac{\underline{H}'P\underline{H}}{\underline{H}'Q\underline{H}}, \tag{65}$$

which is the output signal-to-noise ratio for the general array processor. If Q is the cross-spectral density matrix for uncorrelated, identically distributed noise, then Q = I and

$$\mathrm{SNR} = \frac{\underline{H}'P\underline{H}}{\underline{H}'\underline{H}} \tag{66}$$

is maximized by the same choice of filter vector $\underline{H}$ that maximizes

$$\alpha = \frac{H'RH}{H'H} \, . \tag{67}$$

More generally,

$$\lambda_i = \max_{\underline{H}}\{\alpha\} \quad \text{for } \underline{H}'\underline{E}_j = 0, \quad j = 1,2,\ldots,i-1 \tag{68}$$

where λ_i is the ith eigenvalue of R and $\underline{H} = \underline{E}_i$ is the corresponding eigenvector. Thus a complex signal with L-orthogonal modes and cross-spectral density matrix

$$P = \sum_{i=1}^{L} b_i \underline{E}_i \underline{E}_i' \tag{69}$$

can be extracted from noise by an adaptive realization of the criterion expressed by (68) [36,37]. An example of such a signal in sonar occurs in a multipath environment where the $\underline{E}_i$ vectors are related to the direction vectors for the various arrival path modes. A further example would be to the extraction of frequency diversity signals with random orthogonal modulation processes. In fact, the process suggested above can be seen to be simply an adaptive realization of an L-ary maximum likelihood estimation process in gaussian noise. However, the only a priori knowledge which exists on the signal is that it spans a complex, L-dimensional vector subspace. The adaptive process could be implemented by (a) a constrained, gradient ascent maximization of the instantaneous value of (67) subject to the constraints of (68) [7], (b) an estimation of the cross-spectral density matrix R as in (54) followed by extraction of the eigendata using standard numerical methods or, finally, by (c) a multipass dynamic programming maximization of α at time k, $\alpha(k)$, over the interval $1 \leq k \leq K$.

Suppose that a priori knowledge of the signal exists in the form of a vector parameterization $\underline{\theta}(i,k)$, $1 \leq i \leq L$ and $1 \leq k \leq K$. An adaptive process which estimates the non-stationary process $\underline{\theta}(i,k)$ over the mode index i and time index k is therefore tantamount to signal extraction. Let this parameterization be subject to the inequality constraint $g(\underline{\theta}(i,k)) \leq g_o$ due to, for example, temporal "smoothness" in the signal parameter path. Now, an augmented instantaneous criterion function with $R(k) = \underline{X}(k)\underline{X}(k)'$ is

$$L_i(k) = \frac{|H'(\underline{\theta}(i,k))\underline{X}(k)|^2}{|H(\underline{\theta}(i,k))|^2} + \gamma\, g(\underline{\theta}(i,k)) \tag{70}$$

with the orthogonality constraints

$$\underline{H}'(\underline{\theta}(i,k))\underline{H}(\theta(j,k)) = 0 \quad \text{for } j = 0,1,\ldots i-1. \tag{71}$$

The cummulative criterion function

$$L_i = \sum_k L_i(k) \tag{72}$$

can now be maximized with respect to the sequence $\underline{\theta}(i,k)$ $k = 1,2,\ldots,K$. In (70) γ is an undetermined Lagrange multiplier which reflects the relative importance of the "matching" processing inherent in the first term of (70) and the constraint process expressed in the second term. Maximization of (72) has now been expressed in the form of a numerical two point boundary value problem for which the methods of dynamic programming are applicable. The relationship can be drawn between the process suggested above to an adaptive realization of an L-ary maximum a posteriori (MAP) estimation procedure for gaussian noise.

SUMMARY

An attempt has been made to present an overview of current activities in the application of adaptive techniques to signal processing with emphasis on sensor array processing. The particular emphasis has been on the type of problems encountered in the world of passive sonar. While the coverage has not been complete, sufficient references have been given to serve as a starting point for investigation by the interested reader. The process of minimum variance optimum, adaptive control has been discussed in the context of the discrete sensor array beamforming application for directional signal estimation. Multimodal signals were considered along with the multiple constraints implied in the minimization problem. The implementation issues of minimum complexity, filter realization, algorithm selection and domain of realization (time or frequency) were considered. It was pointed out that insufficient work has been done to offer a "cook-book" procedure for adaptive system design. Finally, an adaptive control approach to a constrained maximization procedure for signal extraction was presented. Alternative implementation procedures which included dynamic programming were suggested for realization of the signal extraction problem by adaptive maximization. Relationships were pointed out between the maximization procedures discussed and maximum likelihood and maximum a posteriori estimation techniques.

REFERENCES

1. Bryn, F., "Optimum Signal Processing of Three-Dimensional Arrays Operating on Gaussian Signals and Noise," J. Acoust. Soc. Am. 34, pp. 289-297, 1962.

2. Mermoz, H., "Filterage Adapte et Utilization Optimale d'une Antenne," Signal Processing with Emphasis on Underwater Acoustics, Proceedings of the NATO Advanced Study Institute, Grenoble, France, 1964.

3. Shor, S.W.W., "Adaptive Technique to Discriminate against Coherent Noise in a Narrow-Band System," J. Acoust. Soc. Am. 39, pp. 74-78, 1966.

4. Widrow, B., et al., "Adaptive Antenna Systems," Proc. IEEE 55, pp. 2143-2159, 1967.

5. Griffiths, L., "A Simple Adaptive Algorithm for Real-Time Processing in Antenna Arrays," Proc. IEEE 57, pp. 1696-1704, 1969.

6. Cox, H., "Array Processing Against Interference," Proceedings of the Symposium on Information Processing, Purdue University, Lafayette, Indiana, pp. 453-464, Apr. 1969.

7. Owsley, N., "A Recent Trend in Adaptive Spatial Processing for Sensor Arrays: Constrained Adaptation," Signal Processing, A NATO Advanced Study Institute, edited by J.W.R. Griffiths et al., pp. 591-604, Academic Press, New York, 1973.

8. Owsley, N. L., Fay, J. W., "Optimum Turbulent Boundary-Layer Induced Noise Suppression with Suboptimum Realizations," to be published in the Journal of the Acoustical Society of America (JASA).

9. Widrow, B., "Adaptive Filters," in"Aspects of Network and System Theory," R. Kalman and N. DeClaris, Eds., pp. 563-587, New York: Holt, Rinehart and Winston, 1971.

10. Huffman, S. D., "Adaptive Linear Estimation Algorithms Applied to Spectral Line Enhancement," Ph.D. Thesis, Duke University, TR-15, August 1978.

11. Lunde, E. B., "The Forgotten Algorithm in Adaptive Beamforming," Proceedings of the NATO Advanced Study Institute on Signal Processing with Emphasis on Underwater Acoustics, Part 1, pp. 411-423, Reidel Publishing Co., Dordrecht, The Netherlands, 1976.

12. Owsley, N. L., "A Frequency Domain Adaptive Beamformer Implementation," Naval Underwater Systems Center, Rpt. 5255, February 1976.

13. Dentino, M., McCool, J., Widrow, B., "Adaptive Filtering in the Frequency Domain," Proceedings of the IEEE, Vol. 66, No. 12, pp. 1658-59, December 1976.

14. Owsley, N., "Summary of Workshop on Adaptive Array Processing," Proceedings of the NATO Advanced Study Institute on Signal Processing with Emphasis on Underwater Acoustics, Part 2, pp. 777-779, Reidel Publishing Co., Dordrecht, The Netherlands, 1976.

15. Ligget, W., "Passive Sonar: Fitting Models to Multiple Time Series," Signal Processing, ec. J.W.R. Griffiths et al., pp. 327-346, Academic Press, New York, 1973.

16. Ibid 14.

17. Frost, O. L., III, "An Algorithm for Linearly Constrained Adaptive Array Processing," Proc. IEEE, Vol. 60, pp. 926-935, August 1972.

18. Yen, N. C., "Ambient Sea Noise Directionality: Measurement and Processing," Naval Underwater Systems Center, New London, Conn., Tech. Rpt. 554, 28 February 1977.

19. Mermoz, H., "Complementarity of Propagation Model Design with Array Processing," Aspects of Signal Processing, Part 2, Giorgio Tacconi, Ed., pp. 463-468, D. Reidel Publishing Co., Boston, Mass., 1977.

20. Cox, H., "Array Processing Against Interference," Proc. of the Symposium on Info. Processing, Purdue University, April 28-30, 1969, pp. 453-464.

21. Pasupathy, S., "Optimum Spatial Processing of Passive Sonar Signals," IEEE Trans. on AES, Vol. AES-14, No. 1, pp. 158-164, Jan. 1978.

22. Lewis, J. B., Schultheiss, P. M., "Optimum and Conventional Detection Using a Line Array," JASA, Vol. 49, No. 4 (Part 1), pp. 1083-1091, April 1971.

23. Pasupathy, S., "Optimal and Conventional Array Processing," Can. Elec. Eng. J., Vol. 3, No. 1, pp. 27-31, 1978.

24. Capon, J., and Goodman, N. R., "Probability Distributions for Estimators of the Frequency-Wavenumber Spectrum," Proc. IEEE 58, pp. 1785-86, DDC AD-723788, 1970.

25. Marcus, M., "Basic Theorems in Matrix Theory," NBS Applied Mathematics Series, 57, pp. 15-18, January 22, 1960.

26. Forsythe, G. E., "Today's Computational Methods of Linear Algebra," SIAM Review, Vol. 9, No. 3, July, 1967.

27. Widrow, B., et al., "Stationary and Nonstationary Learning Characteristics of the LMS Adaptive Filter," Proceedings of the IEEE, Vol. 64, No. 8, pp. 1151-1162, August 1976.

28. Vural, A. M., "A Comparative Performance Study of Adaptive Array Processors," Proceedings of IEEE ASSP Conference, Hartford, Conn., pp. 695-700, April 1977.

29. Makhoul, J., "Linear Prediction: A Tutorial Review," Proc. IEEE, Vol. 63, pp. 561-580, April 1975.

30. Lucky, R., "Principles of Data Communication," McGraw-Hill, New York, 1968.

31. Treichler, J., "The Spectral Line Enhancer - The Concept, and Implementation and an Application," Ph.D. Dissertation, Dept. of E. E., Stanford Univ., Stanford, Calif., June 1977.

32. Burg, J., "Maximum Entropy Spectral Analysis," presented at the 37th Annual Meeting of the Society of Exploration Geophysicists, Oklahoma City, Okla., 1967.

33. Griffiths, L., "Rapid Measurement of Instantaneous Frequency," IEEE Trans. on ASSP, Vol. ASSP-23, pp. 209-222, April 1975.

34. Feintuch, P., "An Adaptive Recursive LMS Filter," Proc. IEEE, Vol. 64, pp. 1622-1624, November 1966.

35. Owsley, N. L., "Wavefront-Array Shape Matching by Adaptive Focusing," Proceedings of the Naval Postgraduate School Workshop on Time Delay Estimation and Applications, Monterey, Calif., May 4-7, 1979.

36. Owsley, N. L., "Spectral Signal Set Extraction," Proceedings of the NATO Advanced Study Institute on Signal Processing with Emphasis on Underwater Acoustics, Part 4, pp. 42.1-42.7, Reidel Publishing Co., Dordrecht, The Netherlands, 1976.

37. Owsley, N. L., "Adaptive Data Orthogonalization," Proceedings of IEEE Conference on ASSP, Tulsa, Okla., April 12, 1978.

ON THE APPLICATION OF RECURSIVE LEAST SQUARES METHODS TO ADAPTIVE PROCESSING

E. H. Satorius[1]

Naval Ocean Systems Center
San Diego, California

M. J. Shensa

Hydrotronics, Inc.
San Diego, California

Least squares criteria have been a recurrent theme in signal processing for years; nevertheless, new algorithms continue to be introduced. In this paper we shall consider a recently developed class of computationally efficient least squares algorithms and their application to adaptive signal processing. Specifically, we shall examine in some detail the so-called lattice least squares algorithms which are recursive in both time and order. These lattice algorithms have a number of important properties

---

[1]Supported by NOSC IR Project ZR94.

Present address: Dynamics Technology, Inc., Torrance, Ca.

ISBN 0-12-514060-6

including rapid initial convergence as well as a computational complexity which depends linearly on the number of filter coefficients (in contrast with the quadratic dependence of conventional implementations). Furthermore, the order-recursive structure allows the dynamic assignment of the particular lattice order (filter length) which may provide the best performance at any instant of adaptation. In light of these properties the lattice least squares algorithms appear quite promising for a wide variety of adaptive signal processing applications.

I. INTRODUCTION

The application of least squares techniques for the estimation of linear systems has been a topic of interest for at least three decades (see e.g., the excellent review article by Kailath [1]). Despite the simplicity of the concept, which ultimately requires only the solution of a set of linear equations (the so-called normal equations), its successful application depends on a wealth of subtleties buried in the mathematical structure. In particular, one is often concerned with efficient computation, numerical stability, possible a priori constraints placed on the system in question, and the treatment of nonstationary data.

In this paper we will be interested in a relatively new class of computationally efficient recursive least squares algorithms which have evolved through the work of Morf, Ljung, Kailath, and others [1-11]. Unfortunately, despite their efficiency, these algorithms are quite complex in structure. As a result their proper application requires a reasonable degree of care and study. A historical parallel might be the fast Fourier transform (FFT). The concept of the discrete Fourier transform (DFT) is relatively simple. Its efficient numerical implementation, the

FFT, is somewhat involved (more precisely, short but sophisticated); and the proper application of the DFT can require a solid understanding of particular details (such as windowing or aliasing).

It is the purpose of this paper to provide a reasonably clear description of the efficient recursive least squares algorithms originally presented in [2-9] and to point out some considerations in their application to adaptive signal processing. It is understood that the user must still expend a reasonable amount of effort to gain familiarity with the variables involved, and many comments (pitfalls?) may not be appreciated until they are experienced. However, hopefully, this account will convince the reader that the dividends are worth the investment.

II. RECURSIVE LEAST SQUARES ALGORITHMS

In order to illustrate the basic properties of the recursive least squares algorithms, let us consider the identification problem depicted in Figure 1.[2] In particular, we wish to model a discrete linear system H by observing its input and output over some time segment, say [0,T]. In general these observations will be corrupted by noise, n_1 and n_2. For simplicity we shall assume G is an all-zero transfer function

---

[2]Although at this point it is more convenient to discuss the recursive least squares algorithms from the point of view of a systems identification problem, we shall ultimately be interested in the output Gx and error e. This will be elaborated upon in Section III. Note also that the parameter t is considered a discrete time index.

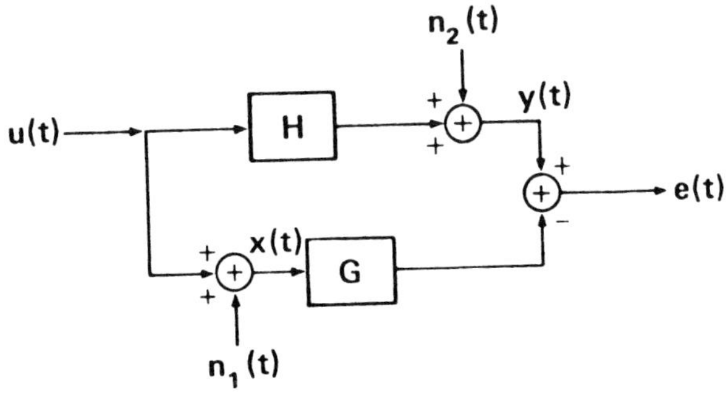

FIGURE 1. A block model for the system identification problem.

$$Gx(t) = \sum_{i=0}^{P} g_i x(t-i) , \tag{1}$$

and the data are scalar functions of time. The more general cases of a rational transfer function model, G, and vector inputs are discussed in [2-4], [6-9].

An obvious candidate for measuring the fit of the model G is the sum of the squares of the output errors over the observation interval:

$$\rho = \sum_{t=0}^{T} e^2(t)$$

$$= \sum_{t=0}^{T} (y(t) - Gx(t))^2. \tag{2}$$

Assuming we have somehow specified P, the order of our model, we choose the coefficients $g_i = (G_P)_i$ so as to minimize the sum of the squared errors, ρ. The resulting estimate of H is the least squares estimate (see, e.g., [12]). In this context G will in general be a biased estimate of H [12], i.e., $E(G) \neq H$ ($E(\cdot)$ denotes expectation) even when H is all-zero and the order P is correct. To insure no bias it is sufficient that H be all-zero of order P, $n_1 = 0$, and n_2 be white noise, independent of x [12].

Taking partial derivatives of ρ with respect to the g_i leads to the normal equations:

$$R_P G_P = F_P \ , \tag{3}$$

where R_P is the P x P matrix

$$(R_P)_{ij} = \sum_{t=0}^{T} x(t-i+1)x(t-j+1) \tag{4}$$

and F_P and G_P are P vectors with

$$(F_P)_i = \sum_{t=0}^{T} y(t)x(t-i+1). \tag{5}$$

If P is large, the inversion of R_P certainly merits special attention; however, as we shall see, the problem is much more extensive.

Often only limited *a priori* knowledge of the order is available. Although we shall skirt the issue of criteria for determining the order (see, e.g. [13]), we note that when a given order is not suitable (e.g., ρ_{min} is unacceptably large), it is necessary to recompute G_P for a larger P. In order to relieve the increased computational burden in this situation, various order-recursive schemes have been proposed. These algorithms, which stem from Levinson's algorithm [14], compute G_{P+1} from G_P thus providing all orders G_p, p = 1,...,P at almost no extra cost.

More precisely, the solution of an arbitrary set of linear equations (for G_P) can be accomplished using several standard methods, e.g., the Gauss reduction or elimination method and the Crout reduction method [15]. These general methods require $P^3/3 + O(P^2)$ operations (multiplications or divisions). By using the square root or Cholesky decomposition which exploit the symmetry and the positive semi-definite properties of R_P, it is possible to reduce this to half the computation ($P^3/6 + 0(P^2)$ operations) [16]. In contrast, the existing order-recursive algorithms are only $0(P^2)$ [5,14] and at the same time supply the solutions G_p for all orders $p \leq P$. It should be pointed out, however, that for large T, the solution of equation (3) does not form the major computational burden. The computation of the R_p matrix requires approximately PT operations [17] which can dominate the computation time if T>>P, as is often the case.

A further source of difficulty arises in an on-line situation in which more data is continually being provided, i.e., T is increasing. Presumably, if the system H is fixed, additional data will result in a better estimate. However, under the above formulation, the incorporation of each additional data point would require the recomputation of R_p as well as the solution of the normal equations. The first algorithms which were developed to deal with this situation were recursive in time but not order. These original implementations (e.g. [18]) required on the order of P^2 operations per time update; however, more efficient methods (fast Kalman) have appeared which reduce this to 0(P) operations per update [2], [3],[19].[3]

Ideally, it would be desirable to have a method which is recursive in both time and order. In this paper we will concentrate on a recently developed class of such algorithms, the so-called least squares lattice algorithms ([4],[6-11]). These algorithms which are both dually recursive (time and order) and numerically efficient (0(P) operations per time update) appear to be one of the most promising candidates for many adaptive signal processing applications.

---

[3] D. Falconer has recently informed one of the authors that several researchers have found adaptive implementations of fast Kalman algorithms to be numerically unstable.

A. Convergence

At this point, we wish to note that there exist other, non least squares algorithms, which are designed to solve the problem of minimizing the expectation of the error (in the limit as $T \to \infty$). Many of these methods, notably the gradient least mean squares (LMS) algorithm [21] and the gradient lattice algorithms [22],[23], are good estimators and computationally more efficient than the recursive least squares algorithm. However, the least squares solution (which is unique regardless of the solving algorithm) enjoys a great advantage inasmuch as it converges much faster. This point requires some clarification.

It can be seen from Figure 1 that the output error e is a function of the data interval T. In a stationary environment, by rate of convergence we shall mean the rate at which the root mean square error $[E(e^2(T))]^{1/2}$ approaches some limit e_∞ (assuming the limit exists). It should be noted, that in the case of least squares this "convergence" has nothing to do with the convergence of the particular algorithm. All such procedures yield the unique solution of the normal equations. However, the solution is a function of the data interval, [0,T]. We may thus examine its behavior as T approaches infinity. G_P considered as a function of T will, in fact, converge to a solution of eq. (3) where

$$(R_P)_{ij} = \lim_{T\to\infty} \frac{1}{T} \sum_{t=0}^{T} x(t-i+1)x(t-j+1) \tag{6a}$$

$$(F_P) = \lim_{T\to\infty} \frac{1}{T} \sum_{t=0}^{T} y(t)x(t-i+1), \tag{6b}$$

providing the limiting R_P and F_P exist.[4] Note that the factor $\frac{1}{T}$

---

[4] In contrast, the convergence of the gradient algorithms must be proven, and, in fact, only holds for restricted values of the algorithms' parameters (see e.g. [24]).

does not affect the solution of equation (3) for finite T. (A more detailed discussion of the convergence behavior of the least squares algorithms may be found in [18].)

A representative example comparing the speed of convergence for three algorithms - the gradient LMS, gradient lattice, and least squares lattice - appears in Figure 2. (For a detailed description including specification of the various input and algorithm parameters the reader is referred to [25],[26].) There is a considerable difference in the three curves. Typically, the maximum convergence rate of the gradient LMS algorithm is limited by the eigenvalue spread of the input correlation matrix [24],[27],[28]. The gradient lattice partially eliminates this spread by approximately orthogonalizing the input data (i.e., orthogonality is approached in the limit as $T \to \infty$ [22],[23],[25]). However, its convergence is slower than the least squares lattice algorithm mainly because of the estimated gradient (gradient noise). In contrast, the speed

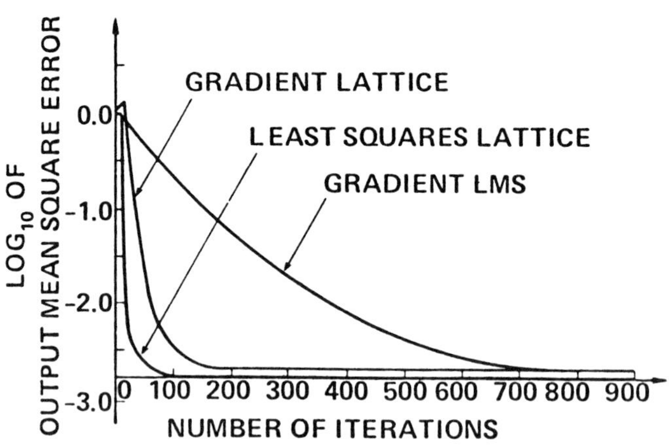

FIGURE 2. The results of a computer simulation comparing the convergence rates of three different algorithms. The conditioning number of the input correlation matrix is 11; for details see [26].

of convergence of the least squares algorithm only depends on the rate at which the estimated correlation matrix (eq. (4)) approaches the true correlation matrix of the input data [18], a limitation which is inherent in any modeling scheme.

B. Adaptive Processing and Windowing

Many of the applications to which one would like to apply least squares techniques (such as speech processing, data equalization, noise cancelling, etc. [19],[23],[29]) require an adaptive model. This can be achieved by limiting T and solving for the least squares parameters using successive blocks of data; i.e., by the use of a moving window. In fact, the desire for such an efficient procedure has been one of the main motivations in seeking time-recursive algorithms.

The particular window used in Section III forgets the past exponentially. More precisely, the expression for the error ρ (eq. (2)) is replaced by

$$\rho = \sum_{t=0}^{T} w^{T-t}(y(t) - Gx(t))^2 \tag{7}$$

where $0 < w \leq 1$. Equation (7) is equivalent to weighting the data x(t) and y(t) by $w^{(T-t)/2}$. The window slides along as T increases (recall that T is the most recent time point at which the model is estimated). If the true system model H is constantly changing, G can only estimate an averaged version of H. The smaller w, the quicker the forget time; that is, the more rapid the adaptation and the less averaging of the true model (although at some point small w will also limit the effective order of G). However, a small value for w will also result in a poorer estimate in the sense that there will be less averaging of the noise.

In addition to w, a fixed window may be imposed on the data. In this paper we shall set $x(t) = 0$ for $t \leq 0$. This results in the so-called prewindowed [7] correlation matrix R_P and is critical in the derivation of the algorithm of Section III. The unwindowed form (which uses values of $x(t)$ for $-P \leq t \leq T$) leads to a more complex algorithm [7]. It is interesting to note that in neither of these cases is R_P Toeplitz. R_P can be made Toeplitz by appropriate windowing (which only uses values of $x(t)$ for $0 \leq t \leq T-P$), however at present there is no corresponding time recursive solution.

C. Stability Considerations

Before presenting and discussing in more detail the recursive lattice least squares algorithm, we wish to mention two other points. In particular implementations, round-off error may require serious consideration. This issue of numerical stability remains a topic for investigation of considerable importance. Secondly, there are applications such as speech processing for which the estimated transfer function, G, or its inverse must be a stable filter (i.e., all its poles must lie within the unit circle). In order to guarantee filter stability, constraints must be put on the minimization problem; or the error criterion, eq. (2), must be modified; or both (see, for example, [23],[30],[31]). With regard to this issue, it should be noted that the particular structure (see Section III) of the least squares lattice algorithm provides a natural environment for the study of numerical and filter stability properties [8],[20],[23],[31-33].

III. A LEAST SQUARES LATTICE ALGORITHM

In this section we present a modified form of the least squares lattice algorithms discussed in [10], [26]. All these algorithms (which utilize the weighted error function (7)) represent corrected versions of the least squares lattice algorithm originally introduced in [9]. The particular version presented here (as well as in [26]) utilizes the residuals: $y(T) - G(T-1)x(T)$ ($G(T-1)$ represents the estimate of G at time T-1) rather than $y(T) - G(T)x(T)$ as intermediate variables. (Note, this only reflects a modification in the algorithm. The problem solved is still the minimization of $\rho(T) = \sum_{t=0}^{T} w^{T-t}(y(t) - G(T)x(t))^2$.) The resulting algorithm lends itself more readily to certain applications such as noise cancelling and decision directed data equalization where $y(t)$ is not observed but must be estimated from $G(t-1)x(t)$.

Figures 3 and 4 contain a flow diagram of the algorithm for a Pth order least squares lattice at time T. We now present the algorithm followed by some hopefully clarifying remarks.

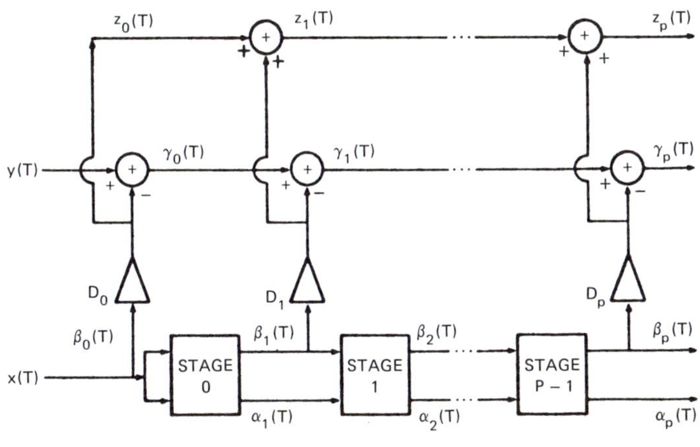

FIGURE 3. Diagram of the least squares lattice structure. The gains D_p are given by $D_p = k_p^{\gamma}(T-1)/\varepsilon_p^{\beta}(T-1)[1-\sigma_{p-1}(T)]$, and the lattice stages are pictured in Figure 4.

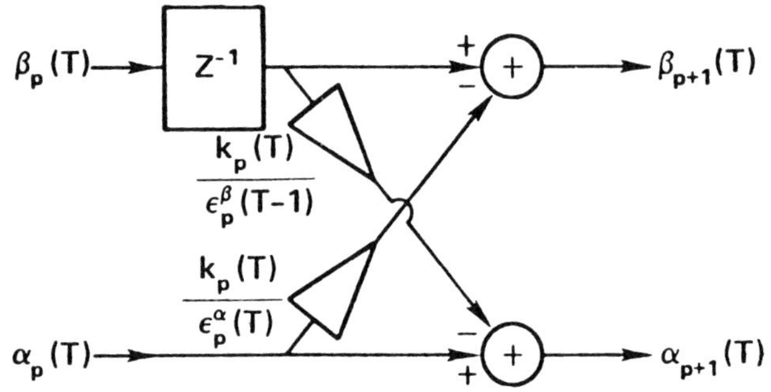

FIGURE 4. Diagram showing the pth stage of the lattice.

The variables:

$$\varepsilon_p^\alpha(T), \alpha_p(T), \varepsilon_p^\beta(T), \beta_p(T), \gamma_p(T) \; ,$$

$$k_p^\gamma(T), z_p(T) \text{ with } p = 0,1,\ldots,P$$

$$\sigma_p(T), k_p(T) \quad \text{with } p = 0,1,\ldots,P-1$$

$x(T), y(T)$ the inputs of Figure 1.

Initialization:

For $p = 0,1,\ldots,P$:

$$\varepsilon_p^\alpha(-1) = \varepsilon_p^\beta(-1) = \delta \tag{8a}$$

$$\beta_p(-1) = k_p^\gamma(-1) = 0 \tag{8b}$$

$$k_p(-1) = 0 \text{ (for } p < P) \tag{8c}$$

Time update:

For $T \geq 1$, the variables at time T are computed from those at time T - 1 by

$$\alpha_o(T) = \beta_o(T) = x(T) \tag{9a}$$

$$\varepsilon_o^{\alpha}(T) = \varepsilon_o^{\beta}(T) = w\varepsilon_o^{\alpha}(T-1) + x^2(T) \tag{9b}$$

$$z_{-1}(T) = \sigma_{-1}(T) = 0 \tag{9c}$$

$$\gamma_{-1}(T) = y(T) \tag{9d}$$

Update the order: for p = 0,...P - 1

$$k_p(T) = w\, k_p(T-1) + \frac{\alpha_p(T)\beta_p(T-1)}{1 - \sigma_{p-1}(T-1)} \tag{10a}$$

$$\alpha_{p+1}(T) = \alpha_p(T) - \frac{k_p(T)}{\varepsilon_p^{\beta}(T-1)}\beta_p(T-1) \tag{10b}$$

$$\beta_{p+1}(T) = \beta_p(T-1) - \frac{k_p(T)}{\varepsilon_p^{\alpha}(T)}\alpha_p(T) \tag{10c}$$

$$\varepsilon_{p+1}^{\alpha}(T) = \varepsilon_p^{\alpha}(T) - \frac{k_p^2(T)}{\varepsilon_p^{\beta}(T-1)} \tag{10d}$$

$$\varepsilon_{p+1}^{\beta}(T) = \varepsilon_p^{\beta}(T-1) - \frac{k_p^2(T)}{\varepsilon_p^{\alpha}(T)} \tag{10e}$$

$$\sigma_p(T) = \sigma_{p-1}(T) + \frac{(\beta_p(T))^2}{\varepsilon_p^{\beta}(T)} \tag{10f}$$

also for p=P:

$$z_p(T) = z_{p-1}(T) + \frac{k_p^{\gamma}(T-1)}{\varepsilon_p^{\beta}(T-1)(1-\sigma_{p-1}(T))}\beta_p(T) \tag{10g}$$

$$\gamma_p(T) = y(T) - z_p(T) \tag{10h}$$

$$k_p^{\gamma}(T) = w\, k_p^{\gamma}(T-1) + \gamma_{p-1}(T)\beta_p(T) \tag{10i}$$

A. Remarks on the Variables

The ε_p^β are the least squares backward prediction errors of order p for p = 0,...,P; i.e.,

$$\varepsilon_p^\beta(T) = \min_b \sum_{t=0}^{T} \left| x(t-p) - \sum_{i=0}^{p-1} b_i^p(T)x(t-i) \right|^2 w^{(T-t)}. \tag{11}$$

Let $\hat{b}_i^p(T)$ be the minimizing coefficients. Then $\beta_p(T)$, the error residual at time T, is given by

$$\beta_p(T) = x(T-p) - \sum_{i=0}^{p-1} \hat{b}_i^p(T)x(T-i). \tag{12}$$

In the stationary case (with w = 1) they are orthogonal in the limit in the sense that $\lim_{T\to\infty} E(\beta_p(T)\ \beta_q(T)) = 0$ for $p \neq q$ [11],[23]. This orthogonality property is related to the gradient lattice formulation where it is responsible for its improved convergence over the LMS gradient method [22],[23],[25].

Similarly, the ε_p^α are the least squares forward prediction errors

$$\varepsilon_p^\alpha(T) = \min_a \sum_{t=0}^{T} \left| x(t) - \sum_{i=0}^{p-1} a_i^p(T)x(t-i-1) \right|^2 w^{(T-t)} \tag{13}$$

and the $\alpha_p(T)$ are the residuals at time T

$$\alpha_p(T) = x(T) - \sum_{i=0}^{p-1} \hat{a}_i^p(T)x(T-i-1). \tag{14}$$

The $\alpha_p(T)$ are not orthogonal. They are introduced as auxiliary variables in the recursion and correspond to the so-called adjoint variables characteristic of many optimization problems.

$\gamma_p(T)$ corresponds to e(T) of Figure 1 and is given by:

$$\gamma_p(T) = y(T) - \sum_{i=0}^{p} \hat{g}_i^p(T-1)x(T-i) , \tag{15}$$

where the coefficients, $\hat{g}_i^p$, solve the original minimization problem (eq. (7)), i.e., the $\hat{g}_i^p$ minimize:

$$\sum_{t=0}^{T} \left| y(t) - \sum_{i=0}^{p} g_i^p(T)x(t-i) \right|^2 w^{T-t}.$$

The variables $z_p(T)$ representing the filter output are given by

$$z_p(T) = y(T) - \gamma_p(T) = \sum_{i=0}^{p} \hat{g}_i^p(T-1)x(T-i) \tag{16}$$

They differ from the least squares estimate of y only in the occurrence of $\hat{g}^p(T-1)$ in place of $\hat{g}^p(T)$. Note that the $\gamma_p(T)$ of equation (15) differ from the pth order least square error residual (e(T) of Figure 1 for a pth order model G) through the same substitution ($\hat{g}^p(T-1)$ for $\hat{g}^p(T)$). The algorithm could have been formulated using $\hat{g}^p(T)$ ([9],[10]), but the present form is more convenient (see explanation at the beginning of this section).

$k_p(T)$ has an interesting interpretation (in the stationary case with w = 1) as the cross-correlation in the limit as $T \to \infty$ between $\alpha_p(T)$ and $\beta_p(T-1)$; i.e., $\lim_{T\to\infty} E(\alpha_p(T)\beta_p(T-1)) = \lim_{T\to\infty} k_p(T)/T$ [9]. (A similar interpretation holds for $k_p^\gamma(T)$ relating $\gamma_{p-1}(T)$ and $\beta_p(T)$.) One can also show that $\varepsilon_p^\alpha(T)/T$ and $\varepsilon_p^\beta(T)/T$ are, in the limit, the autocorrelations of $\alpha_p(T)$ and $\beta_p(T)$, respectively. In fact, in the stationary case (with w = 1) $\lim_{T\to\infty} \varepsilon_p^\alpha(T)/T = \lim_{T\to\infty} \varepsilon_p^\beta(T)/T$, and in the limit of large T, the gains $k_p(T)/\varepsilon_p^\alpha(T)$ and $k_p(T)/\varepsilon_p^\beta(T-1)$, approach the so-called "PAR COR coefficients" [23],[33].

B. Remarks on the Algorithm

Nowhere in the above algorithm do the filter coefficients $g_i^P(T)$ of the model estimate G_P appear. If one only needs G at a single time T (for example when the error e(T) of Figure 1 is sufficiently small) one may compute it from the lattice variables using $O(P^2)$ operations and some additional memory [9-11]. However, if one needs $G_P(T)$ (not just $z_p(T)$ or $\gamma_P(T)$) for repeated values of T, there may be difficulties. In this case, it may be more computationally efficient to use one of the time-only recursive least squares algorithms discussed in Section II (i.e., [18],[19]).

A second related remark regarding the least squares lattice filter concerns any situation where it is desirable to stop updating the filter but to continue processing data. (Such a situation might arise, for instance, in a data equalization application where one first trains an adaptive equalizer algorithm with a reference signal prior to actual data transmission; i.e., preset equalization [18].) One possible approach to this problem is to freeze the algorithm gains at some time T, and then use the resultant structure to filter the data (i.e., freeze all intermediate variables except α_p and β_p - see Figures 3 and 4). Note, however, the resulting filter is only an approximation to $G_P(T)$; i.e., if the original data x(0),...,x(T) is fed back into the frozen lattice, the output in general is not $z_P(T)$ at time T. This is because as the data originally flowed through the lattice, the gains were varying (updating in time). The values of the variables β_p so obtained are therefore different from those which would arise by passing the same data through the lattice with the gains frozen at the particular values they held at time T. On the other hand, if the input data are stationary and the lattice filter coefficients are frozen near convergence, the resulting filter will closely approximate $G_P(T)$ inasmuch as both approach $G_P(\infty)$.

Fortunately, not having the filter coefficients $g_i^P(T)$ readily available is not usually a serious problem since in most applications we are only interested in the output $z_P(T)$ (e.g., in equalization) or the output error $\gamma_P(T)$ (e.g., in noise cancelling), or the frozen lattice is a good approximation to the desired filter (e.g., in speech processing).

Our final remark concerns the initialization. We note that for $T \leq P$ the correlation matrix $R_P(T)$ is singular. This is reflected in the fact that the least squares problem only has a solution for $P \leq T$. To circumvent this fact without altering the algorithm a small positive quantity δ has been introduced (in analogy with the treatments given in [18],[19]). This insures non-singularity of $R_P(T)$ for all T. Although the results appear to be insensitive to δ over a large range of values, if δ is too large it may slow initial convergence. Too small a δ can cause an overshoot which may be handled by placing a limiter on the variables. Alternatively, one may use a different algorithm for T = P ($O(P^2)$ operations) and then begin the update. The authors are currently investigating the subject.

IV. APPLICATIONS

In this section we briefly discuss several potential applications for the recursive least squares algorithm treated in the previous sections.

A. Noise Cancelling

As detailed in the review paper by Widrow, *et al* [29], an important application of adaptive processing is the on-line cancellation of a known noise source from a signal of interest. In the notation of Section III, the "primary" input y(t) contains both signal and noise, and x(t) is the noise "reference" input which is used to cancel noise from y(t). The noise canceller output is then the error output sequence $\gamma_p(t)$ (see equation (15)).

As noted in [29], adaptive noise cancelling has been applied to a wide range of problems including the removal of noise from speech [35] and biomedical applications such as cancelling the maternal ECG waveform in fetal electrocardiography [34]. It can even be used with a single input provided the signal is highly coherent in comparison with the noise [29]. In that case the primary y(t) contains both the signal and noise, while the reference x(t) is a delayed version of y(t); i.e., $x(t) = y(t-\Delta)$. The delay is chosen sufficiently long to decorrelate the noise but not the signal [36-38].

There also arise numerous situations (such as adaptive antenna array processing) where it is desirable to use multiple inputs for noise cancelling. In that case, it is necessary to develop a vector input formulation of the noise cancelling algorithm. The modifications tend to be straightforward for the various adaptive algorithms including the gradient LMS (see eg. [29]), the gradient lattice ([22],[40]), and various least squares recursive algorithms ([2-4], [6],[7],[9],[19]).

We note that in the majority of the above referenced noise cancelling applications, the gradient LMS algorithm has been utilized. More recently, alternative algorithms have been applied to these problems [22],[39]. The recursive least squares algorithms discussed in the previous sections hold promise of providing still improved performance.

B. Speech Processing

Aside from the noise cancelling applications of adaptive processing to speech enhancement [35],[38] there also exist applications in the areas of speech modeling and analysis. In particular, as discussed in the review article by Makhoul [17] (see also the book by Markel and Gray [41]), a rather successful model which has been applied to speech analysis is based on simple linear prediction. In particular, a weighted sum of delayed speech samples, $g_1^p y(t-1) + \ldots + g_p^p y(t-p)$, are used to approximate (in the sense of least squares) the current speech sample y(t). This corresponds to Figure 1 with y(t) equal to the input sampled speech to be analyzed and x(t) = y(t-1). The least squares model G is then the pth order 1-step linear predictor. The coefficients may then be used in an inverse filter to reconstruct a synthesized speech signal of the form G^{-1} s(t). As discussed in [17], the input s(t) may either be a noise source (to produce the unvoiced or fricated sounds such as f, th, s, sh) or a periodic source (to produce sounds such as vowels and nasals).[5] The linear prediction filter coefficients can also be used to analyze the spectral content of the speech [17]. In particular, the spectral estimate $S(\omega) = 1/|1 - \sum_{i=1}^{P} g_i^P e^{-j\omega i}|^2$ (or a scaled version - see [17]) may be computed for $0 \leq \omega \leq \pi$ and analyzed to determine the spectral characteristics of the sampled speech signal.

Typically, in many speech analysis and modeling applications which employ linear prediction models, block data (i.e., non-time recursive) modeling algorithms have been used (such as those discussed in Section II - see, e.g., [5],[17]). More recently, in

---

[5]An interesting application of linear prediction and, in particular, lattice structures to the commercial development of a speech synthesis system can be found in [42],[43].

certain on-line applications, time recursive modeling algorithms have been proposed (see, e.g., [8],[44]). It thus seems likely that future work in speech modeling and analysis will see more application of efficient recursive least squares algorithms.

C. Geophysics

A major problem in geophysics involves the deconvolution of seismic waves [45]. Such waves are generated either by earthquakes or artificial means such as occurs in oil exploration. In particular, by determining the travel times and strengths of the various multipath arrivals information is obtained concerning the layered rock structure and/or source mechanism.

Unfortunately, the seismic traces are typically contaminated by the presence of so-called "multiple reflections" or "reverberations". These slowly decaying wave trains most commonly arise when the seismic sensors are located on the surface of water. For this case, the energy from the seismic source becomes trapped in the water layer thus creating the reverberation.

Among the most popular methods which have been proposed to eliminate the reverberation is the method of predictive deconvolution [45],[46]. This technique models the seismic trace by an all-pole process driven by an uncorrelated sequence. This sequence, which represents the reflection coefficients of the deep rock layers, contains the desired geophysical information.

Specifically, in terms of the variables of Section III, the seismic trace corresponds to $y(t)$, and the input $x(t)$ is a delayed version of y; i.e., $x(t) = y(t-\Delta)$. Ideally, the delay is matched to the fundamental period of the reverberation [45]. The error, $\gamma_p(T)$, then represents the deconvolved trace.

As in speech processing, block data processing methods have been usually applied to the deconvolution problem [46], although more recently time recursive algorithms have been proposed [47].

It would be interesting to also apply the least squares lattice algorithm in Section III to this problem. (With respect to this point, it is interesting to note that lattice forms have recently been considered in geophysics applications (see, e.g., [48]).)

D. Data Communications

Adaptive processing algorithms have been applied to a wide variety of problems in data communications (see, e.g., the review article by Lucky [49]) although perhaps the most common one is data equalization. In data equalization one is interested in eliminating the amplitude and phase distortion which may be introduced by the communication channel. Since in many applications, the channels are time varying, considerable research in the past two decades has been directed towards the design of adaptive equalizers. These equalizers are time-varying filters constructed to continually adjust their coefficients so as to minimize some measure (typically mean square) of the channel distortion. In the context of this paper, the inverse of the channel corresponds to G(T), the modeling criterion is the minimization of ρ (eq. (7)), the transmitted message is y(T) (in decision-directed equalization this is estimated by thresholding the output $z_p(T)$), and the received message is x(T).

Much of the early work in this field (up to 1973) has been described in [49] (see also [27]). More recent work [18],[19],[25],[26],[50] has concentrated on the development of fast converging algorithms with convergence rates insensitive to the amount of amplitude distortion present in the channel. It is with respect to this issue that the recursive least squares algorithms show great promise ([18],[19] but see footnote 2, [26]). Finally, it should be noted that another application of adaptive processing to data equalization (based on linear prediction methods) has recently been developed by Morgan [51]. Clearly, the application of recursive estimation algorithms to communication systems remains a fruitful and active area of research.

E. Other Applications

We briefly mention some additional applications; instantaneous frequency tracking [52],[53]; intrusion detection [54],[55]; acoustic Doppler extraction [56], and, of course, on-line system identification [12],[57]. Considering the pervasive use of least squares techniques in signal processing, we feel confident that the above listing is far from complete.

REFERENCES

1. Kailath, T., "A View of Three Decades of Linear Filtering Theory," IEEE Trans. Inform. Theory, Vol. IT-20, pp. 146-181, Mar. 1974.
2. Morf, M., "Fast Algorithms for Multivariable Systems," Ph.D. Thesis, Stanford University, Stanford, Ca., 1974.
3. Ljung, L., Morf, M., and Falconer, D., "Fast Calculation of Gain Matrices for Recursive Estimation Schemes," Int. Journal of Control, Vol. 27, No. 1, pp. 1-19, 1978.
4. Morf, M., Lee, D., Nickolls, J., and Vieira, A., "A Classification of Algorithms for ARMA Models and Ladder Realizations," Proc. IEEE Inter. Conf. on Acoustics, Speech, and Sig. Proc., Hartford, Ct., pp. 13-19, May 1977.
5. Morf, M., Dickinson, B., Kailath, T., and Vieira, A., "Efficient Solution of Covariance Equations for Linear Prediction," IEEE Trans. on Acoustics, Speech, and Sig. Processing Vol. ASSP-25, No. 5, pp. 429-433, Oct. 1977.

6. Morf, M., "Ladder Forms in Estimation and System Identification," 11th Annual Asilomar Conf. on Circuits, Systems, and Computers, Monterey, Ca., Nov. 7-9, 1977.

7. Morf, M., Vieira, A., and Lee, D., "Ladder Forms for Identification and Speech Processing," Proc. 1977 IEEE Conference Decision and Control, New Orleans, La., pp. 1074-1078, Dec. 1977.

8. Morf, M., and Lee, D., "Fast Algorithms for Speech Modeling," Final Report to Defense Communication Agency, Contract DCA 100-77-C-0005, Information Systems Laboratory, Stanford University, Stanford, Ca., Nov. 1978.

9. Morf, M., and Lee, D., "Recursive Least Squares Ladder Forms for Fast Parameter Tracking," Proc. of the 1978 IEEE Conf. on Decision and Control, Jan. 12, 1979, San Diego, Ca., pp. 1362-1367.

10. Pack, J. D., and Satorius, E. H., "Least Squares, Adaptive Lattice Algorithms," Naval Ocean Systems Center Technical Report 423, San Diego, Ca.

11. Shensa, M. J., "Recursive Least Squares Lattice Algorithms - A Geometrical Approach," Sub. for Publication to <u>IEEE Trans. on Automatic Control</u>.

12. Eykhoff, P., "System Identification: Parameter and State Estimation," New York: Wiley, 1974.

13. Akaike, H., "A New Look at Statistical Model Identification," <u>IEEE Trans. Automat. Contr.</u>, <u>Vol. AC-19</u>, pp. 716-723, Dec. 1974.

14. Levinson, N., "The Wiener RMS (Root Mean Square) Error Criterion in Filter Design and Prediction," <u>J. Math. Phys.</u>, <u>Vol. 25</u>, No. 4, pp. 261-278, 1947. Also Appendix B, in N. Wiener, "Extrapolation, Interpolation and Smoothing of Stationary Time Series," Cambridge, Mass.: M.I.T. Press, 1949.

15. Hildebrand, F. B., "Introduction to Numerical Analysis," New York: McGraw-Hill, 1956.

16. Faddeev, D. K., and Faddeeva, V. N., "Computational Methods of Linear Algebra," San Francisco, Ca.: Freeman, 1963.
17. Makhoul, J., "Linear Prediction: A Tutorial Review," Proc. IEEE, Vol. 63, No. 4, pp. 561-580, April 1975.
18. Godard, D., "Channel Equalization Using a Kalman Filter for Fast Data Transmission," IBM Journal of Research & Development, pp. 267-273, May 1974.
19. Falconer, D. D., and Ljung, L., "Application of Fast Kalman Estimation to Adaptive Equalization," IEEE Trans. Comm., Vol. COM-26, No. 10, pp. 1439-1446, Oct. 1978.
20. Markel, J. D., and Gray, A. H., Jr., "Roundoff Noise Characteristics of a Class of Orthogonal Polynomial Structures," IEEE Trans. Acoust., Speech, and Sig. Proc., Vol. ASSP-23, pp. 473-486, Oct. 1975.
21. Widrow, B., "Adaptive Filters," in "Aspects of Network and System Theory," R. Kalman and N. DeClaris, Eds., New York: Holt, Rhinehart, and Winston, pp. 563-587, 1971.
22. Griffiths, L. J., "An Adaptive Lattice Structure for Noise-Cancelling Applications," in Proc. IEEE Int. Conf. on Acoust., Speech, and Sig. Proc., Tulsa, Ok., April 1978, pp. 87-90.
23. Makhoul, J., "A Class of All-Zero Lattice Digital Filters: Properties and Applications," IEEE Trans. on Acoust., Speech, and Sig. Proc., Vol. ASSP-26, No. 4, pp. 304-314, Aug. 1978.
24. Widrow, B., McCool, J., Larimore, M., Johnson, C., Jr., "Stationary and Non-Stationary Learning Characteristics of the LMS Adaptive Filter," Proc. IEEE, Vol. 64, No. 8, pp. 1151-1162, Aug. 1976.
25. Satorius, E. H., and Alexander, S. T., "Channel Equalization Using Adaptive Lattice Algorithms," IEEE Trans. Comm., Vol. COM-27, No. 6, pp. 899-905, June 1979.
26. Satorius, E. H., and Pack, J. D., "Application of Least Squares Lattice Algorithms to Adaptive Equalization," Sub. for publication to IEEE Trans.on Comm.

27. Gersho, A., "Adaptive Equalization of Highly Dispersive Channels for Data Transmission," Bell Syst. Tech. J., Vol. 48, pp. 55-70, 1969.

28. Shensa, M. J., "Time Constants and Learning Curves of LMS Adaptive Filters," Naval Ocean Systems Center Technical Report 312, San Diego, Ca., Nov. 1978.

29. Widrow, B., et. al., "Adaptive Noise Cancelling: Principles and Applications," Proc. IEEE, Vol. 63, pp. 1692-1716, Dec. 1975.

30. Burg, J. P., "Maximum Entropy Spectral Analysis," Ph.D. Dissertation, Geophysics Dept., Stanford Univ., Stanford, Ca., May 1975.

31. Makhoul, J., "Stable and Efficient Lattice Methods for Linear Prediction," IEEE Trans. Acoustics, Speech, and Signal Processing, Vol. ASSP-25, No. 5, pp. 423-428, Oct. 1977.

32. Itakura, F., and Saito, S., "Digital Filtering Techniques for Speech Analysis and Systems," Proc. 7th Int. Cong. Acoust., Budapest, Paper 25-C-1, pp. 261-264, 1971.

33. Markel, J., and Gray, A. H., Jr., "On Autocorrelation Equations as Applied to Speech Analysis," IEEE Trans. on Audio and Electroacoustics, Vol. AU-21, No. 2, pp. 69-79, April 1973.

34. Cox, J. R., Jr., and Medgyesi-Mitschang, L. N., "An Algorithmic Approach to Signal Estimation Useful in Fetal Electrocardiography," IEEE Trans. Biomed. Eng., Vol. BME-16, pp. 215-219, July 1969.

35. Pulsipher, D., Boll, S., Rushforth, C., and Timothy, L., "Reduction of Nonstationary Acoustic Noise in Speech Using LMS Adaptive Noise Cancelling," Proc. IEEE Int. Conf. on Acoust., Speech, and Sig. Proc., Washington, D. C., April 2-4, 1979, pp. 204-207.

36. Treichler, J. R., "Transient and Convergent Behavior of the Adaptive Line Enhancer," IEEE Trans. Acoust., Speech, and Sig. Proc., Vol. ASSP-27, pp. 53-62, Feb. 1979.

37. Zeidler, J. R., Satorius, E. H., Chabries, D. M., and Wexler, H. T., "Adaptive Enhancement of Multiple Sinusoids in Uncorrelated Noise," Trans. Acoust., Speech, and Sig. Proc., Vol. ASSP-26, No. 3, pp. 240-254, June 1978.

38. Sambur, M. R., "Adaptive Noise Cancelling for Speech Signals," IEEE Trans. Acoust., Speech, and Sig. Proc., Vol. ASSP-26, pp. 419-423, Oct. 1978.

39. Satorius, E. H., Smith, J. D., and Reeves, P. M., "Adaptive Noise Cancelling of a Sinusoidal Interference Using a Lattice Structure," in Proc. IEEE Int. Conf. on Acoust., Speech, and Sig. Proc., Washington, D. C., April 2-4, 1979 pp. 929-932.

40. Griffiths, L. J., "Adaptive Structures for Multiple-Input Noise Cancelling Applications," in Proc. IEEE Int. Conf. on Acoustics, Speech, and Sig. Proc., Washington, D. C., April 2-4, 1979, pp. 925-928.

41. Markel, J. D., and Gray, A. H., Jr., "Linear Prediction of Speech," Berlin: Springer, 1976.

42. Wiggins, R., and Brantingham, L., "Three-Chip System Synthesizes Human Speech," Electronics, pp. 109-116, Aug. 31, 1978.

43. Brantingham, L., "Speech Synthesis with Linear Predictive Coding," Interface Age, pp. 72-75, June 1979.

44. Morgan, D. R., and Craig, S. E., "Real-Time Adaptive Linear Prediction Using the Least Mean Square Gradient Algorithm," IEEE Trans. Acoust., Speech, and Sig. Proc., Vol. ASSP-24, No. 6, pp. 494-507, Dec. 1976.

45. Wood, L. C., and Treitel, S., "Seismic Signal Processing," Proc. IEEE, Vol. 63, No. 4, pp. 649-661, April 1975.

46. Peacock, K. L., and Treitel, S., "Predictive Deconvolution: Theory and Practice," Geophysics, Vol. 34, No. 2, pp. 155-169, April 1969.

47. Griffiths, L. J., Smolka, F. R., and Trembly, L. D., "Adaptive Deconvolution: A New Technique for Processing Time-Varying Seismic Data," Geophysics, Vol. 42, No. 4, pp. 742-759, June 1977.

48. Alam, M. A., "Orthonormal Lattice Filter - A Multistage Multichannel Estimation Technique," Geophysics, Vol. 43, No. 7, pp. 1368-1383, Dec. 1978.

49. Lucky, R. W., "A Survey of the Communication Theory Literature," IEEE Trans. on Inform. Theory, Vol. IT-19, No. 5, pp. 725-739, Nov. 1973.

50. Gitlin, R. D., and Magee, F. R., "Self-Orthogonalizing Adaptive Equalization Algorithms," IEEE Trans. on Comm., Vol. COM-25, No. 7, pp. 666-672, July 1977.

51. Morgan, D. R., "Adaptive Multipath Cancellation for Digital Data Communications," IEEE Trans. Comm., Vol. COM-26, No. 9, pp. 1380-1390, Sept. 1978.

52. Griffiths, L. J., "Rapid Measurement of Digital Instantaneous Frequency," IEEE Trans. Acoust., Speech, and Sig. Proc., Vol. ASSP-23, pp. 209-222, April 1975.

53. Hsu, F. M., and Giordano, A. A., "Line Tracking Using Autoregressive Spectral Estimates," IEEE Trans. on Acoust., Speech, and Signal Proc., Vol. ASSP-25, No. 6, pp. 510-519, Dec. 1977.

54. Ahmed, N., et. al., "On an Intrusion-Detection Approach via Adaptive Prediction," to appear in the IEEE Trans. on Aerospace and Electronics Systems.

55. Hass, R. J., Lenhert, D. H., and Ahmed, N., "On a Microcomputer Implementation of an Intrusion-Detection Algorithm," to appear in the IEEE Trans. on ASSP.

56. Keeler, R., and Griffiths, L. J., "Acoustic Doppler Extraction by Adaptive Linear Prediction Filtering," J. Acoust. Soc. Amer., Vol. 61, pp. 1218-1227, May 1977.

57. Rabiner, L. R., Crochiere, R. E., and Allen, J. B., "FIR System Modeling and Identification in the Presence of Noise and with Band-Limited Inputs," IEEE Trans. on Acoust., Speech, and Sig. Proc., Vol. ASSP-26, No. 4, pp. 319-333, Aug. 1978.

ADAPTIVE POWER PLANT CONTROL: PROBLEMS AND PROSPECTS[1]

Raman K. Mehra
John S. Eterno

Scientific Systems, Incorporated
Cambridge, Massachusetts

Sadashiva S. Godbole

The Babcock and Wilcox Company
Lynchburg Research Center
Lynchburg, Virginia

This paper discusses potential applications of adaptive control theory to problems in electric generating plant operation. A brief discussion of nuclear and fossil-fueled power plant design is included for background information. The problems described include feedwater level control, coordinated plant control during normal and low power operation, control during contingencies or partial

[1]This work was supported in part by DOE Contract ET-78-C-01-3394 and EPRI Contract TPS78-811-1.

ISBN 0-12-514060-6

system failure, nuclear core power distribution control, and fault detection and diagnosis. Although adaptive control is particularly well suited to some aspects of these problems, its application is hindered by several obstacles, both technical and human. The opinions expressed are the result of the authors' recent work in power plant control and a recent survey of nuclear control problems they helped to conduct for the Electric Power Research Institute.

INTRODUCTION

This paper discusses the role of adaptive control in modern power systems with particular emphasis on generating plant control. We will be more concerned with relatively slow (seconds to days) time constants, and will not consider the high-speed automatic voltage and frequency control systems. Our discussion will necessarily be general, since power plants are usually built to order, by several parties, and very few standard designs exist, although recent trends in nuclear units are toward standardization to facilitate licensing.

We will begin by describing nuclear and fossil-fueled (coal and oil) generating stations, and then proceed to discuss generally accepted problems in plant operation. Next, we describe those areas where adaptive control may be expected to help, and finally we discuss the difficulties in actually implementing adaptive controllers.

BACKGROUND: POWER PLANT OPERATION

Nuclear and fossil-fueled power plants operate by using a controlled source of heat to make high-pressure steam which is expanded in turbines, producing work to turn large alternators.

The major difference between nuclear and fossil units is the source of heat used to produce the steam, but other subtle differences exist which significantly affect the detailed control system design. Excellent references on plant operation are available in [4],[6] and [7].

Nuclear Plants

A simplified schematic of a pressurized-water-reactor (PWR) is shown in Figure 1. Several parallel components (e.g., pumps and steam generators) are grouped for clarity. The plant is usually divided, as shown, into primary (reactor coolant) and secondary (steam) circuits for convenience.

The reactor is a pressure vessel containing fissionable, slightly enriched uranium. The rate of reaction (or reactivity) is controlled by the position of neutron-absorbing control rods, the concentration of boron in the coolant, and the average tempera-

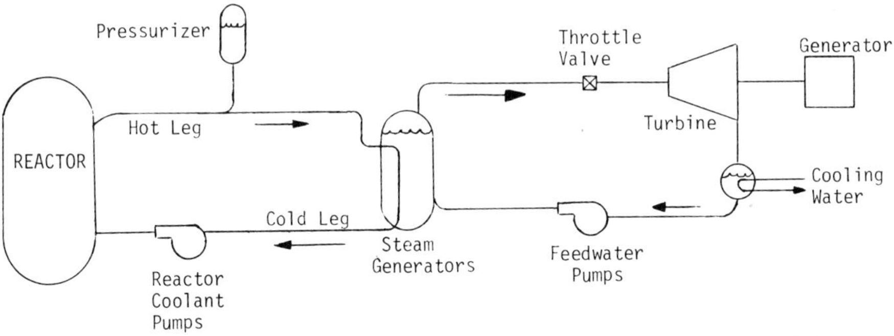

FIGURE 1. Simplified nuclear plant schematic.

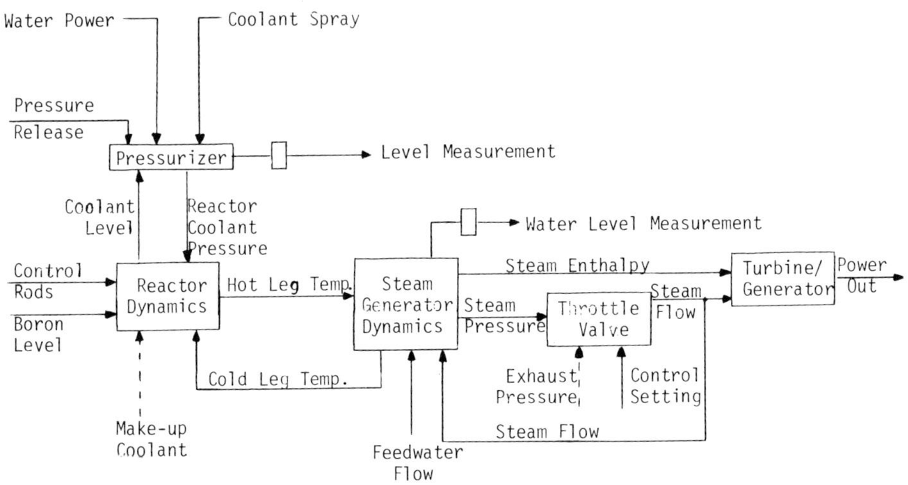

FIGURE 2. Functional inputs and outputs of PWR components.

ture (thus density) of the coolant. Reactivity is also a function of the past history (e.g., 30 hrs.) of power production through long-term poison (neutron absorbing) effects. Control inputs to the reactor may be seen in Figure 2.

The reactor heat is transferred to cooling water (under a pressure of about 2200 psi) which is circulated to a steam generator. The steam generator is a heat exchanger which uses the high temperature reactor coolant (about 600°F) to boil water on the secondary side, producing steam at about 1000 psi and 550°F. The secondary side input to the steam generator is high pressure water, pumped in several stages, from the condenser. The primary side steam generator output is slightly cooled (about 550°F) reactor coolant, which is pumped back to the reactor by the pumps shown. The amount of heat transferred in the steam generator is a complex function of the temperatures, pressures and flows on each side, but the basic inputs and outputs may be seen in Figure 2.

Two basic types of steam generators are used in PWR's. The first is a recirculating design, where only a fraction of the water (secondary side) that is flowing through the generator is turned into steam. The remaining liquid is combined with feedwater and recirculated through the heating section. The second type of generator is a "once-through" design, where all the water is converted to steam in one pass.

Between the reactor and steam generator, in the hot leg, is a large pressurizer which maintains coolant pressure at the desired set point. The pressurizer works by adding or subtracting heat (using electric heaters and water sprays, respectively) to a saturated liquid-vapor mixture, causing a change in the mixture ratio. The pressurizer also serves as a surge tank to accommodate coolant volume changes (with temperature) over a large operating range. The long-term liquid level is controlled by adding or bleeding coolant at the reactor input (cold) leg.

On the secondary side, the steam generator produces saturated steam (recirculation types) or slightly superheated vapor (once-through designs) which is expanded in large turbines to produce power. The flow of steam is controlled by the throttle valve, but the flow is also a function of steam and exhaust (condenser) pressures. The steam path may be much more complex than that shown here, with steam being used to power auxiliary systems, pumps, and feedwater heaters.

The secondary cycle is completed when the turbine exhaust is condensed, using cooling water, chemically treated and then pumped, in several stages, back to the steam generator input.

While the investigation of any problem of interest would require a dynamic model for each of the components discussed above, the basic interactions are visible in Figure 2. The available plant inputs (excluding pressurizer functions) for normal operation are the reactivity controls (control rods and boron concentration), feedwater flow, and throttle valve opening. These variables may be manipulated in several different ways to achieve the desired overall plant response, while satisfying constraints, as discussed below.

Nuclear Control Objectives

The primary control objective is to produce the power demanded by the network while satisfying all plant constraints. Because of the low fuel cost of nuclear units, they are usually run at full power all the time (base loaded) while more expensive plants are brought up to handle the daily peaks and brought down at night when demand drops (load cycling). Nonetheless, during contingencies (e.g., temporary load disruptions) or unusual operating conditions (e.g., a surplus of hydroelectric power which must be used or wasted), it becomes important for nuclear plants to quickly respond to load changes.

To change power output, the steam flow through the turbine must be changed. In general, this may be accomplished by altering the pressure of the steam or the area of the throttle valve; nuclear plants usually change the throttle area. For high frequency load changes, the throttle is controlled by a speed governor on the turbine, thus ensuring a stable line frequency. (This is part of the automatic frequency control that maintains the 60 Hz. line frequency.) At the lower frequency load changes of interest to us, the governor characteristic is changed, by command, to produce different power levels at constant line frequency.

When more power is required and the steam flow is increased, the steam generator energy is depleted. Unless the reactor power is also increased, the steam pressure will drop causing a decrease in steam flow and power output. To actively change the reactor power, the control rods may be withdrawn or the boron concentration decreased. Reactor power is also increased automatically when the steam flow is increased (assuming the feedwater flow is also increased to keep steam generator level constant). The increased secondary flow causes the primary side temperature to drop (more heat is being transferred) which increases reactivity. Although this power transfer arrangement is less efficient than the higher reactor temperature would permit, the phenomenon has a stabilizing effect on plant control. One complication, however, is that this temperature coefficient of reactivity changes with boron concentration and core life.

Apart from the power balances discussed above, the most important plant control tasks involve the feedwater and steam generator systems. The two types of steam generators--recirculating and once-through--require different control strategies. The recirculating flow units try to maintain a constant water level in the generator secondary side, which keeps the primary tubes submerged. This water level control task is similar to that of fossil-fired drum boiler units. Modern controllers use three signals to control this level by changing the feedwater pumping rate: the current water level, the feedwater flow rate, and the steam flow rate. Single input controllers, using only water level measurements, are fooled by the non-minimum phase nature of the problem, which often causes the water level to drop before rising (shrink and swell) when less steam, and therefore water, is needed. The shrink/swell phenomenon in particular is common to both nuclear (recirculating) and fossil (drum) plants.

Once-through steam generators are controlled somewhat differently. In these units, the water (liquid) level is a boundary which moves as a function of load, with low load creating a lower level and more superheating of the steam. The controlled quantity in these generators is steam pressure, which is a function of the heat being added to the fluid, the steam flow rate, and the water flow rate. The water flow is thus controlled to maintain pressure, although a derivative signal (a load indication from the first stage turbine pressure) is also used to anticipate pressure demands [4].

One other local loop control problem that deserves special mention is the control of the spatial power distribution in the core. While the overall plant controller is primarily concerned with the total power produced in the reactor, fuel temperature constraints and the desire to evenly burn the available fuel require that care be taken in the distribution of neutron flux as well as its total value. The instantaneous flux is a function of the arrangement and "worth" of the fuel assemblies, the control rod positions (both full and part-length rods), the boron concentration of the coolant, the temperature distribution of the core, and the past history of the power distribution. [4] and [3] contain thorough descriptions of these dependencies.

The important time constants in core control, ignoring the fastest, passively stable interactions, range from seconds to months. The faster phenomena are controlled by the instantaneous core state (rod positions, temperatures, etc.), while the long term effects are governed by the net fuel burn-up resulting from the short-term strategies. The long-term behavior, in turn, affects the coefficients of the short term processes (e.g., the boron changes with core life affect the temperature coefficient of reactivity). Between the two extremes are the xenon-iodine transients (neutron absorbing fission decay products) with time constants of hours. These dynamics in particular result in spatial power oscillations with periods of 25 to 35 hours. The dominant oscillatory mode is along the vertical axis of the core cylinder.

Radial and azimuthal power imbalances are less severe and are usually controlled by fuel rod shuffling during refueling operations.

Reaching these basic control objectives is made difficult by several sets of constraints on plant states and control inputs. The most important constraints are the fuel temperature and temperature derivative limits, which restrict reactor power changes. Other limits exist on actuator ranges and speeds, turbine component temperatures (especially at low-load steam conditions), and water levels in the pressurizer and steam generators (both high and low limits). Some of these constraints, especially those affecting reactor operation, are enforced by the independent protection (reactor safety) system. It is generally desirable to avoid these limits since safety system shutdowns inevitably require lengthy complicated re-start procedures. Most control systems try to maneuver within their own, tighter limits to avoid this safety system intervention.

Fossil Plant Design

There is a greater variety in fossil plant designs than in nuclear stations, and our description below is limited, therefore, to "typical" characteristics. Our discussion generally applies to coal, oil, and natural-gas fueled boiler-turbine-generator (BTG) sets, although the steam side of other generators (wood, multi-fuel, etc.) is usually similar.

In many respects, fossil plants are more complex and difficult to control than nuclear ones. This is primarily a consequence of the hot combustion flame and the resulting higher steam temperatures. At several places in modern, superheated-steam plants, the steam temperature may approach 1050°F, and an error of only 20°F in temperature control can result in significant pipe damage.

A simplified fossil plant diagram is shown in Figure 3. Fossil plants generally burn the fuel in a large furnace with vertical pipes along the walls. The pipes contain water which is partially vaporized by the time it reaches the top of the furnace. The steam from this water is collected and dried in a drum, which sends the liquid and new feedwater back to the bottom of the furnace to be heated again. Steam from the drum (at, for example, 2300 psi and 655°F) goes through more pipes in the combustion gas path to be superheated (to about 1050°F), and then is sent through the throttle valve to the high pressure turbine.

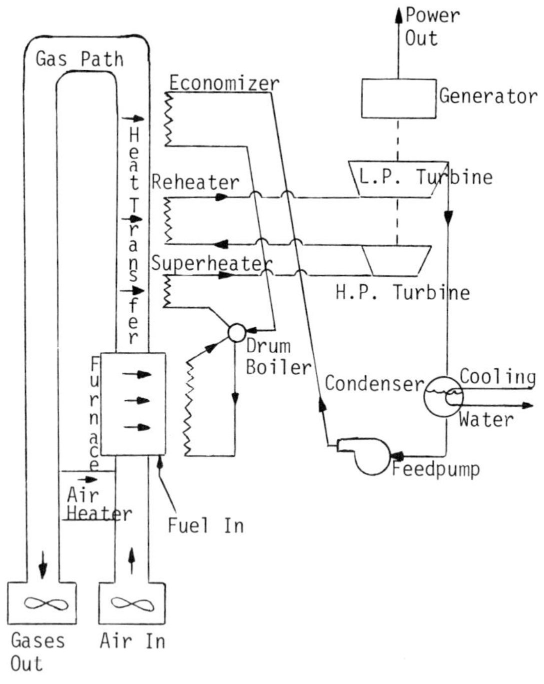

FIGURE 3. Fossil plant schematic.

The high pressure turbine exhaust (e.g., at about 700°F and 610 psi) is sent back to the furnace to be reheated (to the 1050°F pipe limit) and then goes to the intermediate and low pressure turbines. This turbine exhaust is condensed and pumped back to the drum through heaters, called economizers. Many other steam paths are not shown in the figure, but typically steam is drawn from the turbines to heat the feedwater, drive fuel and water pumps, and start other boilers.

There is a great variation in operational plant designs, some using more or fewer superheaters, reheaters, feedwater heaters or other devices to increase efficiency. There are also several different boiler designs, including once-through boilers (analogous to the nuclear once-through steam generators) and super-critical plants, which operate above the critical pressure of water (3206 psia), and only expand the water to steam at the throttle. The primary reason for these variations is efficiency, but the complications often result in poor controllability, with some of the most efficient plants also being the most temperamental.

The furnace in fossil units is designed to ensure complete combustion and almost total heat removal from the combustion product gases. The heat transfer is mostly radiative in the main boiler, where the combustion gas is hottest, shifting to convective transfer in the superheater sections, where the exhaust gas has cooled enough to freeze the corrosive ash suspended in it. Before being exhausted, the gases may be recirculated to control combustion and superheater temperatures.

The combustion dynamics of most fossil units are quite fast, and a new firing rate can be established long before the steam flow has set up to take advantage of it. The major delay in the combustion path is a transport delay in some coal conveyor systems, which can be as long as tens of seconds.

Fossil Control Objectives

Control of fossil plants has much the same overall objectives as those of nuclear units, but the particular details differ. Because of the high superheater temperatures and their closeness to the damage point of the piping, extreme care is taken in temperature control, and strong control action (water spray) is available at the sacrifice of efficiency.

The control of output power in BTG's may be classed as boiler-following, turbine-following, or coordinated. In boiler following mode, the throttle valve controls steam flow, and the boiler is controlled to maintain steam pressure. Thus, when the valve is opened to increase power, the pressure begins to drop and the fuel is increased to maintain pressure. The boiler thus follows the throttle. In turbine-following mode, the reverse happens. The fuel rate is adjusted to produce the power required, and the throttle is controlled to maintain pressure. The best response is available with the third type, coordinated control, where the boiler, throttle and feedwater pumps are all coordinated to control power, water level and steam conditions.

Another distinction is sometimes made in power control, that between sliding and constant pressure techniques. The constant pressure approach maintains steam pressure and reduces power by closing the throttle valve. This approach means that steam is available at low power for sudden load increases. Unfortunately it also means that the steam is "throttled down" at low power, with a significant loss of plant efficiency and possible turbine stress due to temperature drops caused by the throttle. The sliding pressure technique strives to keep the throttle wide open, while controlling power (steam flow) by adjusting pressure (via fuel and water flows). This technique is slower but more efficient than constant pressure control. At least one researcher,

[8], has suggested a dual-mode control, where sliding pressure control is used normally but a constant pressure algorithm is employed during alert conditions, when rapid power changes may be anticipated.

Most other plant controls use local loops to maintain temperature, pressure, or flow setpoints. For instance, air flow is usually controlled as a function of fuel flow, with adjustments made to ensure combustion efficiency, while furnace pressure is controlled (in some cases) independently with exhaust fans or dampers. The plant controller only requests a fuel flow (or heat input), and the local loops do the rest. Predictably, this philosophy results in certain compromises in transient performance, but is entirely adequate for the steady state operation originally envisioned for most plants.

CURRENT CONTROL PROBLEMS

Although power plants perform reasonably well under a wide variety of conditions, there is considerable room for improvement in most areas of operation. As one source phrased it, [1], referring to the need for periodic control tuning, "in fact, this performance has been adequate for system needs but not satisfactory." We believe several areas of plant control are "adequate, but not satisfactory."

Nuclear Control Problems

Several problem areas were identified in our recent survey for EPRI (in collaboration with Babcock & Wilcox under contract no. TPS78-811-1):

Feedwater control is one of the most common problems in nuclear plants, especially during plant transients. The problem occurs in level control in recirculating steam generators and flow control, with level oscillations, in once-through units. The problem apparently is caused by the shrink/swell phenomena discussed above and hydraulic flow instabilities resulting from the interaction of separate control valves and pumps [2]. These instabilities are poorly handled by classic controllers with constant gains. A first step toward solving the problem is to exchange information between operators, vendors, and architect/engineers about currently working (or not working) pump-valve-control system combinations.

A second, broad problem area is coordinated plant control during low-power operation and transients. The basic problems require coordination of available control inputs in a manner tolerant of plant aging characteristics and nonlinearities. Despite the control design difficulties involved, one is struck by the lack of success to date, and the relatively poor performance of most existing controllers. These do an adequate job, but with an undue amount of operator intervention required (especially at low power).

Coordinated plant control during contingencies is another area for improvement. During contingencies (e.g., plant component failure or outside load disruption) very different requirements are placed on the control system than during normal operation. In normal use, "good control" implies tight regulation and smooth response around efficient operating points, but during emergencies, fast response while satisfying all constraints becomes most important. Emergency controllers must be designed to make the best of limited plant capabilities (e.g., minimal bypass flow), and many current controllers perform poorly in this area. Indeed, the limits of plant response are often not known, and the control system rarely exploits all that is available.

The problem of controlling the spatial power distribution in the core (core control) is currently receiving much research interest. This interest is due, in part, to the large core sizes being currently designed, and the increased ability to compute on-line solutions. The problem is more severe in large, natural uranium reactors than in the light water enriched uranium types in the U. S., and some of the more interesting research is being done for these foreign designs [3],[5].

The general problem of information management and human/machine interface design needs much improvement in most plants, especially older units. One of the most important parts of this problem is the timely detection and identification of failed components or operator errors. Power plants in general have many (hundreds of) measurements available, only a few of which are usually used for on-line control or fault detection. A more comprehensive processing of all the information, with full regard for finding the inevitable failed sensors, would be invaluable in improving plant operation and safety.

Fossil Plant Problems

Fossil plant control is generally less sophisticated than that in nuclear plants, although fossil units may be able to respond faster because of fewer constraints on direct heat increase. Nonetheless, better performance is required of fossil units in several areas, and current research interest seems high.

In general, fossil units suffer from the same type of problems as nuclear units. Feedwater level control, coordinated plant response, low-level and automatic-start operation, contingency control, and fault detection and diagnosis are all areas ready for improvement. In addition, in fossil units several control devices, for example air dampers, in the combustion path are subject to extreme conditions and possible failure, often unnoticed until the device is asked to respond. The harsher conditions, higher temperatures, and variety of plants (with consequent lack of mathematical models) often makes fossil control more difficult than nuclear operation.

POTENTIAL APPLICATIONS OF ADAPTIVE CONTROL

Several problems in power plants have aspects that seem to cry out for adaptive control. (It is, however, difficult to say outright that the whole problem requires adaptive control, since most power plant problems are complex enough to require several simultaneous "solutions.") For convenience, we divide the type of problem by time scaling.

At the slowest level, several plant problems are aggravated by plant "aging," or gradual changes in plant performance due to such things as residue buildup in ducts and pipes, constant ongoing maintenance, changes in nuclear core reactivity with burnup, or changes in fossil fuel quality (especially coal). These changes generally degrade system performance far below that available with recently-tuned controllers. One reference [1] lists a degradation in maximum plant response rate from the design goal of 10-15% of full power per minute to 3% with out-of-tune controllers. This condition is tolerated rather than subjecting the plant to the complex (and costly) re-tuning procedure. Clearly automatic, on-line tuning systems would be appreciated.

At a somewhat faster level, daily weather changes and load cycles can noticeably change plant performance. These changes may result in turbine exhaust pressure variations (changing with condenser temperature) or imperfect air-fuel ratio with consequent inefficiency and increased emissions. A controller able to adapt during the day to changing conditions, or passively tolerant of these parameter changes, would be an improvement.

At a faster level still, plant transient response, with its inherent nonlinearities, may be viewed as a challenge to adaptive control design. Ideally, a single controller could be made to operate at low power with one set of parameters and algorithms, gradually change its behavior as power increased, and assume a new characteristic at full power. Current plant controllers were generally not designed for automatic low-level operation and transition, but such a feature is increasingly regarded as being important.

Finally, contingency control, where the plant must respond to a component failure or external disturbance, can easily benefit from adaptive approach. Automatic fault detection, identification, and compensation would do much to reduce the operator burden during critical periods.

PROBLEMS IN THE APPLICATION OF ADAPTIVE CONTROL

Several obstacles have hindered the application of adaptive control (or simply modern control) in power systems in the past and must be addressed by those wanting to change the future. Power plants are highly complex, nonlinear, time varying, constrained systems. In common with much of process control, detailed plant knowledge and accurate mathematical models are seldom available. Control behavior is also often determined by operator actions, and human interfaces become just as important as computational elegance.

Many control problems in plant operation can be easily resolved by hardware design changes or additions. For example, "runback" to low, house load (plant auxiliary systems only) can be accomplished quite quickly if extra bypass piping and condenser capacity are installed to handle a sudden burst of "unused" steam. These additions are quite expensive, however. In general, it is a challenge to the control designer to make the best of whatever capacity already exists, especially in older plants under new operating conditions. In some cases, however, fundamental plant limits are encountered, and only redesign of critical components will yield more performance. One of the problems facing the control designer is the lack of information about where the fundamental plant limits are.

The control and plant failures that are most troublesome are the quiescent ones which only become visible when it is too late to correct for them. An example is a sticking air valve, which is normally open and is suddenly asked to close. There was no indication before it was needed, that the valve would stick. These unexcited failures are practically impossible to find, even with the best adaptive routines, although it may be possible to <u>search</u> for them, through periodic on-line testing. An automated search procedure (self-test) would be more reliable (and less tedious) than an operator procedure, and coupled with advanced detection algorithms, should operate without disrupting normal plant operations.

Finally, two obstacles are related to human resistance to adaptive control. The first is normal (healthy!) conservatism on the part of plant designers. Adaptive control systems are usually hard to analyze, and hard to guarantee the performance of, in an industry (especially nuclear) where guarantees are required.

But much of the resistance to control improvements must rest with the utilities and vendors for tolerating the poor control in existence. Plant controllers <u>do</u> control the plants reasonably well in normal operation, but usually with an undue amount of operator intervention and tuning. A control engineer is struck by the relatively poor controllability of plants in general, and the tolerance for this state by most users. One reason this situation still exists is the lack of obvious direct economic benefit, to the operating utility, of improved controllability. The situation is analogous to aircraft design, where a smooth-handling craft may be no more productive, safe, or reliable than one that requires constant pilot attention. Nonetheless, the smooth handling craft has advantages, not the least of which is its ability to handle unexpected problems, and is universally preferred by pilots. Awareness of the subtle benefits of controllability is increasing, but modern plants still resemble older planes, with much room for improvement.

APPENDIX

As part of our contract with EPRI, we conducted two surveys of control system problems and design approaches. The first survey was sent to people familiar with nuclear control practice and problems; the second to control researchers who, we hoped, would offer solution suggestions. Only the response to the first survey was large enough to consider representative, and the questions and answers from it are discussed below.

The questionnaire was subjective, rather than multiple choice, to permit as much flexibility as possible in the answers. The responses were therefore hard to quantify, although certain trends and similarities were clear. We note that the sample group was not chosen statistically, and no statistical significance is implied in our grouping of the responses. Also, several people responded with letters or reports which we interpreted and tallied for the appropriate questions. The questionnaires in general required a fair amount of interpretation, and we apologize for any misunderstanding of individual responses.

Return Statistics for the Questionnaire

Of the over 250 questionnaires distributed, we received 39 pertinent responses (questionnaires or letters). Thirty-four of these were complete enough to quantify, and these are considered in our summary below. Since several people (and in two cases groups) cooperated on the questionnaires, the 34 responses represent the opinions of 45 identified individuals and two industrial groups. Several NSSS vendors were represented, as well as government research laboratories, university nuclear engineering departments, and European safety inspectors. Most respondents were familiar with pressurized water reactors, but several other types, including boiling water, heavy water (Canadian and Indian reactors), gas cooled and liquid metal fast breeder designs, were also represented.

1. In your opinion, what problems exist in nuclear power plants for which improved control design concepts will increase the overall plant availability and performance?

Several problems were identified in the replies. The problems listed by six or more respondents were:

Core Control: the spatial power distribution in the core.

Load Change Capability: transient response of the entire plant to desired load changes.

Coordinated Control: the supervised control of primary and secondary loops to improve performance and reduce sensitivity to disturbances.

Feedwater Flow/Level Control: the control of water level in the steam generator.

Contingency Control: the ability of the control system to ensure smooth performance in the event of component failure, turbine trip, or loss of lead.

Operator Interfaces: the presentation of information to the operators, control room design, and operator inputs to the control system.

Other problems, listed in three, four or five questionnaires, were:

Control System/Protective System Interaction: the extraction of protective system status or trip margin for use by the control system, also the interlock or inhibiting of operator responses to avoid trip.

Robustness: in the control sense, the insensitivity of the control system to plant (or control) parameter changes.

Auto-Start: automation of the start-up procedures, or improved control of process variables during start-up and low-load operation.

On-Line Estimation: estimation of process states from plant measurements in an on-line, recursive fashion.

Self Diagnosis/Fault Detection: the ability of a control system to diagnose its own failures or to identify plant component failures.

Component Failure: the reduction of common component failures either through redesign or through control change.

2. What control problems are you currently working on?

Generally resembling the problem list above, the current research interests of the respondents (four or more) included:

Core Control

Feedwater Level

Load Change (Plant Transient) Ability

Coordinated Control

State Estimation

Computerized Control (the application of digital computers to plant control tasks)

3. What are the important characteristics of the above problems from a control point of view? (e.g., nonlinear, time varying, equality/inequality constraints on inputs, outputs of functions of inputs, outputs.)

The problems in nuclear plant control were felt to possess (by five or more) the following characteristics:

Nonlinear system

State and control constraints

Time varying system

Multivariable process

Distributed parameter dynamics (space dependency)

Also mentioned by several people, in different contexts, were the presence of disturbances, known and unknown, in the plant.

4. What kind of control system design approach are you going to take? How are constraints handled? How are plant nonlinearity and change with core life handled? How are steady state constant offsets from desired values of outputs eliminated? How are the feedforward and feedback control inputs derived and calculated?

The control design approaches mentioned in the questionnaires included most popular types. Those listed by three or more respondents were:

Modern control (linear state space, time-domain problem formulation)

Classic control (frequency domain)

Combination of modern and classic control

Use of integrators to remove steady state errors

Gain scheduling (gains computed as a function of load)

Parameter changes with core life included in the controller or model

On-line identification of the plant parameters

Other suggested approaches included:

Inverse Nyquist Array

Multivariable Frequency Response

Model Algorithmic Control

Maximum Principle

Heuristic Control

Multistage Mathematical Programming with periodic optimization

5. What are the functions of the control system? (e.g., start-up, normal operation [steady state load regulation, daily load following (planned load changes), spinning reserve (unanticipated load changes)], contingencies (pump failures, sudden loss of electrical load, etc.)).

The control function, or operating modes, of the controllers under development that were listed by three or more people were:

Start-up

Normal operation including load regulation, planned load following, and spinning reserve

Contingency operation

Shutdown

Runback

6. What kind of models are you developing for control system design? (e.g., first principles/heuristic, linear/nonlinear, continuous/discrete, state space/impulse response/ARMA, stochastic/deterministic, order reduction by formal technique (which?) or by knowledge of plant.)

The models that were used for the controllers included many different types. Some confusion arises in interpreting the responses, however, since a number of answers referred to simulation models while others referred to mathematical models used for control design. We intended for this question to refer to the latter type and relate to the control approach of Question 4. Those answers that we believe refer to mathematical models and were listed by five or more people include:

First principles (physical)

Linear

Nonlinear

State space

Discrete

Deterministic

Those models or features listed in two, three, or four replies included:

Continuous

Stochastic

ARMA

Order reduction by knowledge of plant

Heuristic

7. What are the reasons for your selection of this approach over other competing approaches?

The reasons each researcher chose his modeling or control technique were varied. The answers given by three or more people were:

Proven technique

Technique needed by (appropriate to) the system

Evolutionary development

Designer's familiarity with the method

Knowledge of plant used directly (in the physical models)

Computationally efficient

The evolutionary nature of the control approach was mentioned by those who have worked for some time in the area (usually for vendors), and who believe their current control systems do most things well. The methods tend to evolve from those that have worked well before.

8. What are the major simplifications and assumptions in your approach? What are the off-line and on-line calculational requirements? What operator decisions and actions are required?

The answers to this question often referred to specific simplifications used by the designer. The general techniques suggested by four or more respondents were:

Lumped parameters

Reduced order

Linearization

No general trends emerged in the responses regarding calculational requirements or operator actions.

9. What are the tuning requirements? (e.g., field tests and data, data processing, operator inputs, turn around time, etc.)

The tuning requirements of each approach given by two or more people were:

Field tests and data

On-line tuning

Field tuned at start-up testing

Off-line tuning using operating data

Operator-aided tuning

10. Has your technique been successfully applied? Which applications?

The most common answers to this question were:

Yes, pressurized water reactor (PWR)--8 answers

No, not yet--7 answers

Only simulations to date--4 answers

11. When is this project going to start (or when did it start)? When is it going to be completed and how much progress has been made?

Most of the respondents started their projects after 1975 (9 answers), and the second largest group (5) began between 1971 and 1975, inclusive. Five were expecting to finish in 1979, four in 1980, and five in 1981. Nine of the replies indicated continuous work in this area.

12. Do you anticipate any difficulties or problems in carrying out the proposed approach?

The replies to this question were mostly "yes" or "no" (7 answers each). The most common difficulty expected (by 4 respondents) involved inadequate computer capability.

13. Will the final design be implemented on any operating nuclear power plants? What are the types and sizes of the plants and where are they located?

A number of specific implementations were provided. Ten of the replies listed one or more PWR's.

14. How is the final design going to be tested and evaluated? What are the criteria to be used in the evaluation?

The replies to this question generally referred to specific control criteria used for customer acceptance rather than to general control guidelines. As a result, the answers alone were not very useful in developing the criteria we were looking for. The most frequent responses (listed by four or more people) were:

Start-up (commissioning) tests

Normal operation behavior

Control quality and "noise" levels

Also listed were:

Specified acceptance (test) criteria

Qualitative standards

Long test periods

Maneuver core within limits

Operator evaluation on site

Agreement between field tests and simulation

15. In what way would control design increase the overall plant performance and availability?

The respondents (three or more) felt that performance and availability would be improved by

Avoiding trips and reducing the number of shutdowns

Minimizing sensitivity to component failures

Improving regulation (tighter control)

Reducing stress

16. Is the final design going to be implemented with digital or analog hardware? What do you expect the hardware requirements to be?

Twenty-two answers indicated at least a partial digital implementation. Analog controllers were mentioned by 10 people, nine of whom also expected to use digital equipment. Three responses expected "more and more" digital equipment to be used.

17. Any additional comments or information?

Several interesting comments were sent in the replies. One respondent felt that better models led to increased understanding which led to other, indirect improvements such as fault diagnosis. Another mentioned that improved mechanical design as well as better controls are needed to increase coordination and improve performance. One German researcher suggested that plants in the U. S. attempt to minimize operator duties during contingencies, and he referred to the German KTA-Rule 350A: "Limitations" as a model. One reply said that plant reliability could be improved by a diagnosis system, and that operator decisions and actions were important in any such system.

One respondent suggested that network cost/benefit guidelines are needed as a common basis for utility purchase and industrial research and investment. He felt that several improvements to the plant resulted in improved network efficiency, and therefore reduced utility cost, but that such improvements were hard to justify to plant personnel.

Finally, two of the replies not used in the above summary contained comments of interest. One utility believed that greater use of automation in the fuel handling and radioactive waste systems would increase availability and performance. And one respondent felt that accuracy was important but often missing in analysis. He believed that too much faith is put in design ("nominal") specifications or complex, indirect measurements, and that not enough care is taken in accounting for the effects of tolerances and measurement errors.

REFERENCES

1. DOE, Request for Proposal PRDA No. ET-78-D-01-3088(3089), May 30, 1978.
2. EPRI,"Survey of Feedpump Outages," Energy Research and Consultants Corp., EPRI FP-754 Proj. 641, April 1978.
3. Karppinen, J. and Blomsnes, B., "An Application of Optimization Methods to Spatial Control of Nuclear Reactor Cores," Proc. Int. Symp. on New Trends in Syst. Anal., Recquencourt, France, pp. 393-417, 1976.
4. Kerlin, T. W., "Dynamic Analysis and Control of Pressurized Water Reactors, Academic Press, New York, 1978.
5. Khanna, M., Paulksens, J., and Hinchley, E. M., "The Use of Flux Mapping in the On-Line Spatial Control of the 600 MW(e) Candu-PHW Reactor," Proc. 1977 JACC, pp. 375-380, 1977.
6. Solberg, H. L., Cromer, O. C., and Spalding, A. R., "Thermal Engineering," Wiley, New York, 1960.
7. Weedy, B. M., "Electric Power Systems," 2nd Ed., Wiley, Chichester, 1977.
8. Lindahl, S., "Design and Simulation of a Coordinated Drum Boiler-Turbine Controller," Tech. Report LUTFD2, Dept. of Automatic Control, Lund Institute of Technology, Dec. 1976.

NEW DEVELOPMENTS IN IMPROVING POWER NETWORK STABILITY WITH ADAPTIVE GENERATOR CONTROL

E. Irving

Automatic Control Division
Direction des Etudes et Recherches
Electricité de France

The "steady state stability" of the French power transmission system has been greatly improved by supplying additional signals to the voltage regulator. As usual, the control system parameters, including additional signals, have been chosen in order to cover all the possible operating conditions and network configurations. These additional signals lead to an increase in the "steady state stability" domain even with a regulator having constant parameters. Better results have been obtained using several different adaptive control methods. The first method which has been studied and implemented by analog hardware was based on the hyperstability approach and introduced new concepts as series and partial reference models. In order to completely control the time constants of the closed-loop system, a second adaptive control method using a predictor and an implicit reference model has been studied and

ISBN 0-12-514060-6

implemented by micro-processors. Finally, a simpler and more robust new method,using the reference model optimal aim strategy with a very simple predictor, is introduced.

I. INTRODUCTION

This paper discusses several methods of adaptive control applied to automatic voltage control. Among the different control problems in power systems, those related to the stability of power networks seem of great interest for the following reasons.

Studies made some years ago concerning the "steady state stability" (small variations stability) of the French power network as well as its actual performance in certain special conditions revealed a need for an improvement in the power network unit voltage regulation. This fact led to the use of additional stabilizing signals obtained, for example, from the variations of the active power output [1].

Moreover, studies concerning the dynamic behaviour of future power systems show new stability problems appearing. Such stability problems are essentially due to the weak "steady state stability" of future networks resulting from unbalanced developments of unit power and network short-circuit power. In fact, on the one hand, the spinning power of units is increasing faster than the short-circuit power of the network, and on the other hand, nuclear plants can be situated far from the consumption centers. These two facts lead to a weakening of the static stability.

General studies concerning voltage regulation equipped with stabilizing devices show the best values of the parameters to be dependent on the operating conditions of the generator and the network structure. Thus calculating the optimal values of these parameters has to be a compromise which may not be very efficient in certain restricted situations. In order to maintain efficient operation in all situations, it seems of interest to automatically adjust the values of the parameters of the voltage regulation.

Finally, after the French electrical failure of December 1978, the controllers of the network became aware that stability of the system they are in charge of was no longer a future but a present problem. Following the decision which has been taken to equip all the future 1300 MW units with voltage adaptive control [2], the recent events involved another decision: to equip the present most powerful units in the southeast of France with this new type adaptive regulator. The benefit of the last decision has been estimated to be an additional transmitted power available greater than 3000 MW from the southeast to the Parisian area.

We will begin by describing the full mathematical model of the unit and the network and proceed by the simplified model which has been used in the design in order to discuss the practical control problems.

After this, we recall the partial reference model adaptive control method which has been fully described in [2]. This method is, presently, industrially implemented by analog hardware.

Next, we discuss another adaptive control method using a predictor and an implicit reference model which has been studied in order to fully control the time constants of the closed-loop system.

Finally, we describe a third control method, the adaptive reference model optimal aim strategy which has been chosen in order to increase the robustness of the previous method and to reduce its computational cost. Both of the last two methods have been implemented by micro-processors and are to be tested on the E.D.F. micro-network and power grid system.

II. MATHEMATICAL MODELS

An alternating current electrical generator is defined by its nominal variables which are:

V_n, stator nominal voltage; P_n, active electrical power supplied by the unit to the network; Q_n, reactive supplied power; X_q, quadrature reactance which is assumed to be equal to the direct reactance (large units). The nominal power factor angle ϕ_n can be calculated by

$$\cos \phi_n = P_n (P_n^2 + Q_n^2)^{-1/2} \tag{2.1}$$

The nominal impedance is calculated by

$$Z_n = 3\, V_n^2 (P_n^2 + Q_n^2)^{-1/2} \tag{2.2}$$

From the vector representation of Fig. 1, where the voltage generated by the rotor is E and where the unit is supposed linked to an infinite bus of voltage V_r by a linkage reactance X_t, the following relations can be deduced.

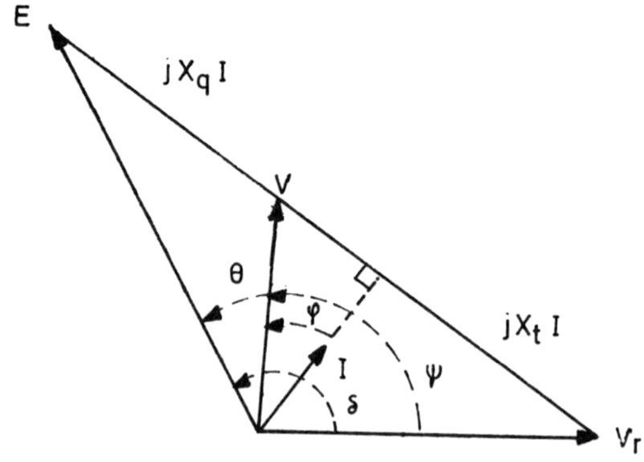

FIGURE 1. Vector representation of alternating voltages.

$$P = 3 V E \sin \theta X_q^{-1} \tag{2.3}$$

$$Q = 3 V (E \cos \theta - V) X_q^{-1} \tag{2.4}$$

These relations are satisfied for any operating conditions defined by V, P, Q where the angle between E and V is called the internal angle given by

$$\cot \theta = \tan \phi + 3 V^2 P^{-1} X_q^{-1} \tag{2.5}$$

and

$$\tan \phi = Q/P . \tag{2.6}$$

Also V_r and X_t are related by

$$X_t = 2 V P^{-1} \cos \phi (V \sin \phi + (V_r^2 - V^2 \cos^2 \phi)^{1/2}) \tag{2.7}$$

Equation (2.7) shows that the variables V, P, Q are not sufficient to define working conditions, so it is necessary to add another variable, for example, X_t. The angle between V and V_r is called the transport angle and the angle between E and V_r is called the total angle. Clearly

$$\delta = \theta + \psi \tag{2.8}$$

The rotation speed of the unit is given by

$$\Omega = p^{-1} d(\delta + 2\pi f_o t)/dt \tag{2.9}$$

where p is the number of pairs of poles and f_o the frequency of the network. From the basic electrical and mechanical equations of the generator, it is possible to deduce the following non-linear dynamical equations

$$d/dt\ \delta(t) = \Omega(t) - \Omega_o \tag{2.10}$$

$$d/dt\ \Omega(t) = J^{-1}(C_m(t) - C_e(t)) \tag{2.11}$$

$$d/dt\ \phi_1(t) = \tau'^{-1}_{d_o}(-\ \phi_1(t)\ B\ I_d(t) + C\ V_f(t)) \tag{2.12}$$

$$d/dt\ \phi_2(t) = \tau''^{-1}_{d_o}(-\ \phi_2(t) + D\ I_d(t) + E\ V_f(t)) \tag{2.13}$$

$$d/dt\ \phi_3(t) = \tau''^{-1}_{q_o}(-\ \phi_3(t) + G\ I_q(t)) \tag{2.14}$$

$$d/dt\ I_d(t) = A^{-1}(-(R_a + B\tau'^{-1}_{d_o} + D\tau''^{-1}_{d_o})I_d(t) - V_d(t) - \Omega(t)\ \phi_q(t)$$
$$-(C\tau'^{-1}_{d_o} + E\tau''^{-1}_{d_o})V_f(t)\ \ \phi_1(t)\tau'^{-1}_{d_o} + \phi_2(t)\tau''^{-1}_{d_o}) \tag{2.15}$$

$$d/dt\ I_q(t) = F^{-1}(-(G\tau''^{-1}_{q_o} + R_a)\ I_q(t)$$
$$-\ V_q(t) + \Omega(t)\phi_d(t) + \tau''^{-1}_{q_o}\phi_3(t)) \tag{2.16}$$

where

$$\phi_d(t) = A\ I_d(t) + \phi_1(t) + \phi_2(t)$$

$$\phi_q(t) = F\ I_q(t) + \phi_3(t)$$

$$C_e(t) = 3/2\ (\phi_d(t)\ I_q(t) - \phi_q(t)\ I_d(t))$$

$$V(t) = (V_d^2(t) + V_q^2(t))^{1/2}$$

$$P(t) = 3/2\ (V_d(t)\ I_d(t) + V_q(t)\ I_q(t))$$

$$Q(t) = 3/2\ (V_d(t)\ I_q(t) = V_q(t)\ I_d(t))$$

$$A = L_d\ \tau'_d\ \tau''_d/\tau'_{d_o}\ \tau''_{d_o}$$

$$B = L_d(\tau'_{d_o} - \tau_d)(\tau'_{d_o} - \tau''_d)\ /\ \tau'_{d_o}(\tau'_{d_o} - \tau''_{d_o})$$

$$C = (M_{af}/R_f)\ (\tau'_{d_o} - \tau_{kd})\ /\ (\tau'_{d_o} - \tau''_{d_o})$$

$$D = L_d(\tau''_{d_o} - \tau'_d)\ (\tau''_{d_o} - \tau''_d)\ /\ \tau''_{d_o}(\tau''_{d_o} - \tau'_{d_o})$$

$$E = (M_{af}/R_f)\ (\tau''_{d_o} - \tau_{kd})\ /\ (\tau''_{d_o} - \tau'_{d_o})$$

$$F = L_q \, \tau''_q \, / \, \tau''_{q_o}$$

$$G = L_q (\tau''_{q_o} - \tau''_q) \, / \, \tau''_{q_o}$$

The numerical values of the previous constants for a 1300 MW unit are

$R_a = 0.977 \; 10^{-3}$ ohm	$\tau'_d = 1.56$ s	$\tau'_{d_o} = 8.323$ s
$\tau''_d = 0.046$ s	$\tau''_{d_o} = 0.071$ s	$L_d = L_q = 2.09 \; 10^{-3}$ henry
$\tau''_q = 0.046$ s	$\tau''_{q_o} = 0.352$ s	$\tau_{kd} = 0.021$ s
$M_{af} = 2.76 \; 10^{-2}$	$R_f = 7.78 \; 10^{-2}$ ohm	$J = 180553 \; m^2$ kg

We deduce that

$$\tau_{id} = A/(R_a + B \, {\tau'_{d_o}}^{-1} + D \, {\tau''_{d_o}}^{-1})^{-1} = 0.08 \text{ s}$$

$$\tau_{iq} = F/(G \, {\tau''_{q_o}}^{-1} + R_a)^{-1} = 0.04 \text{ s}$$

The static equivalent form of equations (2.15) and (2.16) is given by

$$V_d(t) = - R_a \, I_d(t) - \Omega(t) \, L_q \, I_q(t) \tag{2.17}$$

$$V_q(t) = - R_a \, I_q(t) + \Omega(t) \, (L_d \, I_d(t) + M_{af} \, V_f(t) \, / \, R_f) \tag{2.18}$$

with $L_q = F + G$, $L_d = A + B + D$ and $M_{af}/R_f = C + E$.

Now, to obtain the full set of equations, it is necessary to include the reaction of the network, the infinite bus voltage being supposed constant by definition. This reaction can be represented by the following equation

$$\underline{V}(t) = \underline{V}_r + \underline{Z}(t) \, \underline{I}(t) \tag{2.19}$$

$\underline{V}(t)$, $\underline{V}_r$, $\underline{Z}(t)$, $\underline{I}(t)$ are complex variables represented in any system of coordinates. It is thus possible to choose O_q the axis of E and O_d its positive normal as these coordinates. Equation (2.19) can thus be written in the form

$$V_q(t) = V_r \cos \delta(t) - X_t(t)\, I_d(t) \tag{2.20}$$

$$V_d(t) = -V_r \sin \delta(t) + X_t(t)\, I_q(t) \tag{2.21}$$

The previous equations (2.20) and (2.21) complete the (2.15) and (2.16) equations. It can be seen that variations of the linkage reactance X_t may change drastically the dynamical properties of the whole system of equations (2.12) to (2.16). As a matter of fact, the generator which is damped for small values of X_t ($X_t/Z_n = 0.2$) can be oscillatory for intermediate values ($X_t/Z_n = 0.7$) and even be unstable for large values ($X_t/Z_n = 1.2$).

For the design of the adaptive system, it is possible to deduce a more simplified model. For each operating point defined by the parameters V(t), P(t), Q(t) and $X_t(t)$ the following simplified dynamical linearized equations can be obtained if the dynamical aspects of equations (2.13) to (2.16) are neglected.

$$\frac{d}{dt}\begin{bmatrix} \Delta\delta(t) \\ \Delta\Omega(t) \\ \Delta P_e(t) \end{bmatrix} = \begin{bmatrix} 0 & 1 & 0 \\ 0 & -D & -H \\ \alpha\tau^{-1} & \beta\tau^{-1} & -\tau^{-1} \end{bmatrix} \begin{bmatrix} \Delta\delta(t) \\ \Delta\Omega(t) \\ \Delta P_e(t) \end{bmatrix} + \begin{bmatrix} 0 & 0 & -G \\ 0 & H & 0 \\ \alpha\tau^{-1} & 0 & 0 \end{bmatrix} \begin{bmatrix} \Delta V_f(t) \\ \Delta P_m(t) \\ \frac{d}{dt}\,\Delta X_t(t) \end{bmatrix} \tag{2.22}$$

$$V(t) = h_1\, \Delta\delta(t) + h_3\, \Delta P_e(t) \tag{2.23}$$

This simplified model has been fully described in [2]. It should be mentioned that the parameters α, β, γ, τ, G are dependent on V(t), P(t), Q(t), $X_t(t)$. Also, $X_t(t)$ has another effect as a disturbance on the angle $\delta(t)$.

Equation (2.22) is not used for adaptive control design because the total angle $\delta(t)$ is not conveniently measurable. Using (2.22) and (2.23) $\delta(t)$ is eliminated to obtain the following equation

$$\frac{d}{dt}\begin{bmatrix} \Delta V(t) \\ \Delta\Omega(t) \\ \Delta P_e(t) \end{bmatrix} = \begin{bmatrix} a_{11} & a_{12} & a_{13} \\ 0 & -D & -H \\ a_{31} & a_{32} & a_{33} \end{bmatrix} \begin{bmatrix} \Delta V(t) \\ \Delta\Omega(t) \\ \Delta P_e(t) \end{bmatrix} + \begin{bmatrix} b'_{11} & 0 & -h_1 G \\ 0 & H & 0 \\ b'_{31} & 0 & 0 \end{bmatrix} \begin{bmatrix} \Delta V_f(t) \\ \Delta P_m(t) \\ \frac{d}{dt}\Delta X_t(t) \end{bmatrix} \tag{2.24}$$

which is the one used for design.

III. THE PRACTICAL CONTROL PROBLEM

When considering the behaviour of the unit, several control problems must be solved to satisfy the following requirements.

First requirement

In spite of the variations of V(t), P(t), Q(t), $X_t(t)$, the behaviour of the state variables of the unit must be sufficiently damped (static stability).

Second requirement

Control the static value of V(t), since its dynamic behaviour is acceptable if the first requirement is fulfilled.

Third requirement

In spite of the sudden variations of the mechanical power $P_m(t)$, such as the closing down of the unit valve after a network fault, the unit speed $\Omega(t)$ must undergo the least possible variations (transient stability).

Fourth requirement

The control system must work with the same efficiency for different static values of $V_c(t)$ (set point input of $V(t)$) and $P_m(t)$ (linearization requirement).

Fifth requirement

As far as possible, minimize use of the valve to maintain stability because, first, it is not an efficient way to tackle the previous problems from the standpoint of thermal time constants and, second, the reliability of the unit may be endangered (nuclear units).

For these reasons, only the field voltage $V_f(t)$ is used to fulfil the previous requirements. If the measurements of $\Omega(t)$ and $P_m(t)$ are available, the following control law is proposed [2].

$$\begin{aligned} \Delta V_f(t) = {} & k_o\ \Delta V_c(t) - k_1\ \Delta V(t) - k_2\ \Delta\Omega(t) \\ & - k_3\ \Delta P_e(t) + k_4\ \Delta P_m(t) \\ & + k_5\ d\ \Delta P_m(t)\ /\ dt \end{aligned} \tag{3.1}$$

where $k_o\ \Delta V_c(t)$ is a feedforward term including the setpoint input, $- k_1\ \Delta V(t) - k_2\ \Delta\Omega(t) - k_3\ \Delta P_e(t)$ are the feedback terms including the measurable state variables of the controlled system, and $k_4\ \Delta P_m(t) + k_5\ d\ \Delta P_m(t)\ /\ dt$ are feedforward terms including the disturbance input $\Delta P_m(t)$.

IV. EXISTENCE OF A CONTROLLER STRUCTURE

We will show briefly, using the simplified model, that the previous control law has all the necessary degrees of freedom to compensate for any variation of the parameters of the network.

Recall that the simplified model takes the form (neglecting the disturbance due to the network)

$$dy(t)/dt = Ay(t) + Bu(t) \tag{4.1}$$

$$u^T(t) = [\Delta V_f(t), \Delta P_m(t)] \tag{4.2}$$

$$y^T(t) = [\Delta V(t), \Delta\Omega(t), \Delta P_e(t)] \tag{4.3}$$

where A is a 3 x 3 matrix, and B is a 3 x 2 matrix. The transfer function matrix of the model is

$$y(s) = (sI - A)^{-1} B\, u(s) \tag{4.4}$$

Equation (4.4) may be expressed as

$$\Delta(s) \begin{bmatrix} y_1(s) \\ y_2(s) \\ y_3(s) \end{bmatrix} = \begin{bmatrix} B_1(s) & D_1(s) \\ B_2(s) & D_2(s) \\ B_3(s) & D_3(s) \end{bmatrix} \begin{bmatrix} u_1(s) \\ u_2(s) \end{bmatrix} \tag{4.5}$$

where $\Delta(s)$, $B_i(s)$, $D_i(s)$, $i = 1,2,3$, are polynomials of the form

$$\Delta(s) = \Delta_o + \Delta_1 s + \Delta_2 s^2 + s^3 \tag{4.6}$$

$$B_i(s) = b_{i1} + b_{i2} s + b_{i3} s^2 \qquad i = 1,2,3 \tag{4.7}$$

$$D_i(s) = d_{i1} + d_{i2} s + d_{i3} s^2 \qquad i = 1,2,3 \tag{4.8}$$

A general form of a linear controller is given by the equation

$$S(s)\ u_1(s) = \sum_{i=1}^{3} R_i(s)\ y_i(s) + C(s)\ r(s) + T(s)\ u_2(s) \tag{4.9}$$

where r(t) is an external reference input, the term $T(s)\ u_2(s)$ represents a feedforward control, S(s), C(s), T(s), and $R_i(s)$ are polynomials chosen to fulfil the requirements of the practical control problem. Using (4.9) in (4.5) the transfer function of the closed-loop system becomes

$$[S(s)\Delta(s)I - B(s)R^T(s)]y(s) = B(s)C(s)r(s) + [B(s)T(s) + D(s)S(s)]u_2(s) \tag{4.10}$$

where $B(s)^T = [B_1(s), B_2(s), B_3(s)]$

$R(s)^T = [R_1(s), R_2(s), R_3(s)]$

$D(s) = [D_1(s), D_2(s), D_3(s)]$.

This equation is the one used to specify the dynamic properties of the closed-loop. It can be proved that the following choice

$$S(s) = 1 \tag{4.11}$$

$$R_i(s) = k_i \qquad i = 1,2,3 \tag{4.12}$$

give all degrees of freedom for the determination of the roots of the polynomial $\tilde{A}(s)$ defined by

$$\tilde{A}(s) = \det[S(s)\Delta(s)I - B(s)R^T(s)] \tag{4.13}$$

assuming that the polynomials $B_i(s)$ are prime.

So, the first requirement can be fulfilled.

Moreover, the following choice of C(s)

$$C(s) = k_o \tag{4.14}$$

allows the second requirement on static gain to be fulfilled.

Finally, if T(s) is given by

$$T(s) = k_4 + k_5 s \tag{4.15}$$

then it is possible to compensate for disturbances on $y_2(t)$ (the speed of the unit) caused by the variations of $u_2(t)$ (mechanical power $P_m(t)$). So, the third requirement can be fulfilled.

V. EXPLICIT REFERENCE MODEL ADAPTIVE CONTROL METHODS

The idea of the explicit parallel reference model adaptive control method is to adjust the parameters of a controller by comparing the output of the closed-loop system with the output of a reference model which has the desired behaviour of the closed-loop system. The automatic adjustment of the regulator parameters is carried out in such a way that the error between the output of the reference model and the output of the closed-loop system is made as small as possible.

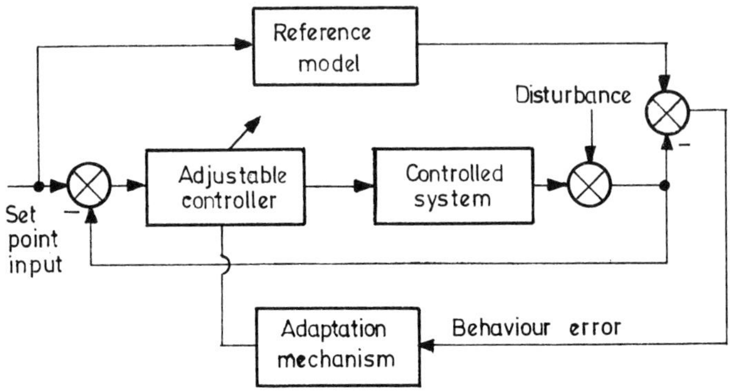

FIGURE 2. Parallel reference model adaptive system.

Figure 2 illustrates such a structure. A basic condition that must be fulfilled by the adaptation mechanism is the stability of the error. Moreover, the convergence of the error must be faster than the equivalent time constant of the reference model, when the set point input is zero, and unmeasurable disturbance (network fault) makes the output of the controlled system deviate greatly from zero. Since the output of the reference model is zero, the adaptation will null the behaviour error very quickly at the price of "violent shocks" to the regulation loop components which are harmful to unit reliability. A way to avoid this and to eliminate disturbances in a better way is to introduce a new behaviour error, such as the output of a reference model connected in series with the previous behaviour error as shown in Fig. 3.

The matrix transfer functions of the two reference models may be chosen as follows:

$$(A_m - sI)^{-1} B_m$$

for the parallel reference model and

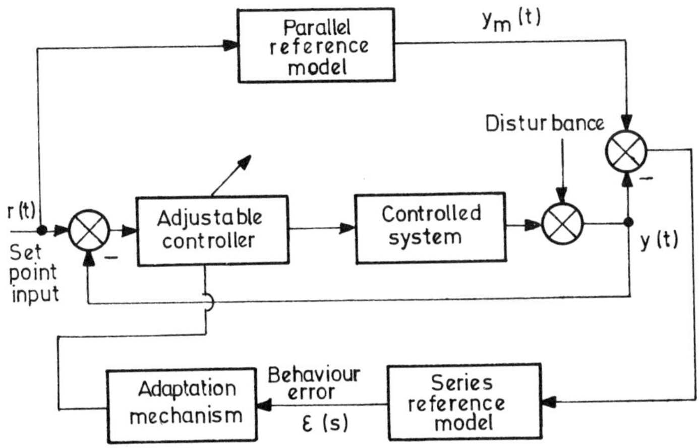

FIGURE 3. Series-parallel reference model adaptive system.

$$(1 + \tau_f s)^{-1} (A_m - sI)$$

for the series reference model. The new behaviour error is then given by

$$\begin{aligned} \varepsilon(s) &= (1 + \tau_f s)^{-1} (A_m - sI) . \\ &\quad ((A_m - sI)^{-1} B_m r(s) - y(s)) \qquad (5.1) \\ &= B_m r_f(s) - A_m y_f(s) + s\, y_f(s) \end{aligned}$$

where $r_f(s)$ and $y_f(s)$ are defined by

$$r_f(s) = (1 + \tau_f s)^{-1} r(s)$$

$$y_f(s) = (1 + \tau_f s)^{-1} y(s)$$

with τ_f being the time constant (small) of a filter needed to obtain a smooth value of the derivative of y(t).

If $\varepsilon(s) = 0$, then

$$y(s) = (sI - A_m)^{-1} B_m r(s)$$

or equivalently

$$\begin{bmatrix} y_1(s) \\ y_2(s) \\ y_3(s) \end{bmatrix} \Delta_m(s) = \begin{bmatrix} B_{m1}(s) \\ B_{m2}(s) \\ B_{m3}(s) \end{bmatrix} r(s) \tag{5.2}$$

Remarks:

Remark 1

If

$$r(s) = 0$$

a sudden disturbance on $y(t)$ will be eliminated in a soft and dynamically controlled way dictated by the following differential equation

$$\Delta_m(p) \begin{bmatrix} y_1(t) \\ y_2(t) \\ y_3(t) \end{bmatrix} = 0 \qquad p = d/dt$$

Remark 2

The method described so far relies on cancellation of plant zeros so that they can be replaced by the zeros of the model transfer function. Such cancellation is accomplished by the generation of compensator poles. If these poles are lightly damped an oscillatory control will result which is harmful to the unit.

As it is impossible to completely null the behaviour error $\varepsilon(t)$, we have proposed [2] to null a "trade off" $Q\varepsilon(t)$ of the error $\varepsilon(t)$ in what was called the partial reference model adaptive control method. The main drawback of the partial reference model adaptive control method is the fact that the dynamical properties (at least concerning poles) of the closed-loop are not maintained constant, even if the trade off error is nulled.

VI. IMPLICIT REFERENCE MODEL ADAPTIVE CONTROL METHODS

Before introducing implicit reference model adaptive control methods, it is noted that explicit reference model adaptive control methods avoid explicit identification of the controlled system. There is no need to assume convergence of parameters which usually requires injecting an extra excitation into the closed-loop system. This excitation, being harmful to unit reliability, would rarely be accepted by people in charge of the plants (especially in nuclear plants).

Implicit model reference adaptive control methods require identification, but do not assume convergence of the adjustable parameters of the identifier. It is only assumed that the output of the identifier converges to the output of the identified system. This does not imply the need for an extra excitation. In short, what we assume now, is convergence of what we may call the "external behaviour error" of the identifier. We call this special kind of identifier a "predictor". For this scheme we can prove convergence of the output of the controlled system to the output of an implicit reference model without assuming extra excitation.

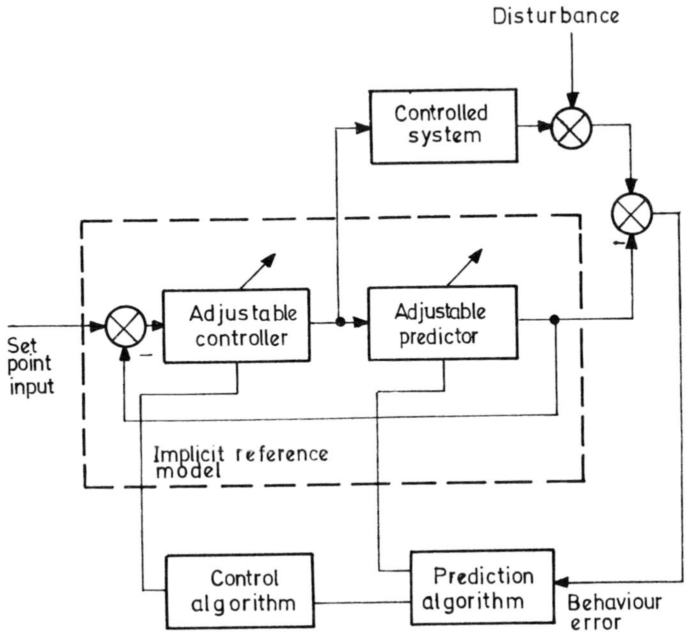

FIGURE 4. Implicit reference model adaptive control system.

Before discussing this proof of convergence, let us examine in Fig. 4 the scheme of implicit model reference adaptive control.

It is seen that the loop composed of the adjustable controller is not closed on the controlled system as is usually done, but on the adjustable predictor. This feature allows a stability proof which is sketched next in two parts. First, we assume that the predictor algorithm is such that the output of this predictor converges to the output of the controlled system. Second, we assume that the controller adjustment algorithm, taking the internal parameters of the predictor into consideration, will give to the adjustable controller such parameters that the closed-loop composed of the adjustable predictor and the adjustable controller will be a stable system with given time constants. Then, this closed-loop is equivalent to a reference model. The output of this implicit reference model is stable. The conclusion of stability of the control system follows.

With this method, we can improve the partial reference model adaptive control method by maintaining the dynamical properties of the controlled system constant. In fact, as the controlled system (the unit linked to the network) may be oscillatory and even unstable, the parallel scheme of the adjustable predictor does not work and it is necessary to use a series-parallel scheme for the predictor.

Implementation of the implicit reference model adaptive control method is much more intricate than the explicit partial reference model adaptive control method. So, instead of analog hardware, a realization by microprocessors (16 bits) has been chosen. We can use the following discrete time simplified model of the unit linked to the network

$$y(t+1) = \phi y(t) + \Gamma u(t) \tag{6.1}$$

$$y^T(t) = (\Delta V(t), \Delta\Omega(t), \Delta P_e(t)) \tag{6.2}$$

$$u^T(t) = (\Delta V_f(t), \Delta P_m(t)) \tag{6.3}$$

where $\Delta V(t)$, $\Delta\Omega(t)$, $\Delta P_e(t)$, $\Delta V_f(t)$, $\Delta P_m(t)$ are the discrete time equivalent of the "variations" defined in the continuous time case (4.10) [2].

Specifics of this implementation may be found in [2]. Suffice it to say that it requires a considerable amount of computation, and it suffers from the drawback of large convergence times for the predictor. A simpler, more robust algorithm is needed. The algorithms discussed next potentially have these properties.

VII. AN ADAPTIVE PREDICTOR ALGORITHM

A predictor algorithm using recursive least squares needs some additional features to avoid difficulties which may arise when it is used in an adaptive system.

Let us examine these particular aspects applied to the predictor of the simplified model of the unit. The simplified model (6.1) can be written in the following form

$$\begin{bmatrix} y_1(t) \\ y_2(t) \\ y_3(t) \end{bmatrix} = \begin{bmatrix} \phi_1 \\ \phi_2 \\ \phi_3 \end{bmatrix} \begin{bmatrix} y_1(t - \Delta t) \\ y_2(t - \Delta t) \\ y_3(t - \Delta t) \end{bmatrix} + \begin{bmatrix} \Gamma_1 \\ \Gamma_2 \\ \Gamma_3 \end{bmatrix} \begin{bmatrix} u_1(t - \Delta t) \\ u_2(t - \Delta t) \end{bmatrix} \tag{7.1}$$

The previous system of equations shows that the predictor problem can be split into three decoupled predictor problems which may be solved using parallel computation. So each predictor must be designed such that its output converges in the quickest way possible to the output of the following model.

$$y_i(t) = \phi_i y(t - \Delta t) + \Gamma_i u(t - \Delta t) \qquad 1 \le i \le 3 \tag{7.2}$$

As the whole simplified model (7.1) may be unstable, it is necessary to utilize a series-parallel predictor of the following form

$$\begin{aligned} \hat{y}_i(t) &= \hat{\phi}_i(t)\ y(t - \Delta t) + \hat{\Gamma}_i(t)\ u(t - \Delta t) \\ &= x^T(t)\ \hat{p}(t) \end{aligned} \tag{7.3}$$

This choice of a series-parallel predictor avoids the stability problem because the equation of the output behaviour error of the predictor has no dynamic dependence on the dynamics of the identified system as in the purely parallel case, see [3].

Let

$$x^T(t) = (y^T(t - \Delta t), u^T(t - \Delta t))$$

$$\hat{p}^T(t) = (\hat{\phi}_i(t), \hat{\Gamma}_i(t))$$

and p(t) be defined by

$$p^T(t) = (\phi_i, \Gamma_i) .$$

p(t) can be shown to satisfy the following state space equation

$$p(t + \Delta t) = p(t) + v(t)$$

$$y_i(t) = x^T(t) \; p(t) + e(t) \tag{7.4}$$

with v(t) the parameter noise and e(t) the measurement noise with the following statistics

$$E(v(t) \; v^T(t)) = \Xi(t)$$

$$E(e^2(t)) = \sigma^2 .$$

The predictor algorithm can simply be a Kalman predictor [4] satisfying the following equations

$$\hat{p}(t) = \hat{p}(t - \Delta t) + (\sigma^2 + x^T(t) \; \Pi(t) \; x(t))^{-1} \; \Pi(t) \; x(t) \; (y_i(t) - x^T(t) \; \hat{p}(t - \Delta t)) \tag{7.5}$$

where

$$\Pi(t + \Delta t) = \Pi(t) + \Xi(t) - \Pi(t)\, x(t)\, (\sigma^2 + x^T(t)\, \Pi(t)\, x(t))^{-1} x^T(t)\, \Pi(t)$$

The need for adaptive control is due to the fact that $\Xi(t)$ may vary from time to time at unknown instants. $\Xi(t)$ being unknown is generally set to zero and the equation (7.5) shows that $\Pi(t)$ will decrease continuously in this case and attain very small values. The small values of $\Pi(t)$ involve too slow an adaptation which is a severe drawback, especially after a sudden variation of the parameters of the controlled system. The well-known remedy which is generally applied in such a situation is the multiplication of $\Pi(t)$ by a factor $\lambda(t)$ called a "forgetting factor" satisfying the inequality

$$1 \leq \lambda(t) \leq \lambda_{max}$$

It has been found efficient to choose $\lambda(t)$ in such a way that the trace of $\Pi(t)$, related to the speed of adaptation, remains constant. Unfortunately, the use of a forgetting factor may result in quasi-singularity of $\Pi(t)$ when the data of the predictor becomes singular i.e. the matrix $W(t)$ defined by

$$W(t) = \sum_{i=0}^{t} x(t-i)\, x^T(t-i)$$

becomes singular.

Once $W(t)$ becomes singular, the matrix $\Pi(t)$ will remain singular even though the data vector $x(t)$ results in a non-singular $W(t)$ at a later time. To avoid this problem the diagonal elements of the matrix $\Pi(t)$ are multiplied by $\lambda(t)$ so that the trace remains a constant and the off diagonal terms by a factor less than $\lambda(t)$ (e.g. .95 $\lambda(t)$). This procedure has been found to work well in different adaptive control problems.

VIII. AN ADAPTIVE REFERENCE MODEL OPTIMAL AIM STRATEGY

The optimal aim strategy, fully described in [5], seems very attractive as a possible control algorithm for the unit voltage adaptive control due to the following reasons. First, its implementation is considerably simpler than the corresponding implementation of the implicit reference model control method, even with a non-linear model of the unit. Second, the objective function to be minimized, the vector "distance" of a predicted point (output of a predictor) to a given target, is compatible with the criterion minimized by a least squares predictor. This compatibility may lead to a more robust algorithm in the transient phases of the predictor. This aspect has been checked numerically for both simplified and physical models used for the simulation of the controlled system.

Unfortunately, the initial formulation of this method did not guarantee that the transient behaviour would be practically bounded. A remedy to this situation has been found. It is to use a series reference model and to change the coordinate of the target at each instant of time in such a way that an equivalent time constant of the controlled system can be defined.

By combining the adaptive predictor with the model reference optimal aim strategy, we obtain what may be called an adaptive reference model optimal aim strategy. This has been checked numerically with success on the full physical model including two different units linked by three lines to a consumer and to an infinite bus. However, the convergence time of the predictor must be smaller than the number of adjustable parameters multiplied by the sampling interval. This is equal to 0.5 seconds in our case. In fact the convergence time may be, because of noise, about 2 seconds. This may be too large in several critical situations (severe network faults).

Other modifications of this approach are possible and are being investigated. One in particular, which uses a non-linear model, offers the possibility of providing very quick adaptation and good control of the unit even for large transients.

IX. CONCLUSION

Several adaptive control methods have been described in this paper. Most of them have been checked successfully in physical representative simulations. Taking advantage of the microprocessor revolution, they give the hope that significant improvements will be realized in the behaviour of the power grid network.

REFERENCES

1. Monville, J. P., "Dispositif permettant d'améliorer la stabilité statique des turbo-alternateurs," Bull. D.E.R. Electricité de France, Série B - réseaux électriques, 1978.
2. Irving, E., et al., "Improving Power Network Stability and Unit Stress with Adaptive Generator Control," Automatica, Vol. 15, pp. 31-46, 1979.
3. Landau, I. D., "Adaptive Control: the Model Reference Approach," M.D.I., New York, 1979.
4. Jazwinski, A. H., "Stochastic Processes and Filtering Theory," Academic Press, New York, 1970.
5. Barnard, R. D., "Optimal Aim Control Strategies Applied to Large-Scale Nonlinear Regulation and Tracking Systems," IEEE Trans. on Circuits and Systems, Vol. CAS. 23 no. 12, December 1976.
6. Bonnard, G., "Micro-reseau." Techniques de I'Ingénieur, art. D 633, January 1973.

ADAPTIVE CONTROL APPLICATION TO AIRCRAFT

E. G. Rynaski

Advanced Technology Center
Calspan Corporation
Buffalo, New York

INTRODUCTION

The adaptive control of aircraft is an idea that has been mostly a glimmer in the eyes of theoreticians and aircraft corporations for approximately 25 years. A large amount of effort was devoted to theoretical and applied research in this area during the latter half of the 1950's and the early 1960's. Among the better known theoretical efforts were those by MIT [1], Autonetics, and Honeywell [2]. A few of the methods were actually mechanized and flight tested, including the application of the Honeywell Adaptive Control System to the X-15 experimental hypersonic research aircraft [3] and more recent application by General Dynamics to the F-111 aircraft [4].

The basic idea was that an adaptive control system should maintain a constant or invariant closed loop dynamic response throughout the entire vehicle flight range. During this time, the parameters of the equations of motions of the vehicle would

ISBN 0-12-514060-6

be unknown and varying "rapidly" over an extremely wide range. The adaptive system, called by some a "universal autopilot", was expected to be applicable without modification to a small light aircraft and to a heavy Goliath transport. For all sorts of reasons, some of which are discussed in this paper, the concept, with one possible exception, did not work as advertised.

Twenty-five years ago, estimates of stability and control parameters for aircraft were based almost completely upon wind tunnel results and heuristically obtained formulae. Because the results were unreliable at best, the adaptive autopilot concept was to have corrected for these large uncertainties. However, more, not less, knowledge of the airframe dynamic characteristics was required for a successful application - perhaps not so much in terms of accuracy of the basic stability and control derivatives of the airframe, but in other areas, such as the nonlinearities of the system, external disturbances like wind shear, turbulence and structural dynamics of the vehicle.

In the early 1960's, the Air Force was actively engaged in a flight test program to try to define what was needed or acceptable to the pilot in terms of the dynamic behavior of an aircraft. This investigation of the cybernetics aspects of the man and his flying machine reaffirmed once again that man is one of nature's most adaptable creatures. The pilot finds a relatively wide range of airplane dynamics to be satisfactory and acceptable. Invariant dynamic behavior for all aircraft over all flight ranges was neither necessary nor particularly desirable [5]. Exact analogies with automobiles are accurate; a driver does not expect nor does he find it necessary to have an automobile respond the same at all speeds, and he does not want identical handling qualities for a sportscar and a large truck. Considerable variations in dynamical behavior are expected and accepted.

As a result of the many negative adaptive control experiences of the past, the concept fell into disfavor in the flight community, and a residual bias still exists.

RECENT ADAPTIVE CONTROL ACTIVITY

The most difficult problems with previous adaptive control systems were probably those involving nonlinearities and structural dynamics. An aircraft has limited control power, and the servos that drive the aircraft control surfaces have rate limits. This means that feedback can be used only to a limited degree to speed up the response of a sluggishly responding open loop system before servo saturation occurs either in position or velocity. One had to be certain that the model would not demand more of the system than it was capable of producing. To guarantee this one had to know both the maneuver limits and the stability and control parameters of the airplane. Those adaptive systems that changed a gain as an integral of a quadratic function of the error would not work because the error could never be driven to zero. System nonlinearities and higher order dynamics guaranteed it. Both of these problems still exist and are compounded by new problems, created by attempts to apply modern control theory.

MODERN CONTROL THEORY

The two primary elements in adaptive control, introduced by modern control theory, are, generally, the concepts of the use of a performance index to specify a control criteria and modern estimation and parameter identification methods. Neither of these ideas have been successfully or adequately applied to the unique requirements of the flight of an airplane.

Improving flying qualities is the primary purpose of aircraft feedback. Flying qualities, discussed in the next section of this paper, are defined in terms of allowable or acceptable regions of poles and zeros of certain transfer functions of the airplane.

Although relationships between closed loop poles of a linear optimal system and the performance index are easily defined, no one has yet been able to establish a relationship between the closed loop transfer zeros and the weighting matrices of a performance index. A relationship of this type, perhaps similar to the root-square locus expression, would be invaluable to the flight control system designer, for he could then establish a direct relationship between a performance index and the resulting flying qualities. This is extremely important to future flight control system design. Single controller flight control systems for recent aircraft such as the F-16, F-17, Space Shuttle, and others contain many compensation networks and other energy storage elements. The order of the closed loop response of these systems has become so high the pilots have had difficulty simply controlling the aircraft on the initial flights. Future control-configured aircraft will require multiple controllers for control functions such as maneuver load control, gust alleviation, flutter suppression and maneuver drag minimization, as well as enhanced flying qualities. Without performance indices tailored to these tasks, the flight control system design problem will be very difficult, if not impossible, to do properly. The need for application has outdistanced applied theory by approximately ten years, the length of time that theory has been substantially deemphasized by traditional sponsors such as DoD and NASA.

The most promising and theoretically complete adaptive systems hypothesize the use of on-line identification as an inherent part of the system. <u>On-line</u> identification, however, is not yet a household word. Experience with <u>off-line</u> identification shows that accuracy is a "maybe" kind of thing, very much dependent upon the vehicle dynamics, the quality and complement of onboard sensors, the input or motions of the airframe and finally, least important, the particular identification algorithm used. The same data can be analyzed for weeks - adjusting such things as noise covariances and instrument biases - and increasingly more accurate results are obtainable. Occasionally, after the first identification has been successfully accomplished, subsequent identifications can be made in one pass, but only over a limited flight range.

For flight control application, a number of fundamental questions of applied theory must be adequately answered, either analytically or by experimental research methods:

1. What degree of identification accuracy is required? Because, as discussed later, the pilot is quite tolerant of a wide variety of airplane dynamic characteristics, a relatively low accuracy may be sufficient, and all dynamic parameters need not be identified to the same level of accuracy. Accuracy requirements of aircraft identification for adaptive flight control application have yet to be defined. If angle of attack, sideslip or Mach number effects are not important, and if the nondimensional stability derivatives are known (and constant), identification of one derivative per degree of freedom of aircraft motion implies knowledge of all stability and control derivatives.
2. Can adequate on-line identification be accomplished using normal pilot or atmospheric excitation to the airplane? This implies a need to know the relationship between the amplitude and frequency content of the pilot control inputs and the resulting parameter identification accuracy. If it is possible to identify the relationship using normal pilot inputs, auxiliary or dither inputs would not be necessary, and the system is likely to be more acceptable to the pilot.

In the following sections, flying qualities and model following are briefly presented to show some of the unique requirements of flight dynamics as they apply to adaptive control. These serve as guidelines that should not be ignored. Failure to adhere to the guidelines can result in an improper application that would reflect detrimentally to adaptive flight control, as well as to adaptive control in general.

FLYING QUALITIES

Usually, the purpose of feedback control for aircraft is to improve or enhance the flying qualities of the vehicle. Therefore, to apply adaptive or any other control to an aircraft without taking into consideration the cybernetics aspects of the problem, makes no sense at all, because flying qualities define the proper match of the man-machine system.

Outlined below is a very brief discussion of the flying qualities aspects of an aircraft that would be of immediate use for adaptive control application. The fundamental differential equations that describe the dynamical motions of an aircraft in the longitudinal-vertical plane of motion are given by

$$\dot{q} = M_q q + M_{\dot{\alpha}} + M_\alpha \alpha + M_{\delta_e} \delta_e \tag{1a}$$

$$\dot{\alpha} = q - L_\alpha \alpha \tag{1b}$$

$$n_z = \frac{V}{g} (\alpha - q) \tag{1c}$$

where q is the change in pitch rate of the aircraft

α is the change in angle of attack of the wing

δ_e is the change in elevator deflection

n_z is normal acceleration

V is velocity of the aircraft with respect to the air mass

$M_q = \frac{\partial M}{\partial q}$, change in pitching moment due to pitch rate change

$M_\alpha = \frac{\partial M}{\partial \alpha}$, change in pitching moment due to angle of attack change

$M_{\dot{\alpha}} = \frac{\partial M}{\partial \dot{\alpha}}$, change in pitching moment due to rate of change of angle of attack

$M_{\delta_e} = \frac{\partial M}{\partial \delta_e}$, change in pitching moment due to change in elevator deflection

$L_\alpha = \frac{\partial L}{\partial \alpha}$, change in lift due to change in angle of attack. The equations (1), of course, neglect many higher order and non-linear effects, and the adequacy of the equations depends upon the individual airframe and the motion range of the airplane. The transfer functions of the motion variables of the equations of motion (1) with respect to an elevator deflection are given by

$$\frac{q}{\delta_e}(s) = \frac{M_{\delta_e}(s + L_\alpha)}{s^2 + s(L_\alpha - M_q - M_{\dot{\alpha}}) - M_\alpha + M_{\dot{\alpha}} L_\alpha} \tag{2a}$$

$$\frac{\alpha}{\delta_e}(s) = \frac{M_{\delta_e}}{s^2 + s(L_\alpha - M_q - M_{\dot{\alpha}}) - M_\alpha + M_{\dot{\alpha}} L_\alpha} \tag{2b}$$

$$\frac{n_z}{\delta_e}(s) = -\frac{V}{g} L_\alpha \frac{\alpha}{\delta_e}(s) \tag{2c}$$

The zero defined by $(s + L_\alpha)$ in Equation (2a) physically means that the airplane must rotate before it can change its flight path, i.e. climb or descend. The response of the aircraft to a step pilot command input depends not only on the damping ratio and natural frequency of the characteristic equation, but also on the value of the zero of the $q/\delta_e(s)$ transfer function. The angle of attack and acceleration response are generally in phase but the pitch rate response leads. A feedback control law can produce a rapid and precise response for pitch rate or acceleration, but the single controller cannot produce both. Basic flying qualities then involve three dynamic quantities of the vehicle response: the zero of the $q/\delta_e(s)$ transfer function, the natural frequency, and the damping ratio of the characteristic polynomial.

Military flying qualities specifications are defined in a document known as MIL-F-8785B(ASG). The basic requirement as shown in

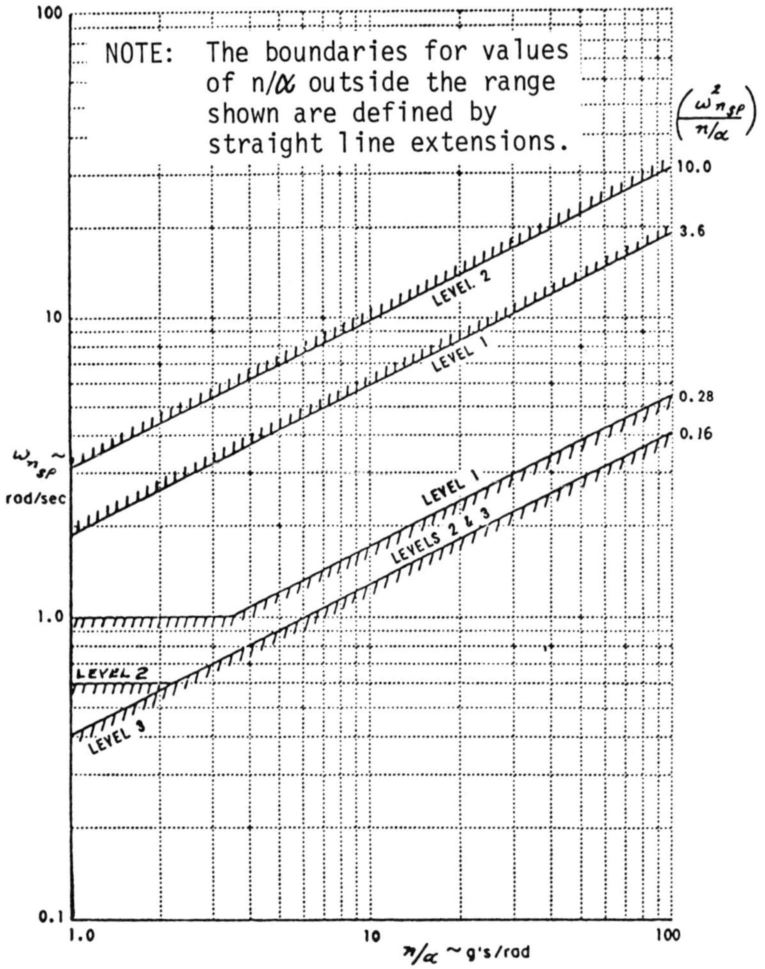

FIGURE 1. Short-period frequency requirements - category A flight phases.

Figure 1 is taken from the specification and is a plot of the aircraft natural frequency and acceleration response sensitivity for tasks that require rapid maneuvering and/or precision tracking. The "levels" 1, 2 and 3 indicate the relative acceptability of the dynamics by the pilot. Damping ratio is specified separately and for best acceptability, level 1, the damping ratio, can be anywhere from $\zeta = 0.35$ to $\zeta = 1.30$. The plot shows that for an aircraft with a particular acceleration sensitivity, the natural frequency may vary by a factor of nearly ten and the damping ratio by a factor of nearly four, and the vehicle dynamics will still be acceptable to the pilot. From this, it can be inferred that every aircraft contains an adaptive control system: the pilot.

Although the pilot is quite tolerant of the range of the natural frequency and damping ratio of his aircraft, he is less tolerant of other factors - particularly time delays and lags. When a pilot injects a command input into his aircraft, he expects the aircraft to start responding to that command very quickly. In fact, if the delay exceeds 100 milliseconds due to computer computation cycle time or other factors, the effect on the pilot will be serious. Equally serious is a non-minimum phase response, in which the aircraft initially responds in the wrong direction before reversing itself. Most of the logic applied to aircraft flying qualities are equally appropriate for motor vehicles as well. If a race driver turned his wheel for a right hand turn and there was a significant delay before starting the turn, or if the vehicle initially moved left before turning right, these effects, if significant, would be intolerable to the driver, as they are to a pilot.

For adaptive control application, it is possible to conclude the following from the flying qualities specification:

- The model dynamics can be allowed to vary over a wide range; therefore,
- The model following system need not follow the model with much precision; however,
- Higher order dynamics delays, digital computer or other transport lag effects and non-minimum phase responses are always strongly detrimental to flying qualities.

In terms of an adaptive control mechanization, the following conclusions may be appropriate:

- Implicit rather than explicit adaptive control may be preferable because imprecise explicit model following will introduce a higher order dynamic response into the aircraft.

- Unless a very rapid digital computer system that produces digital computer lags of less than about 50 milliseconds is used, an analog system is preferable.
- On-line estimation and identification and observer mechanizations must be accomplished with high precision if errors would introduce the effects of higher order dynamic delays into the system response.

In light of what is presently known about aircraft and the very loose flying qualities requirements for aircraft, it can be concluded that a successful adaptive control application would be an analog or very fast digital computer system that allows only gains in the feedback and feedforward part of the control system itself. These gains may be altered by a performance index criteria that contains an implicit model, generally known as the "model-in-the-performance index" method of linear optimal control. The system would generally be structured as shown in Figure 2. Because the

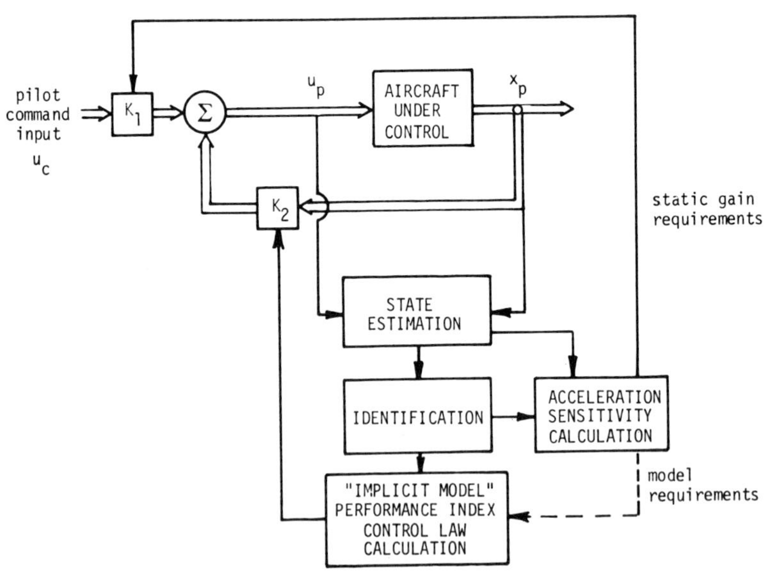

FIGURE 2. Adaptive configuration for aircraft application.

required natural frequency of response (the model) and the pilot command input gain requirements are a function of the acceleration sensitivity, this sensitivity can be easily calculated from the simplified expression

$$\frac{n}{\alpha} = - \left.\frac{n}{\delta_e}\right|_{s.s.} \cdot \frac{-M_\alpha + M_{\dot{\alpha}} L_\alpha}{M_{\delta_e}} \tag{3}$$

where $n/\delta_e \big|_{s.s.}$ can usually be taken to be the value of acceleration obtained approximately 2 sec after a pilot command input.

The advantage of adaptive control as shown in the figure is primarily that adjustments are made independently of the information flow. Changes in the control law are computed without disturbing the system in any way. It is reasonable to expect that the entire adaptive process can be controlled or triggered by the calculation of the acceleration sensitivity. Only when this parameter shows significant change would it be necessary to change the model dynamics in the performance index, or equivalently, change the weighting matrices of a performance index that does not contain a model.

MODEL FOLLOWING

Perhaps the majority of adaptive control systems are based upon the concept of model following in one form or another. In aircraft applications, it has been necessary to investigate and examine the theoretical structures and control laws of model following systems to optimize their suitability for a flight control system.

Two basic situations are possible in the theoretical investigation of model following systems:

1. The number of independent controllers is equal to the number of degrees of freedom of motion of the airplane and,
2. The number of independent controllers is less than the number of degrees of freedom of motion.

The first case is the simpler of the two. The equations of motion of the airplane and of the model are given by the constant coefficient and linear equations of motion:

$$\dot{X}_p(t) = F_p X_p(t) + G_p u_p(t) \tag{4}$$

$$\dot{X}_m(t) = F_m X_m(t) + G_m u_m(t) \ . \tag{5}$$

It is convenient to assume a control law of the form

$$u_p(t) = K_1 X_m(t) + K_2 u_m(t) \ . \tag{6}$$

Substituting Equation (6) into Equation (4) yields

$$\dot{X}_p(t) = F_p X_p(t) + G_p K_1 X_m(t) + G_p K_2 u_m(t) \ . \tag{7}$$

If $G_p K_2 = G_m$ then $G_p K_2 u_m(t) = \dot{X}_m(t) - F_m X_m(t)$ and this can be substituted into Equation (7) to yield

$$\dot{X}_p(t) - F_p X_p(t) = G_p K_1 X_m(t) + \dot{X}_m(t) - F_m X_m(t) \tag{8}$$

and $X_p(t) = X_m(t)$ if the following conditions are satisfied:

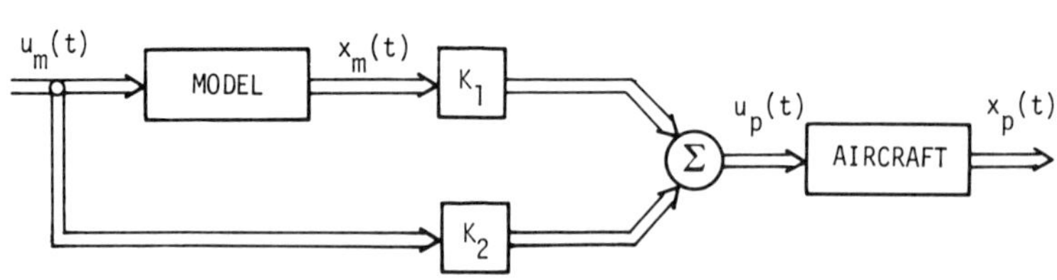

FIGURE 3. Explicit open loop model following system.

$$G_pK_2 = G_m \tag{9}$$

$$-F_p = G_pK_1 - F_m . \tag{10}$$

Because it has been assumed that the number of independent controllers is equal to the number of degrees of freedom of aircraft motion, the gain matrices K_1 and K_2 are exactly given by

$$K_1 = (G_p^TG_p)^{-1}G_p^T(F_m-F_p) \tag{11}$$

$$K_2 = (G_p^TG_p)^{-1}G_p^TG_m . \tag{12}$$

The expressions for the gain matrices K_1 and K_2 are the same for an explicit model following system shown in Figure 3 and the implicit model following configuration shown in Figure 4.

Although theoretically equivalent, the configuration shown in Figure 4 is superior because it is less sensitive; the output will be less prone to wander if the gain matrices are in error. Also, if the gains are not computed exactly, the system of Figure 3 will exhibit a higher order response characterized by the cascading effect of the model dynamics and the airframe dynamics.

The control gains of the model following systems of Figures 3 and 4 are a function of both the plant and the model. It would be

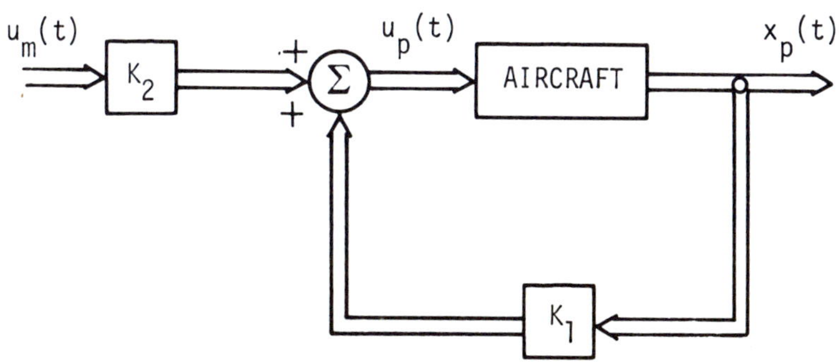

FIGURE 4. Implicit closed loop model following system.

an advantage to define a model following control independent of the model, for then the model could be altered at will without changing the control law. To obtain a control law independent of the model [6], it is only necessary to substitute $\dot{X}_m(t)$ and $X_m(t)$ for $\dot{X}_p(t)$ and $X_p(t)$ in Equation (4) and obtain

$$\dot{X}_m(t) = F_p X_m(t) + G_p u_p(t) \; . \tag{13}$$

Solving for $u_p(t)$ yields the control law

$$\begin{aligned} u_p(t) &= (G_p^T G_p)^{-1} G_p^T \dot{X}_m(t) - (G_p^T G_p)^{-1} G_p^T F_p X_m(t) \qquad (14) \\ &= K_{\dot{x}} \dot{X}_m(t) + K_x X_m(t) \; . \end{aligned}$$

As can be seen, the solution given by Equation (14) is advantageous not only because a nonlinear or time-varying model can be accurately followed, but because a nonlinear or time-varying plant can also be accommodated; the solution given by Equation (14) is entirely general.

A significantly improved model following system is shown in Figure 5 below, where

$$K_{\dot{x}} = (G_p^T G_p)^{-1} G_p^T \tag{15}$$

$$K_x = (G_p^T G_p)^{-1} G_p^T F_p - K_p \; . \tag{16}$$

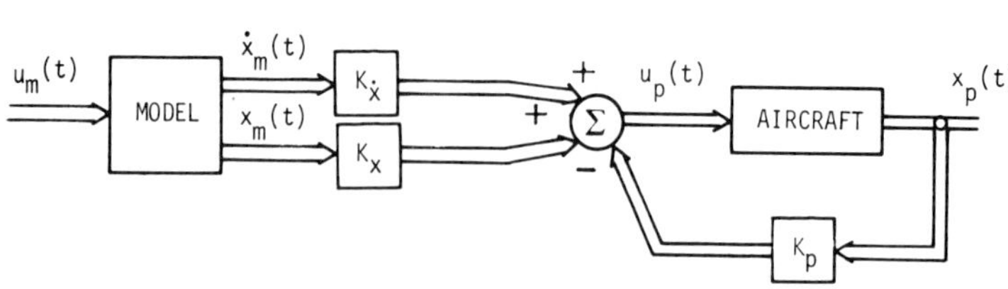

FIGURE 5. Improved explicit model following system.

The feedback around the plant is arbitrary, subject only to the requirement that the characteristic polynomial $\Delta(s) = |Is - F_p + G_pK_p|$ have roots in the left half plane, i.e. the regulator part of the system is stable. From Equation (16) it can be seen that if

$$K_p >> (G_p^T G_p)^{-1} G_p^T F_p \tag{17}$$

then feedback eliminates the need to know the matrix of stability derivatives of the aircraft F_p accurately. The matrix of control effectiveness terms G_p must be identified accurately for any adaptive control configuration.

It is clear that the feedback of an explicit model following system of the type shown in Figure 5 can be designed solely as a regulator using linear optimal control methods to maximize the difference $K_p - G_p^+ F_p$ in order to minimize the sensitivity of the regulator to changes in the plant dynamics. This is a classical example of the use of feedback as Bode originally envisioned.

For the airplane application, it should be noted that the expression $G_p^+ F_p$ implies division of a dimensional stability derivative by a dimensional control derivative. To a large extent, many of the dominant elements of this operation involve stability and control derivatives dimensionalized in exactly the same way. Therefore, many of the elements of the matrix $G_p^+ F_p$ are constants (or nearly so) over a very wide flight range. This fact reinforces the contention that a regulator, most likely with constant feedback gains, can be specified without much difficulty to obtain satisfactory performance for the system.

Limited Control Capability

When the number of controllers on the aircraft is less than the number of degrees of freedom of motion of the aircraft to be controlled, then the control laws for precise model following become a bit more difficult to implement, and problems such as non-minimum phase transfer functions will limit the ability to perform exact model following. To obtain an exact model-following control law, the equations of motion for the aircraft and the model are partitioned as follows:

$$\begin{bmatrix} \dot{X}_1 \\ \hline \dot{X}_2 \end{bmatrix} = \left[\begin{array}{c|c} F_{11} & F_{12} \\ \hline F_{21} & F_{22} \end{array}\right] \begin{bmatrix} X_1 \\ \hline X_2 \end{bmatrix} + \begin{bmatrix} G_1 \\ \hline G_2 \end{bmatrix} u \tag{18}$$

$$\begin{bmatrix} \dot{X}_{1m} \\ \hline \dot{X}_{2m} \end{bmatrix} = \left[\begin{array}{c|c} F_{11m} & F_{12m} \\ \hline F_{21m} & F_{22m} \end{array}\right] \begin{bmatrix} X_{1m} \\ \hline X_{2m} \end{bmatrix} + \begin{bmatrix} G_{1m} \\ \hline G_{2m} \end{bmatrix} u_m \; . \tag{19}$$

Assuming that it is appropriate to model follow to force $X_2(t) = X_{2m}(t)$ to a command input, the partitioning is accomplished such that F_{22} and G_2 are of the same pxp dimension where p is equal to the number of independent plant controllers. For the model, F_{22m} is of dimension pxp and G_{2m} is of pxm dimension.

A solution is obtained in a very straightforward manner if the following control law is assumed:

$$u_p = K_1\dot{X}_{2m} + K_2X_{2m} - K_3X_1 - K_4X_2 \; . \tag{20}$$

Substitution of this control law into Equation (18) yields

$$\dot{X}_2 = (F_{21} - G_2K_3)X_1 + (F_{22} - G_2K_4)X_2 + G_2K_1\dot{X}_{2m} + G_2K_2X_{2m} \; . \tag{21}$$

Examination of equation (21) shows that exact model following requires the term in X_1 to be zero, i.e. $F_{21} - G_2K_3 = 0$ or

$$K_3 = (G_2^TG_2)^{-1}G_2^TF_{21} \; . \tag{22}$$

Substitution of K_3 into Equation (21) and taking a Laplace transform of Equation (21) yields

$$(Is-F_{22} + G_2K_4)X_2(s) = (G_2K_1Is + G_2K_2)X_{2m}(s) \ . \tag{23}$$

Then $X_2(s) = X_{2m}(s)$ if

$$I = G_2K_1 \quad , \quad K_1 = (G_2^TG_2)^{-1}G_2^T \tag{24}$$

$$G_2K_2 = -F_{22} + G_2K_4 \quad , \; K_2 = (G_2^TG_2)^{-1}G_2^T(-F_{22} + G_2K_4) \tag{25}$$

and K_4 is arbitrarily defined, subject only to the restriction that the polynomial equation in s defined by $D_1(s) = |Is - F_{22} + G_2K_4|$ has roots in the left half plane, i.e. it must be stable. A sketch of the resulting model following system is shown in Figure 6 below. As with the configuration shown in Figure 5, the model following control law is independent of the model dynamics. Unlike the configuration of Figure 5, however, feedback from those states that are <u>not</u> to follow the model must be included in the design, while feedback of those states that are to follow the model need not (theoretically) be included in the feedback control law.

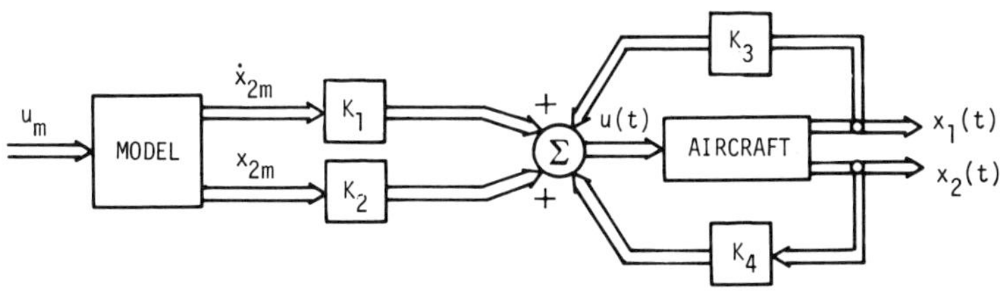

FIGURE 6. Limited control model following system.

Example: Limited Control Model Following

Limited control model following can be viewed as a problem of pole-zero cancellation. In fact, the feedback gain matrix K_3 of Equation (22) implies just that. To show this, the following example is computed using frequency domain rather than the time domain control law of Equation (20).

From Equation (2), the fundamental longitudinal-vertical transfer functions of an aircraft are given by

$$\frac{q}{\delta_e}(s) = \frac{M_{\delta_e}(s+L_\alpha)}{s^2 + 2\zeta\omega s + \omega^2} \tag{26}$$

$$\frac{\alpha}{\delta_e}(s) = \frac{M_{\delta_e}}{s^2 + 2\zeta\omega s + \omega^2} . \tag{27}$$

A model is taken to be of the same form but different order. We shall assume that the model contains actuator dynamics, as

$$\frac{q_m}{\delta_{e_m}}(s) = \frac{M_{\delta_{e_m}}(s+L_{\alpha_m})}{(s+a)(s^2+2\zeta_m\omega_m s+\omega_m^2)} \tag{28}$$

$$\frac{\alpha_m}{\delta_{e_m}}(s) = \frac{M_{\delta_{e_m}}}{(s+a)(s^2+2\zeta_m\omega_m s+\omega_m^2)} . \tag{29}$$

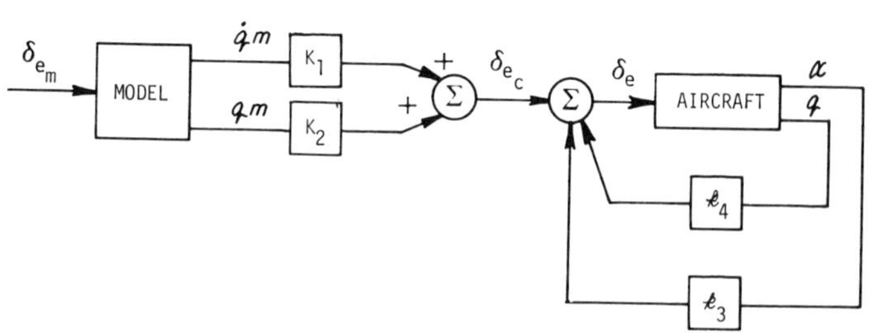

FIGURE 7. Pitch rate model following system.

To calculate a control law that will yield $\frac{q}{\delta_{e_m}}(s) = \frac{q_m}{\delta_{e_m}}(s)$, Equation (20) indicates that the control law should be of the form shown in Figure 7.

The pole-zero cancellation implications of the model-following configuration shown above become clear when the transfer functions $\delta_{e_c}/\delta_{e_m}(s)$ and $q/\delta_e(s)$ are written out as

$$\frac{\delta_{e_c}}{\delta_{e_m}}(s) = \frac{k_1 M_{\delta_{e_m}} s(s+L_{\alpha_m}) + k_2 M_{\delta_e}(s+L_{\alpha_m})}{(s+a)(s^2+2\zeta_m\omega_m s+\omega_m^2)} \tag{30}$$

$$\frac{q}{\delta_{e_c}}(s) = \frac{M_{\delta_e}(s+L_\alpha)}{s^2 + 2\zeta\omega s + \omega^2 + k_4 M_{\delta_e}(s+L_\alpha) + k_3 M_{\delta_e}} . \tag{31}$$

As seen from Equation (30) and Equation (31), the gains k_1 and k_2 allow complete choice of the zeros of the $\delta_{e_c}/\delta_{e_m}(s)$ transfer function while k_3 and k_4 allow complete choice of the poles of the $\frac{q}{\delta_{e_c}}(s)$ transfer function.

The strategy for computation of a model following system then becomes clear. The feedback gains should be set such that one of the poles is at $s = -L_\alpha$ to cancel the zero of the $\frac{q}{\delta_{e_c}}(s)$ transfer function at $s = -L_\alpha$; the second pole is arbitrary. The gains k_1 and k_2 should be calculated to yield zeros of the $\delta_{e_c}/\delta_{e_m}(s)$ transfer function at $s = -L_\alpha$ and at the value of the second pole of the $\frac{q}{\delta_{e_c}}(s)$ transfer function. First calculate the feedback gains k_3 and k_4 such that

$$s^2 + 2\zeta\omega s + \omega^2 + k_4 M_{\delta_e}(s+L_\alpha) + k_3 M_{\delta_e} = (s+L_\alpha)(s+c) \tag{32}$$

where $-c$ is the second arbitrarily chosen pole. Equating coefficients of equal powers of s in Equation (32) yields the two equations

$$2\zeta\omega + k_4 M_{\delta_e} = L_\alpha + c \tag{33}$$

$$\omega^2 + k_4 M_{\delta_e} L_\alpha + k_3 M_{\delta_e} = L_\alpha c \tag{34}$$

or

$$k_4 = \frac{L_\alpha + c - 2\zeta\omega_n}{M_{\delta_e}} \tag{35}$$

$$k_3 = \frac{1}{M_{\delta_e}} (-\omega^2 - L_\alpha^2 + 2\zeta\omega\, L_\alpha) \ . \tag{36}$$

The zeros of the transfer function $\delta_{e_c} / \delta_{e_m}(s)$ are chosen to be at $s = -c$, $s = -L_{\alpha_m}$, as

$$k_1 M_{\delta_{e_m}} s(s+L_{\alpha_m}) + k_2(s+L_{\alpha_m}) = \frac{M_{\delta_{e_m}}}{M_{\delta_e}} (s+c)(s+L_{\alpha_m}) \tag{37}$$

or

$$k_1 M_{\delta_{e_m}} s^2 + (k_1 M_{\delta_{e_m}} L_{\alpha_m} + k_2) s + k_2 M_{\delta_e} L_{\alpha_m} = \frac{M_{\delta_{e_m}}}{M_{\delta_e}} s^2 + \frac{M_{\delta_{e_m}}}{M_{\delta_e}} (L_{\alpha_m} + c) s + \frac{M_{\delta_{e_m}}}{M_{\delta_e}} cL_{\alpha_m} \ . \tag{38}$$

Equating powers of s yields

$$s^2: \quad k_1 M_{\delta_{e_m}} = \frac{M_{\delta_{e_m}}}{M_{\delta_e}} , \quad k_1 = \frac{1}{M_{\delta_e}} \tag{39}$$

$$\begin{aligned} s^1: &\quad k_1 M_{\delta_{e_m}} L_{\alpha_m} + k_2 = \frac{M_{\delta_{e_m}}}{M_{\delta_e}} (L_{\alpha_m} + c), \quad k_2 = \frac{c}{M_{\delta_e}} \\ s^0: &\quad k_2 M_{\delta_{e_m}} L_{\alpha_m} = \frac{M_{\delta_{e_m}}}{M_{\delta_e}} cL_{\alpha_m} \ . \end{aligned} \tag{40}$$

Using the gains k_1, k_2, k_3 and k_4 calculated above, the total input-output transfer function becomes

$$\frac{q}{\delta_{e_m}}(s) = \frac{q}{\delta_{e_c}}(s) \cdot \frac{\delta_{e_c}}{\delta_{e_m}}(s)$$

$$= \frac{M_{\delta_{e_m}}}{M_{\delta_e}} \frac{(s+c)(s+L_{\alpha_m})}{(s+a)(s^2+2\zeta_m\omega_m s+\omega_m^2)} \cdot \frac{M_{\delta_e}(s+L_\alpha)}{(s+c)(s+L_\alpha)}$$

$$= \frac{M_{\delta_{e_m}}(s+L_{\alpha_m})}{(s+a)(s^2+2\zeta_m\omega_m s+\omega_m^2)} = \frac{q_m}{\delta_{e_m}}(s) \quad . \tag{41}$$

Figure 8 shows the resulting exact model following system using the gains calculated above.

The gains shown in Figure 8 are exactly those that would have been obtained using the state space approach defined by the control law of Equation (20). As indicated in Equation (25) and shown by the q feedback in Figure 8, the feedback of the variable that is to follow the model is arbitrary; this gain can be high, low, or zero. If the closed loop plant pole at -c is taken at the origin

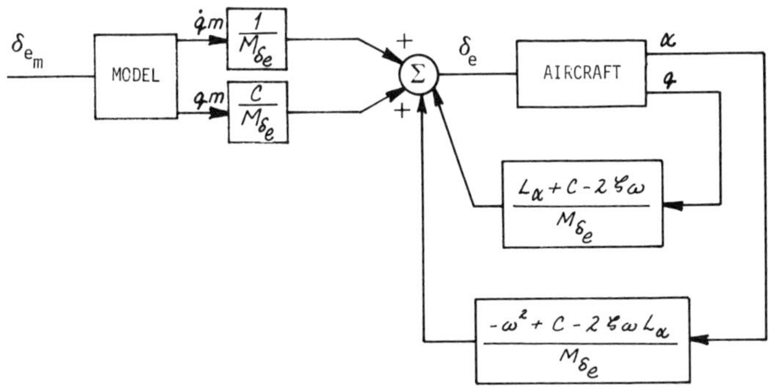

FIGURE 8. Exact pitch rate model following system.

of the s plane, i.e. c = 0, the model output that is to be followed by the plant need not be fed forward from the model. The system shown in Figure 8 is independent of the model order; the additional pole added to the model transfer functions had no effect on the control system design. The important requirement is that the model states and derivatives be available, which is, of course, entirely feasible because the model is explicitly contained within a computer.

If the plant transfer function that follows the model is non-minimum phase, explicit model following is impossible. Feedback around the plant is designed to pole-zero cancel the zeros of the plant transfer function, so model following would require feedback to drive the plant unstable.

For model following systems when the number of controllers is equal to the number of degrees of freedom, there are no non-minimum phase restrictions. For multicontroller systems with fewer control variables than degrees of freedom of motion, model following for systems with non-minimum phase transfer functions is hypothesized to depend upon the feedback gain matrices k_3 and k_4 shown in Equation (21). If feedback gain matrix k_4 can be found to yield a stable closed loop regulator with the constraint of Equation (22), multiple controller model following for non-minimum phase plants may be accomplished.

FINAL REMARKS

The purpose of this paper was to present some guidelines for the application of adaptive control to the aircraft flight control problem. The need for adaptive control in research-oriented aircraft is high now, and will become acute in the not-too-distant future. The need for robust control configurations for operational aircraft is also great, and research should be initiated as soon as possible to try to define those conditions under which a robust, constant control system parameter configuration is sufficient, a gain scheduled system is needed, and a full adaptive system is necessary. Until this is done, the real need for adaptive control systems for aircraft will not be known.

REFERENCES

1. Whitaker, H. P., Yamron, J., and Kezer, A., "Design of Model-Reference Adaptive Control Systems for Aircraft," M.I.T. Instrumentation Lab., Cambridge, Ma., Report R-164, Sept. 1958.
2. Shuck, O. H., "Honeywell's History and Philosophy in the Adaptive Control Field," Wright Air Development Center, Wright-Patterson A.F.B., Oh., Technical Report, January 1959.
3. "Experience with the X-15 Adaptive Flight Control System," NASA Flight Research Center, NASA TN-D-6208, March 1971.
4. Scott, H. Z., "F-111 Self-Adaptive Flight Control System," Proc. Nat. Aerospace Electronics Conf., pp. 459-460, 1967.
5. Chalk, C. R., Neal, T. P., Harris, T. M., et. al., "Background Information and Users Guide for MIL-F-8785B(ASG), Military Specification - Flying Qualities of Piloted Airplanes," AFFDL-TR-69-72, August 1969.
6. Motyka, P., Rynaski, E., and Reynolds, P., "The Theory and Flight Verification of the TIFS Model Following System," AIAA Journal of Aircraft, Vol. 9, No. 5, May 1972.

PERSPECTIVES ON THE APPLICATION OF ADAPTIVE CONTROL TO AIRCRAFT SYSTEMS

Gerhard Kreisselmeier

DFVLR - Institut fuer Dynamik der Flugsysteme
Oberpfaffenhofen
F. R. Germany

This paper tries to outline some of the problems which are relevant for adaptive control applications to aircraft systems, and to indicate in what respect advantages can be expected from adaptive flight control over conventional flight control.

I. INTRODUCTION

In the historical development of aircraft systems, controllers were introduced to automatize straight and level flight, to enable flying by instruments without ground visibility and to simply augment the stability properties, and thus improve the handling qualities of the aircraft. In that period, the development of control systems was very closely related to the development of the flight system itself. This resulted in structurally very simple though very efficient controllers, due to the physical motivation behind it.

ISBN 0-12-514060-6

The further development showed a remarkable extension of the control requirements along with the introduction of new control variables. This led to a large number of relatively simple decentralized single variable control loops for specific subproblems, which are connected through various possible supervisory controllers, the so called autopilot. Although modern control theory concepts have been introduced to flight control since then, much of this physically motivated structure is part of present flight control systems. It appears to be important to recall the strong motivation for these structures, ranging from the physical insight up to the simplicity and flexibility of the realization, when one thinks of introducing new concepts to flight control.

Part of the historical evolution in flight systems is that the region in which the aircraft can operate has been increased significantly, e.g. to high altitudes, high Mach numbers and even to major changes in the geometry of the aircraft. As a matter of fact, increasing the range of flight conditions produced an increase of parameter variations in the dynamic equations of the aircraft. These significant changes in the aircraft dynamics had at least partially to be overcome by means of the control system and, in particular, the controller had to operate satisfactorily for a drastically changing aircraft. Attempts to solve this problem have proceeded in two stages.

Firstly, extensive theoretical and experimental studies have been devoted to obtain a fairly accurate model of the dynamic behaviour of the aircraft for all possible flight conditions. Using this knowledge of the possible parameter variations, the second step then was to design a control system, which is suitably adapted to the flight condition. This was achieved by simply maintaining the earlier control structure but now scheduling the controller gains in a preprogrammed fashion according to the flight condition, for example as a function of the altitude, dynamic pressure, flap positions, etc., which are measured by additional sensors. This gain scheduling resulted in high performance aircraft control, since it involved all the detailed information which was available about the aircraft. Contrary to the expectations of control theorists, application of modern control techniques resulted only in minor performance improvements, if any, in recent years.

To date there are mainly four reasons for the desire to replace the traditional control concept by a more modern (adaptive) concept. Firstly, the complexity of modern flight systems is still increasing, and this makes the number of quantities, with respect to which the gains would have to be scheduled, increasing too. In turn, the burden of the control design as well as of its realization increases to an extent, which may no longer be meaningful nor tolerable. Secondly, the flight region is still being extended, e.g. to high angles of attack and to post stall capability, where at present the knowledge of the aircraft behaviour is much less accurate than it is in the traditional regions. In addition, such parameter changes are no longer slow compared with the aircraft dynamics. Thirdly, it is highly desirable to reduce the complexity and the costs of the control systems, e.g. simply by saving sensors. Last but not least, control theorists feel that recent progress in the theory of adaptive control makes it applicable and well suited to solve flight control problems in an efficient way.

It should be noted in this context, that there are other flight systems such as guided missiles, the control systems of which have been kept less sophisticated compared to aircraft control systems. The main reasons for this are savings of weight and costs, and this ruled out additional sensors and other hardware. For these systems it may even be possible to improve the performance of control by using the same sensor hardware and, in addition, introducing new (adaptive) control concepts.

It is obvious that gain scheduling as used in flight control systems performs a certain adaptation of the controller to the flight condition. Thus, adaptivity is not a new concept in flight control. However, the adaptation is based on extra sensor signals and on predesigned and preprogrammed gain scheduling. Both of them are neither required nor used in a corresponding adaptive control system. A further principal difference between adaptive control and gain scheduling is that in the gain scheduling case the adaptation is dynamically decoupled from the control, whereas in the adaptive control case it is not (consider the gust alleviation control as an example). This indicates that the nonlinearities associated with adaptive control are much more important than those associated with gain scheduling control. Therefore the transition from traditional to new adaptive control structures will not simply solve the parameter variation problem in flight control systems, but will introduce new problems as well. The feasibility of adaptive control for flight systems depends on how it can be "adapted" to the in flight requirements and how its inherent problems can be solved.

In an attempt to give a perspective for the possible application of adaptive control to aircraft systems, this paper discusses some recent progress in adaptive control theory relative to practical flight control problems, and a number of problems which are associated with adaptive control and which may have to be solved prior to or along with practical applications.

II. ON THE STATUS OF ADAPTIVE CONTROL THEORY

If a plant is to be controlled, such as an aircraft, which is to operate over a wide range of <u>known</u> operating conditions, the control theorist is confronted with the question of how to come up with a fairly general concept for solving this type of control problem. Because all the details of a specific application would complicate his analysis, he imbeds the problem in a slightly more general one:

"Given a completely unknown plant, construct a control system, which is capable of adapting its own controller gains suitably to the unknown plant."

Obviously, this adaptive reformulation of the control problem ignores all the information, which is available about the plant in the practical application at hand. On the one hand, removing the necessity of having a priori knowledge is an advantage, but on the other hand, ignoring available information, may be very disadvantageous in view of obtaining the best possible design. This aspect will have to be discussed in more detail later.

There are essentially two different concepts for solving the above adaptive control problem. In the direct adaptive control approach, mainly represented by model reference adaptive control, the controller gains are adjusted directly, i.e. without explicitly identifying the unknown system parameters. As opposed to this, in the indirect adaptive control approach the unknown system parameters are identified first, and from the current parameter estimates the controller gains are adjusted. Without going into mathematical details, the assumptions and properties for both approaches will be discussed now.

The unknown plant generally is assumed to be a linear, time invariant system. Furthermore, at present the most rigorous results are for single-input single-output systems. Therefore the unknown plant can be described in state space form as

$$\begin{aligned} \dot{\underline{x}} &= \underline{A}\underline{x} + \underline{b}u \\ y &= \underline{c}^T\underline{x} \end{aligned} \tag{1}$$

or, equivalently, in the frequency domain as

$$y = \frac{N_p(s)}{D_p(s)} u . \tag{2}$$

A. Model Reference Adaptive Control (Direct Approach)

The basic idea of the model reference concept is to select a suitable reference model and to determine the control in such a way that the transfer behaviour from some command input u_c to the unknown plant output y asymptotically matches that of the reference model, i.e. the transfer behaviour satisfies

$$u_c \rightarrow y \equiv u_c \rightarrow y_m$$

if the adaptation is complete.

It is well-known from linear control theory, that ideal model following can be achieved by means of various different linear control structures. They make use of both feedforward and feedback compensation and mainly differ from the amount of feedback which is employed [4]. But whatever the control structure is, if the output of (2) is to match the output y_m of some reference model

$$y_m = \frac{N_m(s)}{D_m(s)} u_c , \tag{3}$$

then it is necessary that the plant input becomes

$$u = \frac{D_p(s)}{N_p(s)} \cdot \frac{N_m(s)}{D_m(s)} u_c . \tag{4}$$

In (4) the inverse plant transfer function is involved, and stable control is obtained only if $N_p(s)$ is a Hurwitz polynomial.

Therefore the exact model following concept, whether adaptive or not, is limited in its application to minimum phase plants.

In the adaptive model reference approach, feedback compensation with adjustable gains is chosen typically, so as to cancel the plant zeros and generate the model poles, and feedforward compensation then essentially introduces the desired model zeros. Of course, the adaptation of the compensator gains, which is based on the error between the plant and the model output, makes the control law highly nonlinear. Fig. 1 shows the basic control structure.

Only recently, considerable progress has been made in obtaining rigorous proofs of global asymptotical stability for such model reference adaptive schemes (e.g. [7-11]). This means, that the plant output will always converge to the model output, no matter what the plant parameters are. The only restrictions are that the plant must be known to be minimum phase, and that some further knowledge such as the relative degree of the plant (i.e. $\deg(D_p) - \deg(N_p)$) and the sign of the plant gain for high frequencies must be available.

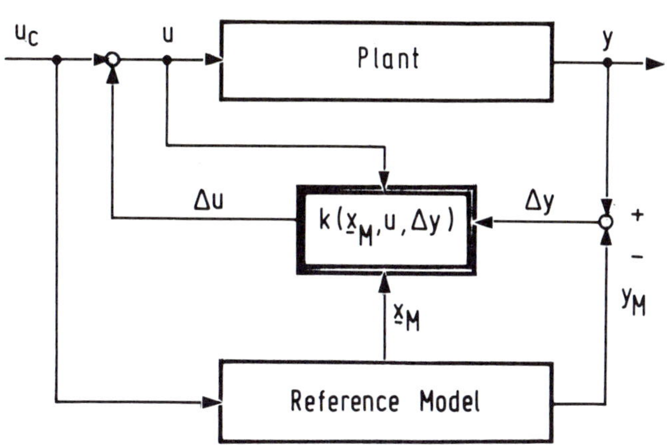

FIGURE 1. Model reference adaptive control structure.

This result is particularly powerful, because the convergence of $y \to y_m$ does not require that the adaptive compensator gains converge, and this in turn does not require that the plant input be "sufficiently exciting" as it would be necessary for identifying the unknown system parameters. This valuable property of the model reference adaptive control approach is due to the fact, that the adaptive laws have been derived by means of the Lyapunov synthesis technique.

It is important to note, however, that the global stability result has been obtained only for the undisturbed plant. This is not surprising, because the design goal and all the corresponding assumptions were to achieve the desired transfer behaviour and did not take into account the disturbance situation. Therefore no global stability results appear to be available so far for the case where unknown disturbances act upon the plant. Nevertheless simulations have indicated that the model reference adaptive controller may well be capable of covering the regulator situation.

From the application point of view making the plant behave as the same model for all operating conditions has certain limitations. For example, an aircraft in the landing approach cannot physically respond as fast as it does in a high speed approach. Therefore the pilot expects a slower response in the first and a faster response in the second situation, which is also part of the handling quality requirements. Application of model reference adaptive control to a simple stability augmentation system thus would require some adaptation of the reference model, too. However, adapting the reference model from measurements of u, y only, would introduce additional nonlinear feedback loops, and the associated stability problem appears not to be covered by the available theory. Adapting the reference model from additional air data measurements and interpreting it as a quasi stationary (i.e. slowly time varying) reference model may be a practically feasible solution to the problem. In conclusion, although model reference adaptive control is applicable to almost arbitrarily unknown plants theoretically, the physics of the plant generally does not permit the use of the same reference model for all (large) parameter uncertainties of the plant in practice.

From practical experience it is also known that exact model matching is more a mathematically tractable concept rather than an actual need. In particular, approximate model matching may be completely satisfactory if it allows the reduction of the necessary control amplitudes and rates by typically fifty per cent.

This is why in the non-adaptive case the model following problem often is solved by means of the linear regulator theory, where a performance index of the form

$$\int_0^\infty [(y-y_m)^2 + \rho\, u^2]dt$$

is minimized. This results in feedforward of the model states together with plant state feedback. By choice of $\rho > 0$ a compromise between model matching accuracy and control effort is determined iteratively, where for $\rho \to 0$ exact model matching would be obtained in the limit. The more detailed design considerations in the non-adaptive linear control case indicate that the adaptive model reference control will be a rough solution to the design problem, generally far from optimum and not an ideal substitute for a detailed design using complete a priori knowledge of the plant.

B. Adaptive Control via Identification and Control Law Synthesis (Indirect Approach)

While model reference adaptive control is based on one specific design goal, namely model following, the adaptive control concept to be discussed now, basically allows arbitrary design goals to be pursued.

The concept is based on an (asymptotic) explicit identification of the unknown system parameters. Using the current parameter estimates as if they were the true values allows the control law to be synthesized on line following any design goal. For example, model following would be one possible application of this concept, where now (in contrast to the model reference adaptive control approach) the reference model can be chosen to depend on the estimates of the unknown system parameters. Also, the design goal can be approximate rather than exact model following. Other applications of the indirect adaptive control concept are (adaptive) pole placement, where the desired pole locations may depend on the unknown system parameter estimates, or the design of state feedback tolerating input saturation, etc.

It is obvious that the main potential of the adaptive control concept using identification plus control law synthesis, is its enormous flexibility in pursuing almost arbitrary design goals in a way which makes it realistic even for large parameter variations. This concept has been applied in many practical applications. However, because it is not based on the Lyapunov synthesis technique, general proofs of global stability do not appear to have been obtained so far.

Recent new results [12],[13] show that a globally stable behaviour can in fact be obtained for a class of adaptive state feedback control laws in connection with an adaptive observer subject to a sufficiently exciting plant input. In this result a fairly general, asymptotic synthesis of an (adaptive) state feedback matrix is based on the current parameter estimates, which are generated by the adaptive observer. An algebraic separation of adaptive state feedback control and adaptive state observation then allows the feedback loop to be closed while assuring global stability. The basic control structure of this approach is depicted in Fig. 2.

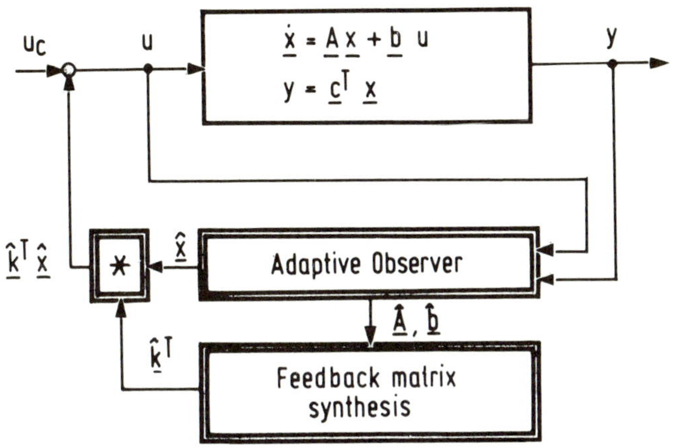

FIGURE 2. Adaptive control via adaptive observation and adaptive control law synthesis.

In contrast to the adaptive model reference approach, no a priori knowledge of the unknown plant other than the dimension of the plant state vector is required, i.e. non-minimum phase plant are also covered. However, proving stability requires that the parameter estimates converge to their true values, which in turn requires that the unknown plant is in some sense "sufficiently excited" by its input. This excitation requirement, which is not present in the model reference adaptive control approach, appears to be inherent to all adaptive schemes, which are based on explicit parameter identification.

As in the model reference adaptive control approach, the global stability result has been established only for the undisturbed plant. Even though the feedback matrix may be designed so as to obtain good disturbance regulation, the convergence of the parameter estimates (in a deterministic sense) will not be obtained for the disturbed plant. But since the convergence of the identification is a basis for the adaptive control concept, it will in general also be a basis for its proof of stability. This is why no global stability results appear to be available so far for the case where unknown disturbances act upon the plant. Nevertheless simulations have indicated that this indirect type of an adaptive controller may also well be capable of covering the regulator situation.

The fact that almost any control law synthesis can be applied to generate the adaptive feedback law offers a potential advantage over the model reference approach, since it allows the control to be a compromise between the control amplitudes and regulation accuracy. Nevertheless it would be rather optimistic to expect that such an adaptive control on the basis of an automatic on-line control law synthesis could be an ideal substitute for a sophisticated off-line control design, which is based on complete a priori knowledge of the plant and a variety of detailed design considerations.

III. SPECIFIC ADAPTIVITY PROBLEMS

A. Excitation

An adaptive control scheme extracts information out of the measured plant input and output signals in order to perform its adaptation with respect to the unknown plant. Of course, inadequate signals can give rise to incomplete adaptation only. Therefore all kinds of adaptive schemes require the system input to sufficiently excite the plant in some sense, in order that convergence of the adaptive process can be obtained. The amount of excitation which will be necessary depends on the adaptive scheme, i.e. depends on how the information is extracted out of the signals.

From the excitation point of view, the model reference adaptive control approach appears to take an outstanding position. This is because the model following error will approach zero whatever the signals are, i.e. independent of any external excitation. Obviously, the feedback signals are generated in such a way that the adaptive scheme, although it may not converge, works so as to satisfy the design goal.

However, thinking of a plant such as an aircraft, which continuously changes its operating conditions, it will be required that the adaptation keeps track of all parameter changes of the plant. This is because undetected adaptation errors can probably not be tolerated, to avoid unpredictable responses in case of a command input. Therefore some sufficient excitation will have to be provided in all types of adaptive control schemes. In a disturbance free situation external test signals may be used, whereas in case of disturbances such as gusts the feedback signal alone provides sufficient excitation.

B. Adaptation Speed

A problem closely related to the excitation problem and of equal importance is the speed of adaptation. It is highly desirable that the controller gains are adapted in a monotonically, non-oscillatory way within a sufficiently short time. Because the adaptation is the result of the available information, it heavily depends on the excitation and, in particular, on the magnitude of the signals. This means, if a step command input is fed into the control system with a non-adapted plant, then the adaptive process is reinitialized, but the way it evolves will fundamentally differ according to the step height. This reveals the high nonlinearity of the adaptive system.

In order that the adaptive speed can be controlled, most of the adaptive schemes contain certain adaptive gains to be chosen by the designer. But to make a systematic choice some principal relationships between the adaptive behaviour and the gain magnitude would be necessary, such as the potential of arbitrarily high speed of adaptation by use of sufficiently high gains.

In the model reference adaptive control approach no such conceptual means appear to have been provided so far for arbitrarily controlling the speed of the adaptive process (nevertheless a suitable choice of the adaptive gains may result in acceptable convergence speeds, see e.g. the paper by Gutmann, Monopoli, Van Allen). In contrast to this, as is well-known, adaptive identification of the unknown plant parameters can be made arbitrarily fast simply by increasing some adaptive gains. Therefore the adaptive control concept using identification plus control law synthesis, allows the adaptation speed to be controlled. It should be noted however, that the adaptation speed still depends nonlinearly on the level and the type of the signals involved, and the desired monotonical behaviour of the adaptive controller gains is not implied.

C. Disturbances

It was mentioned in the previous section that global stability has not been proven for any of the general adaptive control schemes in situations where unknown disturbances act upon the plant. But this is not the only reason why the disturbance regulation may be the most critical mode of operation of the adaptive system. In the absence of disturbances the behaviour of the overall system will typically converge to that of a linear system, which is easy to analyze. However, with disturbances present, convergence does not occur and the adaptive scheme permanently changes the adaptive controller gains, i.e. the whole control system behaves in a completely nonlinear fashion, which is extremely difficult to analyze. Of course, this behaviour depends on the disturbance signal level as well as on the command input signals.

Although disturbances appear to cause a great number of problems theoretically, they have secondary effect, which is very useful. The disturbance measured at the plant output, is fed back through the controller to the plant input, in order to compensate for the disturbance. Thereby the plant input is sufficiently excited in a very natural way. Flight experiments with a simple stability augmentation system showed even better adaptation capabilities in a gusty atmosphere without using external test signals, than it did in a clear air atmosphere, where external test signals were used. Such an effect would be plausible in a case where the feedback increases rather than reduces the disturbance effect. However, this is obviously not the goal of the regulation. Much theoretical work will have to be done to better understand the disturbance case. Some identifiability problems are also involved due to the correlation of the plant input and the disturbance signal.

In case of disturbances a high speed of adaptation is no longer realistic, because the disturbed plant input and output contain misleading information, which might result in large excursions of the adaptive controller gains. Therefore a lower speed of the adaptation appears to be more advisable, which has some low pass filtering effect. Note also, that the disturbance characteristics as well as the disturbance level are unknown and may change with the operating conditions of the plant or even with time, which complicates the situation.

IV. ON ADAPTIVE FLIGHT CONTROL APPLICATIONS

A. Design and Realizability Aspects

Within the decentralized flight control system structure, organized in a hierarchical fashion, certain inner control loops (called stability augmentation systems) are included in addition to the so called autopilot or outer control loops. The inner loops perform high bandwidth control and cope with the parameter variation problem while the outer loops perform various flight path control functions with relatively low bandwidth. So adaptive control applications primarily would have to deal with stability augmentation systems and how they work within the overall control system.

Stability augmentation of an aircraft is more than just control of its basic angles or angular velocities. It involves improvement of the handling qualities of the aircraft, i.e. it is to change the natural modes or the natural transfer function of the aircraft in such a way, that the pilot can more easily handle it. Firstly, this requires detailed knowledge of what the aircraft behaviour should be as a function of the flight condition, and secondly, it requires knowledge of what the aircraft behaviour actually is for all flight conditions. This knowledge is necessary to decide upon what the aircraft behaviour can be made without violating physical constraints. The main design problem thus is to suitably prepare the a priori knowledge and to incorporate it into the adaptive control design.

Using adaptive control with a reference model, the necessity arises to adjust the reference model according to the flight condition. This means that instead of gain scheduling we now have a preprogrammed scheduling of the reference model. The principal difference between the two is that gain scheduling requires exact plant knowledge as well as a complete detailed control design for all flight conditions, whereas scheduling of the reference model only requires knowledge about the dynamic capabilities of the plant.

For the realization of such a control concept, which appears to be suitable for satisfying the handling quality requirements, one could use the model reference adaptive control concept in combination with model scheduling from measured air data, such as altitude and dynamic pressure. Alternatively, identification of the aircraft parameters and synthesis of model following control could be used, which allows the model to be scheduled directly from the estimated aircraft parameters. Either way, it appears that such a concept is realizable on a flight computer, though the procedure is somewhat more complex than traditional gain scheduling. This is particularly true for the lateral motion multivariable case.

Since any realization of adaptive flight control will be on a digital flight computer, discrete versions of the adaptive controllers have to be realized. To be sure that a discrete time control algorithm does not contain (discrete) differentiation, which is not tolerable due to reasons discussed above, it should have a differentiator free continuous time analog. Typically the adaptive algorithms involve vector-vector or matrix-vector operations, which are time consuming on a general purpose computer, due to series data processing. Although it appears that the necessary computations can be done within a reasonable sampling interval, the associated time delay is highly undesirable, since it results in a performance degradation. However this can be overcome by use of special purpose computers, which do this type of operation by parallel processing. The limited accuracy of the computations have also to be paid special attention, and structurally simple algorithms, which are insensitive in this respect, should be preferred. Finally the control computer has its own reliability problems and must be imbedded in a suitable redundancy concept.

The realizability of the adaptive controller also depends on some modeling inaccuracies not discussed so far. Among them are neglected time constants, delays and nonlinearities of the actuator and the structural modes of the aircraft. Also, different from the theoretical assumptions, the aircraft is not linear, especially in high angles of attack. In a classical non-adaptive control design such modeling problems are usually taken care of indirectly, by imposing bounds on the allowable gain and bandwidth in the feedback loop [4]. In an adaptive design it can therefore be expected, that similar considerations might help to improve the realizability. Here not only the direct feedback loop is involved but also the adaptive loop, which also (nonlinearly) contributes to the feedback signal. Imposing bounds on the "gain" and the "bandwidth" in the adaptive loop is particularly difficult because of nonlinearities and the influence of the signal level.

The presence of nonlinearities in the real world may result in instability of the closed loop system, which cannot be excluded theoretically. Therefore the realizability of and the confidence in adaptive control could be increased significantly, if the feedback gains of the controller could a priori be bounded in such a way, that the feedback loop cannot become unstable. Even more desirable would be bounds which guarantee certain minimum (emergency) performance requirements to be satisfied for the whole flight control system. Such bounds can be obtained by using the available knowledge about the aircraft. It remains to be shown how they can be included in theoretical analysis.

Applying adaptive control to aircraft systems has the important advantage that the plant is fairly well-known. This knowledge can be used to make detailed simulations and even optimize certain parameters of the adaptive scheme. Thereby one can make sure that under all realistic circumstances the adaptive system will perform in a satisfactory manner. Note, that for a completely unknown plant there would be much uncertainty about design details as well as the overall performance.

B. Performance Evaluation

The performance of control includes the handling quality improvement and the disturbance rejection capability for all operating conditions. It is highly unlikely that an adaptive control system can perform much better in this respect than traditional highly sophisticated control systems do. (The complete flight envelope can even be covered satisfactorily with a simple constant gain controller, i.e. without gain scheduling [3]). For adaptive control systems the performance of control, in addition, includes the adaptation capabilities and requirements in undisturbed as well as disturbed situations, i.e. the evaluation of the nonlinearity effects. This makes the performance evaluation much more complex.

Although adaptive control is perhaps not likely to improve the performance of control in aircraft systems, it may result in improvements for other flight systems such as guided missiles, where less sophisticated control systems have been used so far.

Efficient design and decreased costs are what make adaptive control particularly attractive. Once the increased complexity, due to the nonlinearity of the adaptive controller, is accepted, it offers a reduction in design costs and considerable flexibility. Firstly, less accurate knowledge about the aircraft is necessary, secondly the adaptive controller saves designing the feedback gains for all flight conditions, it can thirdly tolerate changes of the airframe without redesign, and finally it adapts with respect to all influences, where gain scheduling can take into account only some of them. Furthermore, savings of sensor hardware appear to be possible, since the recent adaptive controllers are typical output feedback controllers and do not require measurements of many state variables.

The overall performance of adaptive control involves performance of both control and its realization costs. Performance improvement over non-adaptive control will have to be shown in practical applications. It cannot be shown solely by arguments.

V. CONCLUSION

This paper tries to outline some of the possible problems which might be relevant for applications of adaptive control theory to aircraft systems. The main objective is that the adaptation concept does not simply solve the parameter variation problem in aircraft control systems, but also introduces new problems. The most important problems appear to be due to the high non-linearity of the control scheme, which have not yet been explored in sufficient detail. For example, the basic stability problem in the presence of disturbance is still an open question, which indicates the theoretical difficulties encountered.

To date aircraft control systems apply simple gain scheduling techniques and thereby effectively solve the parameter variation problem. In view of these sophisticated techniques it is highly unlikely that an adaptive controller can result in much better performance of control. The potential improvements of adaptive control, if any, are therefore visualized more in the direction of saving sensor hardware, requiring less accurate aircraft knowledge (which is of interest when extending the flight range to high angles of attack and post stall) and obtaining more flexibility.

Anyway, much work has to be done to better understand adaptive controllers and simplify their structure for practical applications. To this end, flight experiments might give an important impetus.

REFERENCES

1. Staff of Flight Research Center, "Experience with the X-15 Adaptive Flight Control System," NASA Tech. Note D-6208, 1971.
2. McRuer, D., Askenas, I., and Graham, D., "Aircraft Dynamics and Automatic Control," Princeton University Press, Princeton, New Jersey, 1973.
3. Ball, J. N., and Rynaski, E. G., "Longitudinal Flight Control for Military Aircraft - A Study of Requirements and Design Concepts," Cornell Aeronautical Lab., Final Report No. ID-1757-F-1, P. O. No. L99564, October 1963.
4. Horowitz, I., "Synthesis of Feedback Systems," Academic Press, New York, 1963.
5. Parks, P. C., "Lyapunov Redesign of Model Reference Adaptive Control Systems," IEEE Trans. Auto. Control, Vol. AC-11, pp. 362-367, July 1966.
6. Landau, I. D., "A Survey of Model Reference Adaptive Techniques - Theory and Applications," Automatica, Vol. 10, pp. 353-379, 1974.
7. Monopoli, R. V., "Model Reference Adaptive Control with an Augmented Error Signal," IEEE Trans. Auto. Control, Vol. AC-19, pp. 474-484, Oct. 1974.
8. Narendra, K. S., and Valavani, L., "Stable Adaptive Controller Design - Direct Control," IEEE Trans. Auto. Control, Vol. AC-23, pp. 570-583, Aug. 1978.
9. Feuer, A., and Morse, A. S., "Adaptive Control of Single-Input, Single-Output Linear Systems," IEEE Trans. Auto. Control, Vol. AC-23, pp. 557-570, Aug. 1978.

10. Narendra, K. S., Lin, Y. H., and Valavani, L., "Stable Adaptive Controller Design - Part II: Proof of Stability," Yale University, S & IS Report, No. 7904, April 1979. (To appear in IEEE Trans. Auto. Control, June 1980.)

11. Morse, A. S., "Global Stability of Parameter Adaptive Control Systems," Yale University, S & IS Report, No. 7902R, March 1979. (To appear in IEEE Trans. Auto. Control, June 1980.)

12. Kreisselmeier, G., "Algebraic Separation in Realizing a Linear State Feedback Control Law by Means of an Adaptive Observer," IEEE Trans. Auto. Control, Vol. AC-25, April 1980, (to appear).

13. Kreisselmeier, G., "Adaptive Control via Adaptive Observation and Asymptotic Feedback Matrix Synthesis," DFVLR-Institut für Dynamik der Flugsysteme, Technical Report, June 1979. To appear in IEEE Trans. Auto. Control.

14. Elliott, H., and Wolovich, W. A., "Parameter Adaptive Identification and Control," IEEE Trans. Auto. Control, Vol. AC-24, pp. 592-599, Aug. 1979.

ADAPTIVE FLIGHT CONTROL: A PRAGMATIC VIEW

Gunter Stein

Honeywell Systems and Research Center and
Massachusetts Institute of Technology
Cambridge, Massachusetts

1.0 INTRODUCTION

Adaptive flight control has inspired the imaginations of control researchers and practitioners for over two decades. Yet, it has achieved little success. Few current aircraft use it today and few flight control problems call for it in the future. This paper provides a brief examination of the reasons for this state of affairs. It describes the basic problems which motivate adaptive flight controls, the competing technological solutions, and the current state of the competition. Some upcoming problems with adaptive potential, and the relevance of recent globally stable adaptive concepts are also discussed.

ISBN 0-12-514060-6

2.0 MOTIVATION FOR ADAPTIVE FLIGHT CONTROL

It is popularly argued that adaptive techniques are appropriate whenever uncertainties of a plant model exceed the tolerances offered by ordinary feedback controls. Since aircraft dynamics vary dramatically, the argument goes, flight control is a natural applications area for these techniques. This is, of course, a very simplistic view of the true situation. More realistically, adaptive techniques must be viewed as one of several competing ways to achieve successful control actions in the face of aircraft uncertainties. Whether it offers benefits for a particular application then depends on two primary factors: 1) The nature and magnitude of uncertainties, and 2) alternate methods available to achieve successful control. The first factor determines feasibility. Most adaptive techniques work well for some types of uncertainties but not for others. The second determines viability. Adaptive controls may or may not be competitive with alternate approaches on economic or operational grounds. We consider these factors in a little more detail below.

The Nature of Aircraft Uncertainties

For the aircraft models used to design inner-loop stability and command augmentation systems, uncertainties take two dominant forms. The first consists of unknown parameters in an otherwise known model structure (i.e., coefficients of linearized equations of motion). Their range of uncertainty is due to widely varying flight environments (dynamic pressure, Mach, angle-of-attack) and large configuration variations (c.g. locations, fuel and payload, geometry). However, individual coefficients are strongly interrelated, and only a few must be known accurately in order to design good control laws. As a net result, the important uncertainties reduce to a

State Dynamics:

$$\frac{d}{dt}\begin{bmatrix} q \\ \alpha \\ \delta_e \end{bmatrix} = \begin{bmatrix} M_q & M_\alpha & M_\delta \\ 1 & Z_\alpha & Z_\delta \\ 0 & 0 & -a \end{bmatrix}\begin{bmatrix} q \\ \alpha \\ \delta_e \end{bmatrix} + \begin{bmatrix} 0 \\ 0 \\ a \end{bmatrix}\delta_c + \begin{bmatrix} M_0 \\ g/V \\ 0 \end{bmatrix} + \begin{bmatrix} 0 \\ 1/\sqrt{875V} \\ 0 \end{bmatrix}\sigma_{\omega g}\ \eta_{\omega g}$$

Measurements (discrete):

$$y_k = \begin{bmatrix} q \\ n_z \end{bmatrix}_k = \begin{bmatrix} 1 & 0 & 0 \\ dM_q & dM_\alpha - Z_\alpha V & dM_\delta - Z_\delta V \end{bmatrix}\begin{bmatrix} q \\ \alpha \\ \delta_e \end{bmatrix}_k + \begin{bmatrix} \sigma_g & \eta_{g_k} \\ \sigma_{nz} & \eta_{nz_k} \end{bmatrix} \qquad k = 0,1,2,\ldots$$

Definitions:

q	pitch rate
α	total angle of attack
δ_e	elevator position
δ_c	elevator command
n_z	normal acceleration
$\eta_{\omega g}$	unit intensity white noise
η_{gk}, η_{nz_k}	unit variance white sequences
a	actuator bandwidth (12 r/sec)
d	accelerometer displacement from C.G. (15 ft)
$\sigma_{\omega g}$	vertical gust level (ft/sec rms)
σ_g, σ_{nz}	sensor noise levels
$M_q, M_\alpha, M_\delta, Z_\alpha V, Z_\delta V, V$	dynamic coefficients (functions of c)
$M_0, g/V$	constant trim disturbances (functions of c)

FIGURE 1. Short period aircraft model with systematic uncertainties.

TABLE 1. Model Parameterization

Equation coefficient			Small perturbation parameter	Perturbation parameter uncertainty
M_q	$= -0.23 + 0.028\ M_{\delta o} + c_1$	Mach < 1	c_1	0.065
	$= -0.23 + 0.010\ M_{\delta o} + c_1$	Mach > 1		
M_α	$= (0.61 + c_2)\ M_{\delta o}$	Mach < 1	c_2	0.135
	$= (1.53 + c_2)\ M_{\delta o}$	Mach > 1		
V	$= (200 + c_3)\ \ -M_{\delta o}$	Mach < 1	c_3	31.5
	$= (260 + c_3)\ \ -M_{\delta o}$	Mach > 1		
$(Z_\alpha V)$	$= (53 + c_4)\ M_{\delta o}$	All Mach	c_4	5.0
$M_{\delta o}$	$= c_5$	All Mach	c_5	15.0
$(Z_\delta V)$	$= (7.7 + c_6)\ M_\delta$	All Mach	c_6	0.75
$\bar{q}$	$= (-23 + c_7)\ M_{\delta o}$	All Mach	c_7	2.0

small number of slowly varying parameters. They usually include surface-effectiveness-due-to-elevator, M_δ, sometimes also pitching-moment-due-to-attack-angle, M_α, and occasionally other surface effectiveness coefficients. An example of this form of uncertainty is shown in Figure 1 and Table 1, where the short period dynamics of a fighter aircraft are displayed as third order differential equations parameterized in terms of key aircraft coefficients [1].

The second form of uncertainty consists of unknown and unmodelled dynamics at higher frequencies. These include such effects as structural resonances, unsteady distributed aerodynamics, transport delays in hydraulics and aerodynamics, etc. In principle, the aircraft is an infinite-dimensional dynamic system whose slow-time-scale input/output behavior (say 0.2-10.0 sec) may be approximated by the third order model of Figure 1, but whose fast-time behavior (<0.2 sec) will be quite different. Qualitatively, this type of uncertainty is most easily visualized in the frequency domain. An example is given in Figure 2 which shows a typical transfer function Bode plot for the nominal model in Figure 1 together with a superimposed envelope of possible transfer functions for the true aircraft. The true aircraft could correspond to any reasonably smooth function within the shaded envelope.

The key fact to recognize is that any viable flight control law, whether adaptive or not, must tolerate these relatively unstructured high frequency uncertainties as well as the structured (parameterized) uncertainties in Figure 1. Most successful control systems achieve this required tolerance by limiting their bandwidths to avoid unstructured uncertainties and by adjusting their control parameters within this bandwidth to compensate for the structured variations. This is the primary reason why explicit identification techniques have been so common in flight-proven adaptive systems. One simply identifies certain key parameters and adjusts the control law according to a "schedule" which improves performance subject to explicit bandwidth limits. Such limits are less easily imposed with

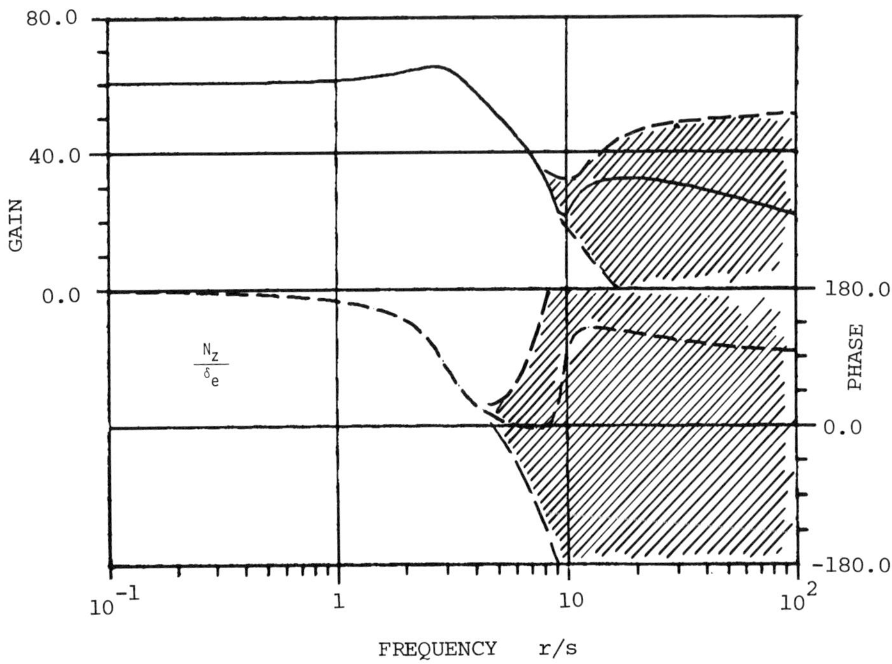

FIGURE 2. Model frequency response with high frequency uncertainties.

implicit adaptive schemes, although careful choice of the adjustable parameters and their permissible range of variation can achieve the desired effect. Certain other proposed adaptive concepts, on the other hand, appear to rely on very special high frequency characteristics to achieve their adaptive properties and would thus prove unacceptable for flight control. More will be said about these concepts later.

Competitors

While some adaptive techniques are thus feasible from the point of view of handling the types of uncertainties prevalent in flight control, they must compete with other ways of doing the same thing. Two principal competing approaches are 1) to use low-gain flight control laws which can tolerate the full range of uncertainties directly, or 2) to provide external airdata for control law adjustment. The first approach is generally ruled out on the basis of performance. Low-gain systems capable of flying the entire flight envelope are unable to satisfy performance requirements for most of today's high performance aircraft and will be even less acceptable for future aircraft incorporating relaxed-static-stability and other control-configured-vehicle (CCV) concepts. The second approach is more successful. It exploits the strong correlations which exist between externally measurable airdata variables (dynamic pressure, $\bar{q}$, Mach number, angle-of-attack, α) and unknown aircraft parameters. On the fighter in Figure 1 for example, M_δ is given almost exactly by a simple linear function of $\bar{q}$ [1]. Thanks to these correlations, control law gains which are originally computed as functions of M_δ, M_α, etc., can be approximated in terms of $\bar{q}$, Mach, and/or α. Although not as good as the original functions, such approximated "airdata schedules" are adequate to meet performance requirements in most current flight control problems.

3.0 THE ADAPTIVE/AIRDATA TRADEOFF

Given these alternatives, the competition among flight control design techniques reduces to a tradeoff between the economics, reliability, and operational features of airdata-scheduled systems versus adaptive systems. This tradeoff has passed through an interesting historical transition. It dates back to the early

1960's when flight technology pushed vehicles such as NASA's X-15 to the edge of space and high performance fighters with large flight envelopes entered military arsenals. Sufficiently accurate and reliable airdata sensors were difficult to build and maintain then, and adaptive techniques appeared to have a strong upper hand. The X-15 research aircraft flew with an operational adaptive system [2], the production F-111 has one [3], and extensive adaptive control tests were conducted with an F-101 flight program [4], with X-20 hardware simulator evaluations [5], and with several F-4 flight programs [6-8]. Despite these various applications, however, the technology did not gain full acceptance. It suffered a particularly serious psychological setback with the unfortunate loss of an X-15 for which the control system, at least by implication, has inherited some blame.

While these developments took place, of course, airdata sensing technology did not stand still. Driven by the same basic need to provide tolerance for uncertainties in flight control, manufacturers of airdata sensors improved their products in accuracy, maintainability, and cost. These improvements were quite successful and have completely reversed the tradeoff. Airdata-scheduling became the preferred approach in most flight control systems designed in recent years. Simply stated, our control engineering predecessors lost the competitive battle to the sensor builders.

As often happens, however, this battle refused to stay settled for long. Another generation of adaptive control advocates reopened the competition in the mid-70's with a collection of new technological arguments. The main arguments are these:

(1) The evolution of improved adaptive control design theories: New developments have appeared in identification theory (e.g. maximum likelihood methods for linear systems [9],[10], in stability analysis and verification of adaptive algorithms (e.g. Lyapunov re-design [11], hyperstability [12], self-tuning [13]) and in approximation of dual control solutions

(e.g. various parallel structures [14-16]). Ostensibly, these improvements will alleviate the marginal performance/reliability problems of earlier adaptive designs.

(2) The trend toward digital control: Flight controllers are now being mechanized in on-board digital computers. Ostensibly, this makes more sophisticated adaptive control concepts practical.

(3) The trend toward redundancy: Future flight-critical control systems will need dual, triple, or even quad redundant mechanizations to insure safety of flight. Airdata-scheduled approaches will then ostensibly suffer the added cost and complexity of multiple airdata sensing systems, and they will also face the difficult problem of finding suitable probe locations for adequate data quality, channel tracking (for effective redundancy management) and dispersion (to avoid common hazards).

(4) The trend toward increased performance: Several anticipated flight control applications appear to require control laws which cannot be adequately approximated by airdata schedules, i.e. they depend on systematic aircraft uncertainties which are not well correlated with $\bar{q}$, Mach or α. Prominent examples include active control of flexure/flutter and coupled control modes. More will be said about these applications later.

These arguments received enough attention in the flight control community to motivate some new analytical design studies and flight experimentation. In the U. S., for example, modest adaptive control efforts have been carried out by NASA as part of the F-8C DFBW (Digital Fly-by-Wire) program [17] and the VALT (Vertical Approach and Landing Technology) program [18]. The F-8C investigation is currently in the post-flight analysis stage, and a brief summary is provided below.

4.0 ADAPTIVE CONTROL RESEARCH WITH THE F-8C

Over the past several years, NASA has been conducting a flight research program to demonstrate digital fly-by-wire flight control technology [17]. The test aircraft is an early 1960's vintage F-8C fighter whose mechanical control linkages and analog augmentation systems were removed and replaced with triply redundant AP-101 digital computers connected directly to the surface actuators. Pilots command the aircraft only through these computers. No direct mechanical connections from stick or pedals to the surfaces remain. By flight control standards, the computers themselves are powerful machines with large memory, high speed, and floating point capability. This offers a unique environment to evaluate some of the new pro-adaptive arguments cited above. Hence, adaptive control was included as a part of the F-8C's flight test agenda.

Candidate "modern" adaptive control concepts for the F-8C were developed by several academic and industrial organizations. These included several explicit identification schemes as well as implicit schemes using model reference, self-oscillation, and multiple-model principles. Detailed descriptions, analyses, and simulator evaluations of the chief candidates can be found in Reference [19]. Based on analytical and simulation evaluations, one candidate was selected for flight evaluation. This was an explicit scheme using a maximum-likelihood-derived algorithm to identify M_δ, M_α and airspeed. The M_δ-estimate was then used to schedule a single gain in each axis of a simplified, bandwidth-limited, linear-quadratic-optimal control law [1]. This system completed flight evaluation in the latter half of 1978, and its performance is documented in Reference [20].

Without going into the details of these various references, the F-8C experience can be summarized by the following major observations. First, the claimed improvements in theoretical adaptive control design techniques did not emerge strongly enough to alter the adaptive/airdata competition. In fact, most sophisticated "modern" concepts failed to survive simulator evaluation with performance ratings high enough to justify competition in flight. The one concept which did reach flight performed as well as airdata-scheduled controls only in the presence of sufficiently large test inputs.

This leads to the second major observation. All flight-proven adaptive systems including this latest one on the F-8C appear to require some test input activity to achieve adequate performance. The required levels can be exceedingly low, yet still be discernable and objectionable to pilots. This is highlighted in Table 2, which shows test signal levels and associated pilot comments from a repre-

TABLE 2. Test Signal Levels in F-8C Flights

Test signal level	Pitch rate deg rms	Normal Acceleration g rms	Elevator deg rms	Pilot comments
0.3	0.056	0.01	0.013	•Test signal undetectable
1.0	0.075	0.01	0.047	•Occasionally noticeable. Can feel light accelerations, less than light turbulence.
1.5	0.125	0.013	0.072	•Noticeable, objectionable if present for a long period of time. •Test signal is repetitious, not random like turbulence. •Nose boom motion detectable.

sentative F-8C adaptive flight. The significance of the test signal requirement is that it threatens to offset any potential redundancy management advantages which adaptive controls might offer in multi-channel flight control implementations.

The third observation is that the F-8C program's results to date say little about the potential of adaptive controls in situations where airdata-scheduled controls do not work. The test aircraft can be controlled very, very well with airdata-scheduling [21], and adaptive approaches have to struggle to do as well. It is perhaps no accident that the best accuracy to which the F-8C's M_δ parameter could be identified from input/output records with reasonable disturbances and test input is approximately equal to the accuracy needed to design good control laws--namely 10-20% (established theoretically and demonstrated in flight [20]). One may conjecture that if greater accuracy were, in fact, required for good control, then it would also be achievable from input/output data. This remains, of course, to be investigated.

In light of the above observations, it appears that even in the post-F-8C era, the competitive battle with airdata is still lost.[1] The implication for adaptive flight control designers is that we must look toward a different set of problems for potential applications of our technology--problems which go beyond the capabilities of airdata. Two major candidates are briefly described below.

5.0 POTENTIAL ADAPTIVE PROBLEMS

There are several problem areas within flight control for which airdata does not provide acceptable solutions. In very high speed flight such as re-entry, for example, it is not possible to deploy

[1] It is interesting to note that the VALT program, using a CH-47 helicopter as its testbed, is coming to similar conclusions [22].

the probes needed for airdata sensing. In small missile applications, it is uneconomical to pay the cost of these sensors. In still other applications, the airdata alternative simply does not work because airdata variables are not well correlated with the key uncertain parameters of the design model. Two examples in the latter category are coupled modes and flexure/flutter control. Both occur in future manned aircraft applications and offer potential for adaptive solutions. Brief descriptions are given below.

Coupled Modes

A major performance trend in modern fighters is the so-called "direct force control" feature in which several surfaces are deflected in coupled fashion to produce pure lift or side force without the usual pitching or yawing moments. These features enhance maneuverability of the aircraft. In order to implement them successfully, it is necessary to know several surface effectiveness parameters quite well. An illustration of this requirement is given by the F-8C's experimental "direct lift" mode, designed for another phase of the overall F-8C DFBW program. This mode uses elevator, δ_e, and symmetric ailerons, δ_a, to achieve pure normal forces. The ideal control law is the following

$$\begin{bmatrix} \delta_e \\ \delta_a \end{bmatrix}_{cmd} = \begin{bmatrix} M_{\delta e} & M_{\delta a} \\ Z_{\delta e} & Z_{\delta a} \end{bmatrix}^{-1} \begin{bmatrix} M \\ F \end{bmatrix}_{cmd},$$

where M_{δ_i}, Z_{δ_i} denote moment and force coefficients of the surfaces and M and F are moment and force commands. The problem with this control law is that it cannot be accurately scheduled with dynamic pressure measurements because these define only a common scale factor of all four elements in the surface coefficient matrix, i.e.

$$\begin{bmatrix} M_{\delta e} & M_{\delta a} \\ Z_{\delta e} & Z_{\delta a} \end{bmatrix} = \bar{q} \begin{bmatrix} c_{11} & c_{12} \\ c_{21} & c_{22} \end{bmatrix}$$

The relative coefficients, c_{ij}, are not well correlated with airdata and must be obtained from other sources. Depending upon the amounts of uncertainty in the c_{ij}'s and the quality of decoupling desired, adaptive techniques may provide viable solutions to this scheduling problem.

Flexure/Flutter Control

A second major trend in modern aircraft is toward lighter structures and/or higher operating speeds. Both tend to aggravate the problem of structural dynamic interactions with the airstream. When these interactions produce lightly damped vibration modes in the airframe, they are usually called "flexure". When the vibrations are negatively damped (unstable), they are called "flutter". All aircraft exhibit flexure and flutter phenomena at sufficiently high airspeeds. In effect, they must obey speed limits in order to avoid excessive structural fatigue and catastrophic failure. These speed limits impose fundamental constraints on aircraft performance.

In recent years, the possibility of extending aircraft speed limits with active control of structural vibrations has received serious consideration. Demonstration flights have been made with active flexure and flutter controls on a B-52 [23], [24], and various wind tunnel tests have been performed with experimental wings [25]. To date, these demonstrations have been carefully selected "point designs", intended to operate at a few fixed flight conditions. No full flight envelope system has yet been developed. The reason for these restrictions is that reliable design models for flexure/flutter are very complex and unavailable in all but a few cases.

Flexure/flutter models are complex because they involve many modes of structural vibration interacting with the airstream on a time scale where steady aerodynamics no longer apply. Typically, the time scale ranges from 0.005 to 0.10 seconds. Hence, referring back to Figure 2, flexure/flutter models must attempt to describe the aircraft transfer functions accurately in the frequency range 10-200 Hz. The functions have a high level of detail in this range with several resonance peaks, different damping characteristics (some may be neutrally stable or even unstable), and different mode sensitivities. All of these characteristics change with flight condition, aircraft geometry, and payload or store configuration.

A qualitative picture of a flexure/flutter transfer function is given in Figure 3. Note from this figure that the flexure/flutter model must again be viewed as having two types of uncertainties:

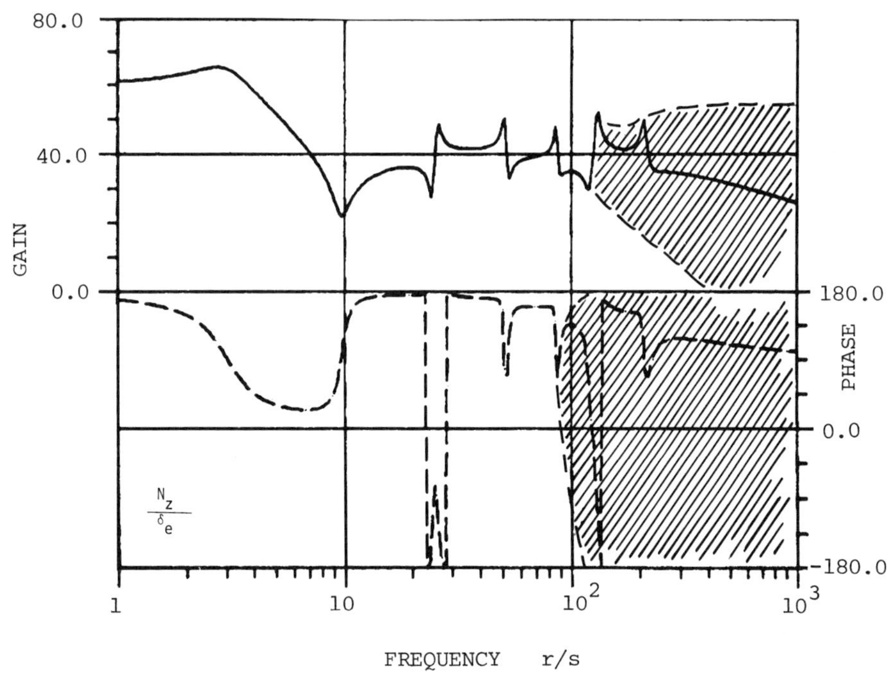

FIGURE 3. Model frequency response with flexure.

(1) Structured (parameterized) uncertainties, now consisting of the "rigid body" parameterization in Table 1 plus mode frequency, damping, and shape parameter for several aeroelastic vibration modes.

(2) Infinite-dimensional unstructured uncertainties at frequencies beyond the model's range of validity.

Viable control system designs must again tolerate both.

Active control systems for flexure/flutter typically attempt to augment the damping of one or more of the critical vibration modes described above. Their bandwidth must therefore exceed these mode frequencies yet fall below the remaining unstructured uncertainties. Control laws for this purpose are generally constructed in two stages [26]:

(1) Sensor choice and placement, in which outputs from several sensors along the wing and fuselage are combined to provide effective "mode sensors" for the modes of interest with little interference from neighboring modes, and,

(2) control synthesis, in which simple compensators are designed to damp vibrations using the effective mode sensor signals.

For such control laws to operate properly over the flight envelope, the sensor combination and compensators must be readjusted as model parameters change. No reliable means have been found to date to achieve this readjustment. Airdata schedules fail because airdata variables are only partially correlated with flexure/flutter parameters. Adaptive techniques may provide the sole alternative.

While adaptive techniques appear to be applicable, however, they are a long way away from being accepted as feasible. Some early studies have been underway to examine their feasibility for flutter problems associated with wing/store vibrations in fighters [26], [27]. These show promising results under certain restrictive assumptions (e.g. the sensor choice problem is solved separately and signs of mode sensor transfer functions are assumed unknown). More analysis, more wind tunnel testing, and eventual flight tests are required to substantiate these early findings.

6.0 THE RELEVANCE OF MORE RECENT ADAPTIVE THEORY

It is perhaps important at this point to emphasize that the adaptive concepts considered for the wing/store flutter studies cited above, and indeed for all the studies and flight experiments discussed so far, are ad hoc, heuristic in nature. They are based on notions such as forced separation of identification and control, model reference adjustment rules with incomplete stability proofs, self-excited limit cycling, etc. These must be contrasted with certain very recent concepts which are backed by rigorous global stability proofs [28], [29]. The latter references set out explicitly to devise adaptive concepts which stabilize arbitrary time-invariant linear plants (in a specified class) and, moreover, achieve specified model-following objectives. Theoretical attacks on this problem are clearly important in their own right, and successful solutions would hold great significance for flight control--particularly for the flexure/flutter problems outlined above. To date, however, available results fall well short of the required generality. The class of systems for which globally stable controllers have been devised in [28] and [29] consist of finite-dimensional scalar plants

$$g(s) = k\,\frac{N(s)}{D(s)}$$

with

(i) known sign of the gain k

(ii) stable numerator polynomial, N(s)

(iii) known bound on the degree of the denominator, D(s)

(iv) known relative degree:

$$\text{degree}\,[D(s)] - \text{degree}\,[N(s)] = m$$

It is evident from Figures 2 and 3 that these restrictions are quite unreasonable for flight control. Our models are infinite dimensional, relative degree is meaningless, minimum-phase properties are often violated, and even the sign of gains is frequently unknown (e.g. flexure/flutter problems). Hence, to use these recent concepts for control design requires the same "act of faith" that adaptive flight control engineers have learned to accept with the more traditional heuristic concepts.

From a frequency domain viewpoint, it is also interesting to note that the assumptions required above actually imply that the high frequency characteristics of the plant are well known--namely,

$$g(s) \rightarrow \frac{k}{s^m} \quad \text{as } s \rightarrow \infty$$

with only the magnitude of k uncertain. This is completely at odds with flight control plants for which uncertainties become more severe at higher frequencies. Moreover, if such systematic high frequency behavior were, in fact, valid for a minimum phase plant, then ordinary root locus construction rules can be used to show that a number of feedback controllers will stabilize the plant. For example,

$$u = -\frac{k\left[\frac{s}{a} + 1\right]^{m-1}}{\left[\frac{s}{k^2} + 1\right]^{m-1}} y$$

with $a > o$ and k sufficiently large.

It is evident that this controller takes advantage of arbitrarily wide bandwidth to achieve its stabilizing action. It, therefore, has no practical value. Its existence prompts the conjecture, however, that the controllers in [28] and [29] employ the same artificial mechanism, though perhaps more subtly disguised in their more complex structures and mathematical derivations. If true, this conjecture would greatly diminish the chances that the present restrictive class of globally stabilizable systems will ever be broadened to include problems of practical significance.

Certainly these questions contribute additional impetus for continuing and broadening these current lines of research.

7.0 CONCLUSION

This paper has provided a brief examination of the role of adaptive methods in flight control. The main point made is that for conventional flight control problems, adaptive control is the losing alternative in a historical competition with explicit airdata-scheduling. The outcome of this competition remains unchanged even after new flight test evaluations of "modern" adaptive concepts in recent years. The implication for adaptive flight control designers is that we must look for other non-conventional problem areas to find potential applications for our techniques. Two possibilities were described, namely coupled control modes and flexure/flutter control. It remains to be demonstrated, however, that adaptive controls can provide viable solutions for these problems. Finally, the relevance of recent globally stable adaptive concepts was examined in the context of flight control. As they stand today, these concepts were seen to handle too narrow a class of systems to be applicable. More research is required and encouraged to broaden this class.

REFERENCES

1. Stein, G., Hartmann, G. L., and Hendrick, R. C., "Adaptive Control Laws for F-8C Flight Test," IEEE Trans. Auto Control, Oct. 1977.
2. Lindahl, J., McGuire, W., and Reed, M., "Advanced Flight Vehicle Self-Adaptive Flight Control System," Air Force Dynamics Lab., Research and Technology Div., Air Force Systems Command,

Wright-Patterson Air Force Base, Ohio, Wright Air Development Div. TR60-651, Pts. I-VI, 1964.

3. Johnson, D. E., "Flight Control Systems Properties and Problems," Vol. II, Block Diagram Compendium, NASA CR-2501, February 1975.

4. Boskovich, B., and Kaufman, R., "Evolution of the Honeywell First Generation Adaptive Autopilot and Its Applications to F-94, F-101, X-15, and X-20 Vehicles," J. of Aircraft, July-August 1966.

5. McDonald, E. H., and Farris, J. A., "The X-20 Flight Control System Development," Systems Engineering Group, Research and Technology Div., Air Force Systems Command, Wright Air Force Base, Ohio, SEG-TDR-64-8, 1964.

6. Kisslinger, R. L., and Wendl, M., "Survivable Flight Control System," Interim Report No. 1 - Studies, Analysis and Approach, Wright-Patterson AFB, AFFDL-TR-71-20, May 1971.

7. Hendrick, R. C., et. al., "Design Criteria for High Authority Closed Loop Primary Flight Control Systems," Wright-Patterson AFB, Ohio, April 1965.

8. Shipley, P. P., et. al., "Self Adaptive Flight Control by Multivariable Parameter Identification," AFFDL-TR-65-90, Wright-Patterson AFB, Ohio, April 1965.

9. Mehra, R. K., Stepner, P. E., and Tyler, J. S., "A Generalized Method for the Identification of Aircraft Stability and Control Derivatives from Flight Test Data," Joint Automatic Control Conference, August 1972.

10. Iliff, K. W., and Taylor, L. W., "Determination of Stability Derivatives from Flight Data Using a Newton-Raphson Minimization Technique," NASA TN D-6579, 1972.

11. Parks, R. C., "Liapunov Redesign of Model Reference Adaptive Control Systems," IEEE Trans. Auto. Control, July 1966.

12. Landau, I. D., "Hyperstability Concepts and Their Application to Discrete Control Systems," JACC, 1972.

13. Astrom, R. J., and Wittenmark, B., "On Self Tuning Regulators," Automatica, Vol. 9, 1973.

14. Stein, G. and Saridis, G. N., "A Parameter Adaptive Control Technique," Automatica, Vol. 5, pp. 731-739, 1969.

15. Tse, E., et. al., "Wide-Sense Adaptive Dual Control of Stochastic Nonlinear Systems," IEEE Trans. Auto. Control, Vol. AC-17, pp. 98-108, April 1973.

16. Deshpande, J. G., Upadhyay, T. N., and Lainiotis, D. G., "Adaptive Control of Linear Stochastic Systems," Automatica, Vol. 9, pp. 98-108, 1973.

17. Szalai, K. J., et. al., "Design and Test Experiments with a Triply Redundant Digital Fly-by-Wire Control System," AIAA paper No. 76-1911, Aug. 1976.

18. Downing, D. R., Bryant, W. H., and Ostroff, A. J., "Flight Test of a VTOL Digital Autoland System along Complex Trajectories," AIAA Guidance and Control Conf., Aug. 1979, Paper No. 79-1703.

19. "Mini-Special Issue on the F-8C Program," IEEE Trans. Auto. Control, Oct. 1977.

20. Hartmann, G. L., Stein, G., and Powers, B., "Flight Test Experience with an Adaptive Control System Using Maximum Likelihood Parameter Estimation Techniques," AIAA Guidance and Control Conference, August, 1979, Paper No. 79-1702.

21. Hartmann, G. L., Hauge, J. A., Hendrick, R. C., "F-8C Digital CCV Flight Control Laws," NASA Contractor Report, NASA CR-2629, February 1976.

22. Hartmann, G. L., and Stein, G., "Digital Adaptive Control Laws for VTOL Aircraft," Proc. Decision and Control Conference, San Diego, Jan. 1979.

23. Burris, P. M., and Bender, "Aircraft Load Alleviation and Mode Stabilization (LAMS) B-52 System Analysis, Synthesis and Design," Wright-Patterson AFB, AFFDL-TR-68-161, November 1969.

24. Roger, K. L., Hodges, G. E., and Felt, L., "Active Flutter Suppression--A Flight Demonstration," J. Aircraft, June 1975.
25. Huang, C., Winther, B. A., and Mills, G. R., "Demonstration of Active Wing/Store Flutter Suppression Systems," AFFDL-TR-78-65, June 1978.
26. Triplett, W. E., et. al., "Preliminary Design of Active Wing/ Store Flutter Suppression Systems for Military Aircraft," AFFDL-TR-74-76, Aug. 1974.
27. Harvey, C. A., Stein, G., and Felt, L. R., "Adaptive Control of Wing/Store Flutter: A Feasibility Study," AIAA Structural Dynamics and Material Conf., April 1979, Paper No. 79-0788.
28. Feuer, A., and Morse, A. S., "Adaptive Control of Single-Input Single-Output Linear Systems," IEEE Trans. Auto. Control, August 1978.
29. Narendra, K. S., and Valavani, L. S., "Stable Adaptive Controller Design - Direct Control," IEEE Trans. Auto. Control, August 1978.

TOWARD ADAPTIVE CONTROL OF LARGE STRUCTURES IN SPACE

Mark J. Balas

Electrical & Systems Engineering Department
Rensselaer Polytechnic Institute
Troy, New York

C. Richard Johnson, Jr.[1]

Department of Electrical Engineering
Virginia Polytechnic Institute & State University
Blacksburg, Virginia

On-line adaptive control is essential for Large Space Structures (LSS) where the modal parameters are poorly known, due to modeling error, or changing, due to variable configurations. It is especially important that such adaptive controllers produce stabilizing controls during adaptation due to the small damping present in LSS. However, any such controller must be based on a

[1]Support by NASA Grant No. NAG-I-7.

ISBN 0-12-514060-6

reduced-order model of the LSS. The spillover from the unmodeled residuals, as well as the modeling error, can deteriorate the performance of the adaptive controller and, if uncompensated, this spillover can defeat the whole purpose of the adaptive control.

This paper investigates adaptive control for LSS using direct and indirect schemes and points out the mechanisms whereby observation spillover can upset the stability of the controller. The framework for nonadaptive control of LSS is reviewed and many of the generic problems of adaptive LSS control are pointed out within this framework. These generic problems must be overcome for successful operation of adaptive LSS control.

1.0 INTRODUCTION

This paper deals with the basic problems inherent in adaptive control of large space structures (LSS), such as satellites and spacecraft, where the structural parameters are poorly known or slowly time-varying.

With the advent of the Space Shuttle Transportation System, it has become possible to conceive very large spacecraft and satellites which could be carried into space and deployed, assembled, or manufactured there. Such LSS would serve a variety of civilian and military needs [1],[2], including electrical energy generation from the solar power satellite - a structure nearly the size of Manhattan Island - to be constructed in space and operated in earth geosynchronous orbit [3]. The control technology needs for such LSS have been discussed in a variety of articles, e.g., [4],[5], and the developing LSS control theory and technology has been surveyed, for example, in [6-10].

The size of these structures, their low rigidity, and the small damping available in lightweight construction materials combine to make LSS extremely mechanically flexible. In theory,

LSS are distributed parameter systems whose dimension is infinite; however, in practice, their dynamics are usually modeled by large scale systems based on approximate elastic mode data. Active control schemes for LSS are often required to meet stringent requirements for their shape, orientation, alignment, and pointing accuracy. Such active control is limited by the capacity of the on-board control computer, the modeling inaccuracy in current finite element computer codes for analyzing structural dynamics, and available control devices (actuators and sensors); therefore, the controller must be based on some reduced-order model (ROM) of the LSS.

Fundamental problems of LSS control include:

(1) selection of appropriate modes to control for desired system performance;

(2) development of ROM for analysis and controller design;

(3) computation of system model and control parameters;

(4) controller design with multiple distributed actuators and sensors;

(5) the number and location of sensors and actuators for efficient control;

(6) the effect of, and compensation for, residual (unmodeled) modes and modeling error on the closed-loop system performance;

(7) adaptive and self-tuning controllers for LSS with poorly known or changing parameters and configurations.

Item (7) is the basic topic of this paper but it must be considered in the context of the other items with which it is completely intertwined.

The need for adaptive control in LSS arises because of ignorance of the system and changing control regimes. The former is due to (a) ignorance of the system structure and order, and (b) ignorance of the system parameters; the latter occurs because of changing configuration of the LSS. Changes in configuration may be due to construction in-space, thermal distortion, or reorientation of subsystems, e.g., rotating solar panels or sunshields; these changes usually produce slowly time-varying parameters. Ignorance of the LSS system structure and order is due to the fundamental problem of modeling a distributed parameter system, e.g., faulty physics, reduced-order models, and ignored nonlinearities; this means that the order of the ROM is lower than that of the actual LSS. Ignorance of the system parameters, while directly related to the system structure, is due to the inherent modeling error present in even the best structural analysis computer codes and to the limitation of testing huge, lightweight LSS on earth; this produces constant but poorly known system parameters. There is a very clear need for an adaptive LSS control methodology that can begin with the best available computed parameters and self-tune its way toward the correct parameters while stably controlling the LSS and, possibly, adapting to variable configurations.

A great variety of adaptive control schemes exists for lumped parameter, small scale systems [11]; in particular, model reference adaptive methods have achieved a great amount of success in producing stable, convergent adaptive controllers, and adaptive observers for systems whose structure is known and whose parameters are constant but poorly known or slowly time-varying, e.g., [12-25]. Adaptive schemes may be direct, i.e., the available control parameters are directly adjusted (adapted) to improve the overall system performance, e.g., [25],[26], or indirect, i.e., the system parameters are identified (based on the assumed system structure) and the control commands are generated from these parameter estimates as though they were the actual values, e.g., [20],[24],[27].

The abundance of adaptive control methods is overwhelming and an an understanding of the interrelationships and structural commonality of these methods is desperately needed; see, e.g.,

[28],[29], for some beginnings in this direction. Furthermore, the use of such methods on distributed parameter or large scale systems, like LSS, is greatly limited by the ROM problem - the adaptive scheme must be based on a ROM of the actual system and, hence, the order of the model is, and must remain, substantially lower than the controlled system. In addition, it seems essential that the LSS adaptive controller provide a stabilizing control in such highly oscillatory systems as LSS.

This paper develops a framework for LSS adaptive control problems and points out generic problems in the use of the most natural direct and indirect adaptive approaches. In other forms, these problems will haunt every use of adaptive control on LSS and must be solved before the valuable benefits of adaptive control can meet the needs of this new application area. A few preliminary attempts at adaptive control for specific distributed parameter systems or LSS have been made in [30-36]; also, for the corresponding parameter identification problem for distributed parameter systems, see [37].

2.0 NONADAPTIVE LSS CONTROL

Following [6], the LSS may be described by the partial differential equation:

$$m(x)u_{tt}(x,t) + D_0 u_t(x,t) + A_0 u(x,t) = F(x,t) \tag{2.1}$$

where u(x,t) represents (possibly, a vector of generalized) displacements of the structure Ω off its equilibrium position due to transient disturbances and the applied force distribution F(x,t).

The mass distribution m(x) is positive and bounded on Ω. The internal restoring forces of the structure are represented by A_0u where A_0 is an appropriate differential operator with domain $D(A_0)$ defined in a Hilbert space H_0 with inner product $(\cdot,\cdot)_0$. In most cases, A_0 has discrete spectrum, i.e.,

$$A_0\phi_k = \omega_k^2\phi_k \tag{2.2}$$

where ω_k are the <u>mode frequencies</u> of vibration and $\phi_k(x)$ are the <u>mode shapes</u>. The damping term D_0u_t is generated by an appropriate A_0-bounded differential operator and may represent gyroscopic damping as well as the very small ($\sim$1/2% critical) natural damping expected in the LSS.

The applied force distribution is given by

$$F(x,t) = F_C(x,t) + F_D(x,t) \tag{2.3}$$

where F_D represents external disturbances and F_C represents the control forces due to M actuators:

$$F_C(x,t) = B_0f = \sum_{i=1}^{M} b_i(x)f_i(t) \tag{2.4}$$

where b_i are the actuator influence functions (usually point devices) and f_i are the control commands. Observations are produced by P sensors:

$$y = C_0u + C_0'u_t \tag{2.5}$$

where $y_j(t) = (c_j,u)_0 + (c_j',u_t)_0$ for $1 \leq j \leq P$ with c_j being the position sensor influence functions and c_j' the velocity sensor ones (usually point devices).

The state variable form of (2.1) and (2.3)-(2.5) is obtained by taking

$$v(x,t) = [u(x,t), u_t(x,t)]^T$$

in $H \equiv D(A_0^{1/2}) x H_0$ with energy norm:

$$||v||^2 = (mu_t, u_t) + (A_0^{1/2}u, A_0^{1/2}u) \tag{2.6}$$

This produces

$$\begin{cases} v_t = Av + Bf;\ v(0) = v_0 \\ y = Cv \end{cases} \tag{2.7}$$

where we consider the disturbance-free case ($F_D \equiv 0$) and define $B \equiv [0\ B_0]^T$, $C \equiv [C_0 C_0']$ and $A \equiv \begin{bmatrix} 0 & I \\ -A_0 & -D_0 \end{bmatrix}$. This distributed parameter system is very oscillatory in the sense that the semigroup $U(t)$ generated by A has very little damping:

$$||U(t)|| \leq M_0 e^{-\varepsilon t} \text{ for } t \geq 0 \tag{2.8}$$

where $\varepsilon \geq 0$ and small and $M_0 \geq 1$.

The desired performance of the actively controlled LSS greatly affects the design of the controller. Many desirable properties of the active structure can be obtained with constant feedback gains applied to the system state $v(x,t)$; such solutions arise for regulator problems and stabilization (pole placement) problems for LSS. However, the full (infinite dimensional) state v is never available from a distributed parameter system; only the P sensor outputs y are available.

Implementable controllers for LSS (and most distributed parameter systems) must be based on finite dimensional on-board control computers which process the sensor outputs y and produce control commands f; thus, a reduced-order model (ROM) of the system (2.7) must be used for the controller design. A ROM can be obtained by projecting the system (2.7) in H onto an appropriate finite dimensional subspace H_N; the projections P (onto H_N) and Q (onto the residual subspace) are usually, but not always, orthogonal. Let $v_N = Pv$ and $v_R = Qv$ and, from (2.7), we obtain:

$$\dot{v}_N = A_N v_N + A_{NR} v_R + B_N f \tag{2.9}$$

$$\dot{v}_R = A_{RN} v_N + A_R v_R + B_R f \tag{2.10}$$

$$y = C_N v_N + C_R v_R \tag{2.11}$$

where $A_N = PAP$, $A_{NR} = PAQ$, $B_N = PB$, etc. The terms $B_R f$ and $C_R v_R$ are called control and observation spillover; the terms $A_{NR} v_R$ are called model error. The ROM for this system is given by (2.9) and (2.11) with $A_{NR} = 0$ and $C_R = 0$:

$$\begin{cases} \dot{v}_N = A_N v_N + B_N f \\ y = C_N v_N ; v_N(0) = P v_0 \end{cases} \tag{2.12}$$

The ROM state v_N and the residual state v_R form the true system state v with total energy $||v||^2$ given by:

$$\begin{cases} v = v_N + v_R \\ ||v||^2 = ||v_N||^2 + ||v_R||^2 \text{ (if projection is orthogonal)} \end{cases}$$

All implementable controller designs based on any ROM must be evaluated in closed-loop with the actual LSS (2.7), and it is in this evaluation that the effects of model error and spillover due to the residuals become apparent.

If the actual mode shapes ϕ_k are known, the modal ROM is a sensible choice:

$$H_N = \{sp\ \phi_1 \cdots \phi_N\}$$

and the model error terms A_{NR} and A_{RN} become zero. Of course, any collection of modes could be used; usually, the most easily excited or critical ones will be chosen. However, in many cases, the partial differential operator A is too complex to provide closed-form mode shapes. Instead finite element approximations of the mode shapes are computed (e.g., via NASTRAN), and these approximate mode shapes can be used to form the ROM for controller design; note that some model error is present when these approximations are used. Henceforth, we will assume the actual mode shapes are available to simplify the discussion but much of our analysis remains valid for approximate mode shapes and other types of ROM.

Modern modal control (MMC) for LSS, as developed in [38], uses the modal (or approximate modal) ROM and develops a controller consisting of a state estimator based on the ROM and a constant gain control law:

$$\begin{cases} \dot{\hat{v}}_N = A_N\hat{v}_N + B_N f + K_N(y-\hat{y}) \\ \hat{y} = C_N\hat{v}_N, \hat{v}_N(0) = 0 \end{cases} \tag{2.13}$$

and

$$f = G_N\hat{v}_N \tag{2.14}$$

This controller design requires the ROM (A_N, B_N, C_N) to be controllable and observable for the calculation of control and estimator gains G_N, K_N. These conditions, in modal terms, provide insight into the number and location of actuators and sensors. From [38], (A_N, B_N, C_N) is controllable and observable, when position sensors are used $(C_0' = 0)$, if and only if

(1) $\min(P,M) \geq$ max mode frequency multiplicity in the ROM

(2) each sub-block of B_N and C_N associated with a mode frequency ω_N of multiplicity α_N must have rank at least equal to α_N.

Similar results hold for other types of sensors, e.g., velocity, acceleration, or mixtures of types [38],[39]. These results are easy to interpret in terms of the mode shapes, e.g., if no repeated frequencies exist, then the above result says that a single actuator and sensor, not necessarily collocated, will do the job as long as neither is located at any of the ROM mode shape zeros. Since LSS have many symmetries and rigid body modes, it is not often that a LSS control problem will have a controllable observable ROM with only one pair of devices; this has consequences for the adaptive control problem to be discussed later.

Let the estimator error $e_N \equiv \hat{v}_N - v_N$ be defined and, from eqs. (2.9)-(2.11) and (2.13)-(2.14), obtain

$$\dot{v}_N = (A_N + B_N G_N) v_N + B_N G_N e_N \tag{2.15}$$

$$\dot{e}_N = (A_N - K_N C_N) e_N + K_N C_R v_R \tag{2.16}$$

$$\dot{v}_R = B_R G_N v_N + B_R G_N e_N + A_R v_R \tag{2.17}$$

This shows the effect of spillover on the closed-loop system: even though $A_N + B_N G_N, A_N - K_N C_N$, and A_R are stable, the closed-loop system need not be stable. <u>When either control or observation spillover is absent</u> $(B_R = 0$ or $C_R = 0)$, <u>then stability is assured;</u>

otherwise, spillover causes pole-shifting and can induce instabilities [38],[40]. Bounds on the destabilizing effect of observation and control spillover were produced in [38] and can be extended to the case where some model error and small nonlinearities are present [41]. Such bounds give an indication of how much spillover the closed-loop system can tolerate.

A variety of methods have been suggested to reduce spillover [6]. One obvious way would be to prefilter the sensor outputs, with a bandpass filter, to substantially reduce observation spillover. This alleviates the worst pole-shifting problem; bounds on system performance with control spillover alone can be found in [42]. Note that the post-filter of the controller outputs could do the same job by reducing control spillover; this interchange of filter and controller is possible due to linearity and time-invariance. The trade-off with this means of reducing spillover is that the prefilter introduces phase distortion which can have a destabilizing effect of its own. Therefore, a very high order filter may be required to keep the phase distortion acceptable; phase-locked-loop quadrature filters may be another solution [43]. Even in nonadaptive LSS control, the spillover and model error problem is a fundamental one.

Finally, we should note in this section that digitally implemented controls would be based on discrete-time versions of the distributed parameter system (2.7). One such version is obtained by using a uniform time step Δt:

$$\begin{cases} v(k+1) = \Phi v(k) + Ef_k \\ y(k) = Cv(k) \end{cases} \tag{2.18}$$

where $\Phi \equiv U(\Delta t)$ and $E \equiv E_0 B = \int_0^{\Delta t} U(\tau)\, d\tau\, B$ and the control command is a constant f_k over the interval $(k-1)\Delta t \leq t < k\,\Delta t$. Other versions of this could be obtained with nonuniform time steps. When the ROM procedure of projecting onto the subspace H_N is used, we obtain the discrete-time versions of (2.9)-(2.11):

$$v_N(k+1) = \Phi_N v_N(k) + \Phi_{NR} v_R(k) + E_N f_k \tag{2.19}$$

$$v_R(k+1) = \Phi_{RN} v_N(k) + \Phi_R v_R(k) + E_N f_k \tag{2.20}$$

$$y(k) = C_N v_N(k) + C_R v_R(k) \tag{2.21}$$

When H_N is the modal subspace, the above become:

$$v_N(k+1) = \Phi_N v_N(k) + E_0 B_N f_k \tag{2.22}$$

$$v_R(k+1) = \Phi_R v_R(k) + E_0 B_R f_k \tag{2.23}$$

$$y(k) = C_N v_N(k) + C_R v_R(k) \tag{2.24}$$

where (2.22) is the same as that obtained by directly discretizing the ROM in (2.12); if the exact mode shapes are not available, these two discretizations may yield different results. In addition, the sampling process can alias residual modes and increase observation spillover and the zero-order hold process can spread-out the control command signal spectrum and, hence, increase control spillover by increasing the energy in the residual mode spectrum; this has been observed and investigated in [44]. Therefore, <u>the time discretization is a very important factor in the design of implementable LSS controllers</u>.

3.0 TOWARD ADAPTIVE CONTROL OF LSS

In order to design MMC, or other controllers, for LSS as proposed in the previous section, we must have knowledge of the ROM parameters (A_N, B_N, C_N). These parameters are obtained from modal data; they are the mode frequencies for A_N and the mode shapes at actuator and sensor locations for B_N and C_N, respectively. This data is required for three reasons:

(1) to determine controllability and observability of the ROM and, hence, to help locate control devices effectively;

(2) to design control and estimator gains;

(3) to use in the state estimator's internal model.

However, we have noted in Sec. 1.0 the sources of error for this data; consequently, a need arises for an adaptive version of the MMC of Sec. 2.0.

The most logical and reasonable procedure to obtain adaptive controllers for a LSS seems to be the following:

Procedure for Adaptive LSS Control

(a) choose a "nice" reduced-order model (ROM); a modal ROM would be the obvious choice;

(b) use your "favorite" lumped parameter adaptive control scheme;

(c) design the adaptive controller as though the ROM were the actual LSS to be controlled, i.e., ignore the unmodeled residual part of the structure;

(d) use this adaptive controller in closed-loop with the actual LSS and hope for the best.

There is nothing wrong with following this procedure as a best first guess - in a way, there is little else that one can do to produce an implementable adaptive LSS controller.

In some cases, spillover is sufficiently small or enough other mathematical structure is present in the system, e.g., a high level of damping in the distributed parameter system [31] or low level of performance required from the controller (increased damping via direct velocity feedback) [32], to allow the adaptive controller to operate. However, these situations are rare with LSS and one would not like to count on the "generosity of nature" or the temporary suspension of Murphy's Law as part of the above design procedure. Consequently, we would add the following items to that procedure:

(e) analyze computer simulations of higher-order models of the LSS in closed-loop with the adaptive controller based on the lower-order ROM (e.g., simulate more modes than you plan to control);

(f) investigate the specific mathematical mechanisms whereby the residual (unmodeled) part of the LSS couples into a given adaptive control scheme (e.g., find out where and how spillover affects the adaptive controller);

(g) obtain mathematical results on the amount of spillover and/or model error that can be tolerated in the closed-loop system and still provide adequate adaptation and control;

(h) develop spillover and model error compensation schemes to augment the adaptive controller when the residuals cannot be tolerated (e.g., when the conditions of (g) are not satisfied);

(i) recheck (g) with this compensation in the closed-loop system.

We believe that, within the basic framework of LSS control as described in Sec. 2.0, this Augmented Procedure (a)-(i) will go a long way toward revealing the problems of adaptive LSS control (and indeed, most other adaptive control situations where control must be based on a ROM) and will help to focus needed attention on these crucial issues. For example, although Step (e) would be done most likely at some point in the system development phase of any project as complex as the construction and operation of a LSS, often it is done much too late and the design is "set in concrete (or in this case, graphite-epoxy)"; the other steps (f)-(i) may not be done at all. Yet, ignoring the effects of the residual unmodeled LSS can produce some disastrous behavior in the adaptive controller; this was pointed out quite clearly in the LSS example in [36].

When an adaptive scheme is applied to such a lightly damped, oscillatory system as a LSS, the stability of the closed-loop system during adaptation is a necessity; therefore, we do not view convergence and (global) stability results as luxuries for adaptive LSS control and shall only consider appropriate those lumped parameter adaptive schemes for which such results are available. However, even a globally stable adaptive scheme may prove to be unstable when it is used in closed-loop with the actual LSS instead of the ROM on which it was based. This is not a failure of the adaptive scheme; it is a failure to satisfy the mathematical hypothesis of the stability result associated with the scheme.

In the rest of this section, we shall study the use of two very general adaptive schemes which seem to illustrate the problems and potential of adaptive LSS control: the indirect schemes of [23],[24], which use an adaptive observer and operate in continuous time, and the discrete-time, direct or indirect, schemes which are based on an Autoregressive Moving Average (ARMA) model of the controlled plant, e.g., [25-27]. These approaches represent a good cross-section of available lumped parameter adaptive control schemes which have been shown to possess the desired stability properties. We emphasize that the point of this section is not to criticize or slander these schemes; rather, we mean to point out where the hypotheses of their stability results are violated, and must be modified, when we attempt to use them on LSS. We feel that consideration of these approaches within the context of the

Augmented Procedure for Adaptive LSS Control (a)-(i) will illustrate the generic difficulties in the application of existing, well-behaved, lumped parameter adaptive control schemes to LSS.

3.1 Multivariable Systems Converted to Scalar Systems

The results of many stable adaptive schemes, e.g., [13],[14], [23],[24], are limited to a single actuator and/or a single sensor; yet, we have seen in Sec. 2.0 that most LSS control problems will involve multiple actuators and sensors. One way to deal with this (although, admittedly it has its drawbacks) is to convert the controllable observable LSS problem via output feedback into one that is controllable and observable from a single actuator and/or single sensor; this can be done with almost any output feedback gains [45-47]. These gains would have to be based on the best available calculated ROM data and the designer must hope that they will continue to do their job during adaptation.

The output feedback modifies the original system (2.9)-(2.11) to become:

$$\dot{v}_N = (A_N+B_NH_NC_N)v_N + (A_{NR}+B_NH_NC_R)v_R + b_Nf \tag{3.1}$$

$$\dot{v}_R = (A_{RN}+B_RH_NC_N)v_N + (A_R+B_RH_NC_R)v_R + b_Rf \tag{3.2}$$

$$y = c_N^Tv_N + c_R^Tv_R \tag{3.3}$$

where H_N is the output feedback gain matrix, b_N, b_R, c_N, c_R are vectors, and f,y have been renamed. Let $A_N + B_NH_NC_N$ be $\bar{A}_N$, etc. and we have that the new ROM $(\bar{A}_N, b_N, c_N^T)$ is a controllable, observable single-input, single-output system and (3.1) and (3.2) become:

$$\dot{v}_N = \bar{A}_N v_N + \bar{A}_{NR} v_R + b_N f \tag{3.4}$$

$$\dot{v}_R = \bar{A}_{RN} v_N + \bar{A}_R v_R + b_R f \tag{3.5}$$

3.2 Indirect Adaptive Controller Design

We apply the design of the adaptive controller in [24] directly to the ROM consisting of (3.4) and (3.3) with the assumption, for now, that

$$\bar{A}_{NR} = 0 \text{ and } c_R^T = 0.$$

The control law is given by

$$f(t) = g_N^T \hat{v}_N(t) + f_c(t) \tag{3.6}$$

where g_N is a <u>constant</u> gain vector, f_c is a "sufficiently rich" external signal (more about this later), and $\hat{v}_N$ is derived from the following adaptive observer (or state estimator):

$$\dot{\hat{v}}_N = F\hat{v}_N + gy + hf; \quad \hat{v}_N(0) = \hat{v}_0 \tag{3.7}$$

where F is an arbitrary, stable matrix and g, h are unknown parameter vectors. The appropriate <u>matching conditions</u> are:

$$\begin{cases} F + G^* c_N^T = \bar{A}_N \\ h^* = b_N \\ \hat{v}_0 = \hat{v}_N(0) \end{cases} \tag{3.8}$$

where g^*, h^* are constant.

Let $p_0^* \equiv [g^{*T} h^{*T} v_N(0)^T]^T$ and note that

$$\dot{p}^* = \tilde{F}p^*; \; p^*(0) = p_0 \tag{3.9}$$

where $\tilde{F} \equiv \text{diag}\,[0\;0\;F]$ and, when $g = g^*$, $h = h^*$, we have from (3.8) and (3.7):

$$v_N(t) = [M(t) I_N] p^*(t) \tag{3.10}$$

where $M(t) \equiv \int_0^t e^{F(t-\tau)} [I_N y(\tau) I_N f(\tau)] d\tau$ and we have used the fact that h, f are scalars.

Now (3.7) can be rewritten as

$$\hat{v}_N(t) = [M(t) I_N] p(t) \tag{3.11}$$

$$\dot{M}(t) = FM(t) + [I_N y(t) I_N f(t)] \tag{3.12}$$

$$M(0) = 0 \tag{3.13}$$

where $M(t)$ is as defined in (3.10), and we have yet to produce the adaptive law to generate $p(t)$. This adaptive law is given by the following:

$$\dot{p}(t) = \tilde{F}p(t) - \alpha(t)[R(t)p(t) + r(t)] \tag{3.14}$$

where $p(0) = p_0$ is arbitrary and

$$\begin{cases} \dot{R}(t) = -qR(t) - \tilde{F}^T R(t) - R(t)\tilde{F} + [M(t) I_N]^T c_N c_N^T [M(t) I_N] \\ R(0) = 0 \end{cases} \tag{3.15}$$

and

$$\begin{cases} \dot{r}(t) = -qr(t) - \tilde{F}^T r(t) - [M(t)I_N]^T c_N y(t) \\ r(0) = 0 \end{cases} \tag{3.16}$$

where $\alpha(t)$ is the adaptive gain and the constant q is chosen to exceed twice the absolute value of the real parts of the eigenvalues of F. The adaptive gain is chosen so that

$$\alpha(t) = \gamma + ||\dot{\mu}(t)|| \tag{3.17}$$

where

$$\dot{\mu}(t) = -\lambda\mu(t) + (N/2)^{1/2}(|y(t)| + |f(t)|) \tag{3.18}$$

with λ positive and $F + F^T \leq 2\lambda I_N$.

3.3 Convergence Results: What Goes Wrong?

All of the above is exactly as stated in [24] where it is also shown in Appendix I and II that

$$\mu(t) \geq ||M(t)|| \tag{3.19}$$

and, with f(t) sufficiently rich in frequencies, there is a t_1 such that

$$R(t) \geq \rho I_N > 0 \text{ for all } t \geq t_1 \tag{3.20}$$

In addition, it is shown in Appendix III that

$$R(t)p^*(t) + r(t) \equiv 0 \tag{3.21}$$

This is very crucial to the stability results of [24] and it is here that observation spillover (i.e., the fact that we are using a ROM) appears - (3.21) is not valid when $c_R \neq 0$; however,

$$R(t)p^*(t) + r(t) = \Delta_R(t) \tag{3.22}$$

where $\Delta_R(t) \equiv \int_0^t e^{-(q + \tilde{F}^T)(t-\tau)} [M(\tau)I_N]^T c_N c_R^T v_R(\tau) d\tau$

Let $e_N(t) \equiv \hat{v}_N(t) - v_N(t)$ and $\Delta p(t) \equiv p(t) - p^*(t)$; then

$$e_N(t) = [M(t)I_N]\Delta p(t) \tag{3.23}$$

and

$$\Delta\dot{p}(t) = [\tilde{F} - \alpha(t)R(t)]\Delta p(t) - \alpha(t)\Delta_R(t) \tag{3.24}$$

Consider $V(t) \equiv \Delta p(t)^T \Delta p(t)$ and we obtain:

$$\dot{V}(t) \leq -\gamma\ \rho\ V(t) - 2\gamma\ \Delta_R(t)\Delta p(t) \tag{3.25}$$

where $V(0) = \Delta p_0^T \Delta p_0$ and $\Delta p_0 = p_0 - p_0^*$.

This follows [24] except for the additional term in (3.25); also note that

$$||e_N(t)||^2 \leq V(t)[1 + \mu(t)] \tag{3.26}$$

Let $\Delta v_N(t) \equiv v_N(t) - v_N^*(t)$ and $\Delta v_R(t) \equiv v_R(t) - v_R^*(t)$ where v_N^* and v_R^* represent the ideal states of (3.4)-(3.5) when the parameters are exactly known:

$$\dot{v}^*_N = (\bar{A}_N + b_N g_N^T) v^*_N + \bar{A}_{NR} v^*_R + b_N f_c \tag{3.27}$$

$$\dot{v}^*_R = (\bar{A}_{RN} + b_N g_N^T) v^*_N + \bar{A}_R v^*_R + b_R f_c \tag{3.28}$$

with $v^*_N(0) = v_N(0)$ and $v^*_R(0) = v_R(0)$.

When the implementable adaptive control law (3.6) is used, we obtain:

$$\Delta\dot{v}_N = (\bar{A}_N + b_N g_N^T)\Delta v_N + \bar{A}_{NR}\Delta v_R + b_N g_N^T e_N \tag{3.29}$$

$$\Delta\dot{v}_R = (\bar{A}_{RN} + b_N g_N^T)\Delta v_R + \bar{A}_R \Delta v_R + b_R g_N^T e_N \tag{3.30}$$

and we have the following result:

THEOREM 3.1: Assume

(1) $c_R = 0$
(2) $\bar{A}_N + b_N g_N^T$ stable
(3) $\bar{A}_R$ stable
(4) $\bar{A}_{NR} = 0$
(5) $f_c(t)$ sufficiently rich (i.e., it has at least 3N/2 distinct frequencies)

Then there is a $\delta > 0$ such that for all $||\Delta p_0||^2 \leq \delta$:

(a) $e_N(t)$ is bounded and (eventually) vanishes with an arbitrary exponential rate
(b) $\lim_{t\to\infty} \Delta v_N(t) = 0$
(c) $\lim_{t\to\infty} \Delta v_R(t) = 0$.

Therefore, even though the closed-loop system with the adaptive controller is highly nonlinear, it is stable while the adaptation is taking place. In particular, (1),(3), and (4) are satisfied if there is no observation spillover ($C_R = 0$) and no model error ($A_{NR} = 0$) and some damping in the residuals (i.e., A_R is stable).

Proof: This result follows from the results of [24] because $\Delta_R(t) \equiv 0$ in (3.22) when $c_R^T = 0$. The stability of (3.29)-(3.30) is determined by $\bar{A}_N + b_N g_N^T$ and A_R when $\bar{A}_{NR} = 0$. Also, if $C_R = 0$, then $\bar{A}_{NR} = A_{NR}$, $\bar{A}_R = A_R$, and $c_R^T = 0$ since it is a row vector of C_R. #

Note that the stability of the system with adaptive control is determined by that of (3.27)-(3.28) - the ideal case where the parameters are known and the external signal f_c is applied. This is natural, since adaptation cannot take place without f_c present; however, after adaptation, we would most likely want to turn off f_c. In addition, we could choose the f_c signal so as not to excite the residual frequencies whenever sufficient spectral separation is present. Still, we would need to turn on f_c now and then, in order to "tune-up" the controller.

3.4 Spillover Compensation for the Indirect Adaptive Controller

The above result is merely a slight extension of the results of [24] to a special case of the adaptive controller based on a ROM instead of the full system. However, it does suggest that some form of compensation should be used to eliminate the observation spillover. Such compensation must be essentially independent of the parameters of the ROM; yet, most methods of spillover reduction require knowledge of the ROM (and some residual) parameters.

One approach to spillover compensation already suggested in Sec. 2.0 is prefiltering the sensor outputs to remove or greatly reduce the term $C_R v_R(t)$. Such prefiltering can be achieved with low-pass or band-pass filters when the ROM frequencies are known and separated from the residual frequencies. However, the modal frequency data is part of the poorly known parameter data.

In an attempt to resolve this predicament, we could try using phase-locked loop (PLL)-based filters with the center frequency of each loop tuned to the best approximation available of the corresponding ROM data. The PLL will adapt itself until it tracks the actual mode frequency, and after "lock-on," it will behave as a narrow band-pass, linear filter which tunes out the observation spillover from the other frequencies [43]. Of course, sufficient spectral separation must be present, the calculated values of the ROM modal frequencies must be sufficiently good, the distortion introduced by the filter must be sufficiently small, and the adaptive controller must not shift the poles around too much. Thus, the PLL prefilter is not a panacea! But, it might work to reduce spillover and, if it does, it has the added advantage that its output could also reveal better estimates of the modal frequencies; this would take some of the load off the adaptive observer. If the adaptation mechanism causes too much pole-shifting, the ROM frequencies could be excited via f_c and identified in open-loop by the PLL filters before the adaptive controller is turned on. Note that modal frequency data is usually better known, via computer approximation, than modal shape data; hence, this approach might not be unreasonable. Another possible, but untried, approach to spillover compensation might be an adaptive version of the orthogonal filter in [48].

Note that some prefiltering (and postfiltering) always takes place due to the bandwidth limitations of the sensors (and actuators). Whether this can be used to advantage in adaptive LSS control is still a matter for speculation, e.g., [33].

A problem that arises with the use of the results in [24] and their modification to LSS is that the constant feedback gains g_N^T must be calculated, in advance, to stabilize $\bar{A}_N + b_N g_N^T$. It would be better if these gains were adapted along with the parameters in the observer. Of course, after adaptation has taken place, they could be recalculated from the "tuned-up" parameters, but, in some cases, the adaptation phase is never over, e.g., slowly-varying parameters. Other approaches could be used for adaptive pole-placement, e.g., [20],[21], but these also have their limitations.

3.5 ARMA-Gettin'

Many discrete-time adaptive control schemes depend on an Auto-Regressive Moving Average (ARMA) representation of the plant in discrete-time, e.g., [25-27]:

$$y(k+N) = \sum_{r=1}^{N} \alpha_r y(k+r-1) + \sum_{r=1}^{N} \beta_r f(k+r-1) \tag{3.31}$$

for some N and appropriate matrices α_r, β_r. What the ARMA says is that, after N time steps, the present output is only related to the past N outputs and inputs. Existence of an ARMA is directly related to the finite dimensionality of the plant (N is usually that dimension) and is obtained using the Cayley-Hamilton theorem for matrices. For LSS, only a <u>quasi-ARMA</u> can exist; these were considered in detail in [49]. From the Appendix of [49], we obtain the quasi-ARMA for the LSS (2.18) or (2.19)-(2.21):

$$y(k+N) = \sum_{r=1}^{N} \alpha_r y(k+r-1) + \sum_{r=1}^{N} \Gamma_r E_N f(k+r-1) + R(k) \tag{3.32}$$

where

$$R(k) = C_R v_R(k+N) + \sum_{r=1}^{N} \Delta_r v_R(k+r-1)$$

$$\Delta_r = \Gamma_r \Phi_{NR} - \alpha_r C_R$$

and Γ_r is defined in the Appendix of [49]. Since R(k) = 0 when $C_R = 0$ and $\Delta_r = 0$, we have the following result:

THEOREM 3.2: When the observation spillover (C_R) and the model error term (Φ_{NR}) are both zero, the quasi-ARMA (3.32) is a true ARMA for the LSS (2.18), and any stable adaptive scheme based on this ARMA will be globally stable when used in closed-loop with the actual LSS (2.18).

When the rather stringent hypothesis of Theo. 3.2 is not satisfied (as it may not be in practice), any adaptive LSS control scheme based on the quasi-ARMA (3.31) must ignore R(k) in order to be implementable. However, R(k) is the term where the residual effects - spillover and model error - enter the scheme and can cause instability. Again, as in Sec. 3.4, prefiltering or other compensation might be tried in an attempt to reduce or eliminate this term.

4.0 CONCLUSIONS

In an attempt to point out the crucial issues and generic problems associated with adaptive control of large aerospace structures (LSS), we have reviewed the framework for nonadaptive LSS control (Sec. 2.0) and, within this framework, have proposed a general procedure, based on reduced-order models (ROM) of the LSS, for obtaining and assessing the problems of adaptive LSS controllers (3.0). In addition, we have taken a closer look at the use of certain well-known, lumped parameter, stable adaptive control schemes in this procedure. Taking these schemes as representative of the basic ideas present in all lumped parameter adaptive control approaches, we have obtained corresponding LSS adaptive controllers and found the following generic problems associated with adaptive LSS control:

(1) LSS are distributed parameter or large scale systems; therefore, the plant dimension is always larger than the dimension of the adaptive controller, which must be based on a ROM;

(2) LSS control must often be done with more than one actuator and sensor; conversion of multivariable to scalar systems via output feedback introduces problems of its own (e.g., stability of the residuals);

(3) LSS control problems are often non-minimum phase due to noncollocated actuators and sensors;

(4) Interaction of the residuals with the adaptive controller may negate the stabilizing properties of the controller due to observation spillover; this interaction is much worse due to the nonlinear nature of adaptive control;

(5) Methods of spillover compensation for LSS often require knowledge of the ROM (and some residual) parameters - the very data that are poorly known;

(6) The adaptation mechanism may shift the closed-loop frequencies around; this counteracts the benefits of any prefiltering unless sufficient spectral separation is maintained;

(7) Indirect adaptive controllers need sufficient excitation from an external signal f_c; however, this signal may substantially excite the residuals.

(8) Discrete-time adaptive controllers can only be based on quasi-ARMA rather than strict ARMA representations of the LSS; this may negate the stability properties of such a controller.

Stable adaptation is essential for such highly oscillatory systems as LSS, yet our preliminary stability results, Theos. 3.1 and 3.2, both require that observation spillover be somehow completely eliminated before it reaches the adaptive control logic;

certainly, this is not an easy thing to do in general! Perhaps, global stability is too much to ask for LSS adaptive control because observation spillover will always be present to some degree in LSS control. However, it seems reasonable to hope for the development of spillover bounds to give some idea of regions of stability for the successful operation of LSS adaptive control. Some comparison should be made between stable adaptive controllers based on ROM and stable, robust control schemes, e.g., [50],[51].

In closing, we would like to say that it is not our intent to present a gloomy picture for the application of adaptive control to LSS. In fact, the need for adaptive control in LSS is already becoming quite clear, and recognition of this need comes, for a change, at an appropriate time - before any LSS have been built and put into space. However, the development of adaptive control for LSS will not take place overnight and will not be done by one or two people. Consequently, what we have tried to stress here for the interested researcher are some of the fundamental problems that arise and the basic steps which need to be taken toward the goal of successful adaptive control of LSS. In the long run, we have high hopes for the success of this endeavor and we expect that adaptive control theory will profit by its association with large aerospace structures, as well.

REFERENCES

1. "A Forecast of Space Technology 1980-2000," NASA SP-387, Jan. 1976.
2. "Large Space Structures - Challenge of the Eighties," Special Section in AIAA Astronautics and Aeronautics, 24-59, Oct. 1978.

3. Glaser, P., et al., "First Steps to the Solar Power Satellite," IEEE Spectrum, 52-58, May 1979.
4. Dahlgren, J., and Gunter, S., "Pointing and Control Technology Needs for the Future Automated Space Systems," presented at AIAA Conf. on Large Space Platforms, Los Angeles, Ca., Sept. 1978.
5. Herzberg, R., et al., "Dynamics and Control of Large Satellites,"AIAA Astronautics and Aeronautics, 35-39, Oct. 1978.
6. Balas, M., "Some Trends in Large Space Structure Control Theory: Fondest Hopes; Wildest Dreams," presented at JACC, Denver, Co., June 1979; to appear in revised form in IEEE Trans. Autom. Contr.
7. Meirovitch, L., and Öz, H., "An Assessment of Methods for the Control of Large Space Structures," presented at JACC, Denver, Co., June 1979.
8. Gran, R., and Rossi, M., "A Survey of the Large Structures Control Problem," presented at 18th IEEE CDC, Ft. Lauderdale, Fl., Dec. 1979.
9. Likins, P., "Attitude Control of Nonrigid Spacecraft," AIAA Astronautics and Aeronautics, May 1971.
10. Likins, P., "The New Generation of Dynamic Interaction Problems," AAS Rocky Mtn. G & C Conf., Keystone, Co., Mar. 1978.
11. Asher, R., Andrisani, D., and Dorato, P., "Bibliography on Adaptive Control Systems," Proc. IEEE 64, 1226-1240, 1976.
12. Narendra, K., and Kudva, P., "Stable Adaptive Schemes for System Identification and Control, Parts I and II," IEEE Trans. Syst., Man and Cybernetics SMC-4, 542-560, 1974.
13. Narendra, K. and Valavani, L., "Stable Adaptive Observers and Controllers," Proc. IEEE 64, 1198-1208, 1976.

14. Narendra, K., and Valavani, L., "Stable Adaptive Controller Design - Direct Control," IEEE Trans. Autom. Control AC-23, 570-583, 1978.
15. Lindorff, D., and Carroll, R., "Survey of Adaptive Control Using Lyapunov Design," Int'l. J. Control 18, 897-914, 1973.
16. Landau, I. D., "Model Reference Adaptive Systems - A Survey - What is Possible and Why?," ASME J. Dyn. Systems, Measurement, and Control, 119-132, 1972.
17. Landau, I. D., "A Survey of Model Reference Adaptive Techniques - Theory and Applications," Automatica 10, 353-379, 1974.
18. Carroll, R., and Lindorff, D., "An Adaptive Observer for Single-Input Single-Output Linear Systems," IEEE Trans. Autom. Control AC-18, 428-435, 1973.
19. Nuyan, S., and Carroll, R., "An Adaptive Observer and Identifier with an Arbitrarily Fast Rate of Convergence," presented at IEEE CDC, 1976.
20. Nuyan, S., and Carroll, R., "Adaptive Identification and Pole Placement for Stable Plants," Univ. of S. Carolina, Report No. EE-CACL-TR-77-1, 1977.
21. Kraft, L. G., "Controllable Form State Variables Obtained from an Arbitrarily Fast Adaptive Observer," presented at IEEE CDC, 1976.
22. Kreisselmeier, G., "Adaptive Observers with Exponential Rate of Convergence," IEEE Trans. Autom. Control AC-22, 2-8, 1977.
23. Kreisselmeier, G., "The Generation of Adaptive Law Structures for Globally Convergent Adaptive Observers," IEEE Trans. Autom. Control AC-24, 510-513, 1979.
24. Kreisselmeier, G., "Algebraic Separation in Realizing a Linear State Feedback Control Law by Means of an Adaptive Observer," to appear IEEE Trans. Autom. Control.

25. Monopoli, R., "Model Reference Adaptive Control with an Augmented Error Signal," IEEE Trans. Autom. Control AC-19, 474-484, 1974.

26. Johnson, Jr., C. R., and Tse, E., "Adaptive Implementation of One-Step-Ahead Optimal Control via Input Matching," IEEE Trans. Autom. Control AC-23, 865-872, 1978.

27. Goodwin, G., Ramadge, P., and Caines, P., "Discrete Time Stochastic Adaptive Control," to appear IEEE Trans. Autom. Control.

28. Johnson, Jr., C. R., "Input Matching, Error Augmentation, Self-Tuning, and Output Error Identification: Algorithmic Similarities in Discrete Adaptive Model Following," presented at 18th IEEE CDC, Ft. Lauderdale, Fl., Dec. 1979.

29. Landau, I. D., "Model Reference Adaptive Control and Stochastic Self-Tuning Regulators - Towards Cross-Fertilization," presented at Workshop on Adaptive Control, Univ. of Illinois, Urbana-Champaign, Il., May 1979.

30. Amouroux, M., Babary, J.-P., and El Jai, A., "Adaptive Open Loop Control for a Class of Distributed Parameter Systems," Automatica 14, 603-607, 1978.

31. Hamza, M., and Sheirah, M., "A Self-Tuning Regulator for Distributed Parameter Systems," Automatica 14, 453-463, 1978.

32. Benhabib, R., Iwens, R., and Jackson, R., "Active Vibration Control of a Flat Plate Using Model Reference Adaptive Techniques," presented at 2nd Symp. on Dynamics and Control of Large Flexible Spacecraft, Blacksburg, Va., June 1979.

33. Potter, J., and Ginter, S., "A New Concept in Adaptive Structural Control," presented at 2nd Symp. on Dynamics and Control of Large Flexible Spacecraft, Blacksburg, Va., June 1979.

34. Montgomery, R., and Johnson, Jr., C. R., "The Dual Momentum Control Device for Large Space Systems: An Example of Distributed System System Adaptive Control," presented at 12th Asilomar Conf. on Circuits, Systems and Computers, Pacific Grove, Ca., Nov. 1978.

35. Johnson, Jr., C. R., and Montgomery, R., "A Distributed System Adaptive Control Strategy," IEEE Trans. on Aerospace and Electronics AES 15, 601-612, 1979.

36. Johnson, Jr., C. R., "On Adaptive Modal Control of Large Flexible Spacecraft," presented at AIAA G & C Conf., Boulder, Co., Aug. 1979; to appear in revised form in J. Guidance and Control, 1980.

37. Goodson, R., and Polis, M. (eds.), "Identification of Parameters in Distributed Systems: A Symp. of the Am. Auto. Control Council," JACC, Austin, Tx., June 1974.

38. Balas, M., "Feedback Control of Flexible Systems," IEEE Trans. Autom. Control AC-23, 673-679, 1978.

39. Hughes, P., and Skelton, R., "Stability, Controllability, and Observability of Linear Matrix-Second-Order Systems," presented at JACC, Denver, Co., June 1979.

40. Ginter, S., and Balas, M., "Attitude Stabilization of Large Flexible Spacecraft," presented at AIAA G & C Conf., Palo Alto, Ca., Aug. 1978.

41. Balas, M., "Control of Flexible Structures in the Presence of Certain Nonlinear Disturbances," presented at JACC, Denver, Co., June 1979.

42. Balas, M., "Modal Control of Certain Flexible Dynamic Systems," SIAM J. Contr. and Opt. 16, 450-462, 1978.

43. Gustafson, D., and Speyer, J., "Linear Minimum Variance Filters Applied to Carrier Tracking," IEEE Trans. Autom. Control AC-21, 65-73, 1976.

44. Ginter, S., "Attitude Control of Large Flexible Spacecraft," M.S. Thesis, Mass. Inst. of Tech., Cambridge, Ma., Sept. 1978.

45. Davison, E., and Wang, S., "Properties of Linear Time-Invariant Multivariable Systems Subject to Arbitrary Output and State Feedback," IEEE Trans. Autom. Control AC-18, 24-32, 1973.

46. Heymann, M., "On the Input and Output Reduceability of Multivariable Linear Systems," IEEE Trans. Autom. Control AC-15, 563-569, 1970.
47. Ding, C., Brasch, F., and Pearson, J., "On Multivariable Systems," IEEE Trans. Autom. Control AC-15, 96-97, 1970.
48. Skelton, R., and Likins, P., "Orthogonal Filters for Model Error Compensation in the Control of Non-rigid Spacecraft," J. Guidance and Control 1, 41-49, 1978.
49. Balas, M., and Johnson, Jr., C. R., "Adaptive Control of Distributed Parameter Systems: The Ultimate Reduced-Order Problem," presented at IEEE CDC, Ft. Lauderdale, Fl., Dec. 1979.
50. Gutman, S., and Leitmann, G., "Stabilizing Feedback Control for Dynamic Systems with Bounded Uncertainty," presented at IEEE CDC, 1976.
51. Nikiforuk, P., Gupta, M., and Tamura, K., "The Design of a Signal Synthesis Model Reference Adaptive Control System for Linear Unknown Plants," ASME J. Dyn. Syst., Meas., and Control 99, 123-129, 1977.

ADAPTIVE CONTROL IN THE PROCESS INDUSTRIES

Pierre R. Bélanger

Department of Electrical Engineering
McGill University
Montreal, Canada

In this paper, we review the status of adaptive control methods in process control. The particular characteristics of process control problems are discussed and compared to those of electromechanical systems. Model reference and enforced separation methods are compared, and some practical aspects are discussed. A brief review of applications is given, followed by a discussion of difficulties and suggestions for further research.

1. INTRODUCTION

Adaptive control has been with us since the 1950's, at least. For the most part, however, it has been the province of experience and art, not to say tinkering. The last decade has seen the development of sound theories of adaptive control. If it is true that we must crawl before we can walk and walk before we can run, we are surely now well out of the crawling stage.

ISBN 0-12-514060-6

We now have several theoretical approaches at our disposal. The self-tuning regulator [1] has already been used in several applications, and the model reference approach [2-4] has also enjoyed some measure of success.

Coinciding with the theoretical developments in control, the economies of the West have been faced with the need for increasing productivity and reducing their use of scarce resources: control affords a means of helping to fulfill both requirements. It is fortuitous that, at the same time, the cost of computing has been dramatically decreasing, making possible the use of increasingly more sophisticated control algorithms.

In this paper, we shall be concerned with the applications - and applicability - of adaptive control techniques to process control. To begin, we shall describe the features of a process control problem that distinguish it from other control problems. Next, we shall review briefly some of the available methods for adaptive control, after which we shall examine some of the practical considerations associated with these methods. Some of the applications of adaptive control will be reviewed, and the results examined in a practical light. Finally, we shall mention some of the difficulties, technical and others, associated with adaptive control and we shall try to point useful research directions for alleviating them.

2. CHARACTERISTICS OF PROCESS CONTROL

Since this paper specifically addresses process control, one would think that there is something about process control that makes it different from the control of electromechanical, electrical power, biological or societal systems. We shall make comparisons here mostly with electromechanical systems: biological systems have much in common with process control and societal systems are not within the realm of competence of the writer.

The first striking difference occurs right at the modelling stage. Electrical phenomena are well described by the simple models of circuit theory. Mechanical systems are usually quite adequately modelled by Newton's laws, although distributed behaviour such as vibration modes is admittedly a source of difficulty. The dynamics of hydraulic components are also well understood, so it is usually possible to obtain, from first principles, an adequate model of an electromechanical system.

This is not the usual situation in process control. To be sure, it is not difficult to write good models of systems involving the flow of fluids, but even then some parameters (e.g. the fiber retention on the wire of a paper machine) will be yielded only through experiment. Many processes involve complex reactions or transfer operations that almost defy modelling. It is often the case that the oldest industries (e.g. cement, paper, dyeing) are the ones where the basic physics and chemistry are the least well-known. On the whole, existing models in process control are steady-state models; they are much easier to derive, often by simple energy and mass balances, and the economic payoff in the optimization of steady-state operating point is sizeable. It is fair to say that dynamic models are possible in process control, but they usually require a great deal of effort by very skilled personnel.

The second difference is in the models themselves. Electromechanical systems often yield models of relatively high dimension ($\sim$30-50) with several highly underdamped poles usually related to bending or vibration modes. On the other hand, dynamic models for process control are often partial differential equations, but they do have one saving grace: the dynamics are almost always well damped, so that the essential behaviour can be retained by models of relatively low order. This is in contrast to the electromechanical models where the highly underdamped modes at least must be retained for proper representation. On the other hand, it will usually be necessary to include some delay along with a low-order lumped model: the traditional model structure in process control is first-or-second-order system plus delay.

A third distinction, of some significance in the context of adaptive control, is the source of the perturbing signals. In many electromechanical systems, the main external signals are command input, for example the commands of the pilot in an aircraft. Most process control systems, on the other hand, are regulators; there are changes in the reference, or set-point, but these are relatively infrequent. On the whole, the principal perturbing signals are disturbances, or upsets, which are sometimes measured but more often are not.

Fourthly, although one expects to do some trimming of controller parameters in an electromechanical system, the system must often be functional with the designed settings (e.g. an aircraft control system). In process control, controller parameters are estimated only well enough to purchase the right controller model: tuning is an essential part of the job. The overwhelming popularity of PID controllers is at least partly explained by the fact that three parameters is about the most that can be juggled in a manual tuning procedure. Despite what was said in the previous paragraph, tuning is almost exclusively done by set point changes. This at least ensures good stability margins; disturbance spectral content and process dynamics variations are handled implicitly by tuning tightly (larger bandwidth, wider process variations).

A fifth distinction, a major one, is in the instrumentation. Electrical and mechanical quantities can be measured by a variety of standard means. The sensors can be selected to accommodate the control system design, and are limited only by economics and, possibly, reliability. In process control, the universal complaint is the inability to measure key process variables. For example, burning zone temperature and rutile composition are two such variables in a T_iO_2 kiln; yet the optical pyrometer cannot provide a reliable absolute temperature measurement, and rutile composition cannot be measured on-line. Processes are often wet, dirty, hot dusty, corrosive and generally hostile to the sophisticated electronics of the aerospace world.

Finally, there is a difference in the need for control. In an aircraft or space vehicle, good control is an absolute necessity: no control, no flight. Most processes have a history that antedates Maxwell and Watt, let alone Bode and Nyquist. They can get along with nothing more than good process operators. Of course, today's processes operate with tolerances that make control mandatory; it still remains that the process industries do not emphasize control as much as, say, process development. This is understandable: the customers are buying the product and no one seems to be complaining too much about quality.

3. ADAPTIVE CONTROL METHODS

There are two basic approaches to adaptive control. The first approach, model reference, attempts to make the input-output behaviour of the controlled plant identical to that of a preselected model. In principle, it is not necessary to identify the plant because the control objective is externally specified: it is possible to act directly on the controller parameters. One may, of course, identify the plant and apply a model-following control, but this is usually not done. Developments up to 1974 are surveyed in Landau [2], and the papers by Feuer and Morse [3] and Narendra and Valavani [4] represent the latest published results.

The other approach might be called the enforced separation approach, since it uses plant parameter estimates to calculate the control law. Identification is necessary here, because the control objective, usually the minimization of a quadratic performance index, depends on the plant. The self-tuning regulator [1] belongs to this class, as does the "adaptive state regulator with adjustable plant model" proposed in [2].

The early work in model reference control was characterized by ad-hoc adjustment rules, validated by simulation. In more recent work, the parameter adjustment rules are selected according to stability theory. In all cases, the adaptation proceeds from the difference between the outputs of the plant and of the model. Auxiliary signals are generated from the plant input and output. Depending on the assumptions concerning the plant, the generation of these signals may require a positive real transfer function involving some of the plant parameters. The adjustment rule is usually of the form

$$\dot{\underline{\Theta}}_i = -\Gamma\underline{\omega}(t)e(t)$$

where Γ is a matrix, $e(t)$ is the output error and $\underline{\omega}(t)$ is the set of auxiliary signals.

In [10] Martin-Sanchez presents an application to a distillation column of a scheme that is at least partially of the model reference class [27]. The identification is by model reference, where the model is adjusted to fit the plant. The derivation of the control introduces a novel twist. The desired output to be matched by the plant is derived by passing the input through a "driver block", i.e. a reference system, but where the measured plant outputs are used instead of the reference system outputs. For example, if the reference system is

$$z(t) = .8z(t-1) + v(t-1)$$

the actual past output y(t-1) is used instead of z(t-1) to generate z(t), the desired output. The plant control u(t) is then used to make the predicted output equal to the desired one, so that the scheme is reminiscent of some sort of generalized minimum-variance control. Note that this method handles discrete plants with delays.

The model reference approach has at least two drawbacks in the context of process control.

First, current continuous-time theory applies only to lumped models, and does not admit time delays. Lumped discrete systems may include delays. Work on discrete systems has been reported in [5-9]. Since the delay in terms of sampling periods is often known (e.g. transport lag), it is possible that theory can accommodate it. There still remains the problem of right-half plane zeros, very frequent in process dynamics. This appears insurmountable since a controller would need to cancel them by unstable poles to fit the reference model.

A second drawback is the fact that model reference theory works on the basis that the plant output is entirely due to the mapping of the plant input into the plant output by the plant input-output map. Practically, this means that most of the output must be due to the response to a known command input. It is difficult to see how the theory could be fit to regulation, especially since the work to date has been almost entirely deterministic; one exception is Landau [5] but that work is concerned with identification only.

Despite these disadvantages, the theory has been used in process control [10] and could be useful to tune controllers for set point changes.

The separation method is not without blemish, but it does not suffer from the drawbacks of the first method. Non-minimum phase dynamics are not a problem since the closed-loop optimum is shaped by the plant dynamics rather than by force fitting to a model. Regulation problems are easily handled, if a stochastic method is used for identification.

We shall focus our attention here on self-tuning regulators (STR), since they have had the lion's share of adaptive control applications in the process industries. The plant is characterized by [1].

$$A(q^{-1})y(t) = B(q^{-1})u(t-k) + C(q^{-1})e(t) \tag{1}$$

with A, B and C nth-order polynomials in the delay variable q^{-1}, and where $a_o = c_o = 1$.

An equivalent characterization is the predictor form.

$$\begin{aligned} y(t+k) = {} & \alpha_1 y(t) +\ldots+ \alpha_n y(t-n) + \beta_o[u(t) + \beta_1 u(t-1) \\ & + \beta_{n+k}u(t-n-k)] + \varepsilon(t+k) \end{aligned} \tag{2}$$

where β_o may either be identified along with the other parameters or fixed a priori. If $C(q^{-1}) = 1$ in Eq. 1 then $\varepsilon(t)$ in Eq. 2 is a moving average of order k.

The practicality of the separation method depends strongly on the simplicity of the control law calculation. The control law is calculated according to the latest parameter estimates which implies a new calculation at every sampling instant. Eq. 2 affords a very simple means of calculating the minimum variance control law, if $C(q^{-1}) = 1$ in Eq. 1:

$$u(t) = \frac{-1}{\beta_o}[\alpha_1 y(t) +\ldots+ \alpha_n y(t-n) - \beta_1 u(t-1)..-\beta_{n+k}u(t-n-k)] \tag{3}$$

This control law is stabilizing if the system is minimum phase. Several recourses are available if it is not. We can solve the Riccati equation, probably not as formidable a job as it may seem, but nevertheless cumbersome. The solution can also be obtained by spectral factorization, as in [11]. In [12], Wittenmark suggests that the problem be circumvented simply by increasing the delay k.

Clarke and Gawthrop [13] suggested a generalized performance index

$$J = E\{[P(q^{-1})y(t+k) - R(q^{-1})W(t)]^2 + [Q(q^{-1})u(t)]^2\} \tag{4}$$

where P, R and Q are polynomials in q^{-1} and w(t) is a command input. If P = 1, R = 0 and Q = ρ, then it is shown that, for an open-loop stable system, the control law is stable for ρ sufficiently large.

In all applications so far, the identification appears to have been carried out by the method of recursive least squares, i.e., assuming $C(q^{-1}) = 1$. A discount factor (or forgetting factor) λ is usually employed, so that the identification minimizes

$$I(t) = \sum_{i=0}^{t} \lambda^{t-i} \varepsilon^2(i) \tag{5}$$

The practical effect of using $\lambda < 1$ is to allow tracking of the parameters: if $\lambda < 1$, the identifier "closes up" asymptotically, and the estimates are frozen.

Since separation is really an ad hoc scheme, it is necessary to study convergence. Astrom and Wittenmark [14] established the rather remarkable result that, if the STR converges at all, the parameters converge to values such that

$$\begin{aligned} E[y(t+\tau)y(t)] &= 0, \quad \tau = k, k+1, ..k+n-1 \\ E[y(t+\tau)u(t)] &= 0, \quad \tau = k, k+1, ..2k+n \end{aligned} \tag{6}$$

which guarantees minimum variance control.

In [15], Ljung further analyzed the convergence, with reference to a set of auxiliary differential equations. Basically, it is known [1] that: i) if the model and system have the same structure, then the true parameter values are a stable convergence point; ii) if the structure is correct and $C(q^{-1}) = 1$, the convergence is global; iii) if $C(q^{-1}) = 1$ in the model and $1 + c_1 q^{-1} + c_n q^{-n}$ in the true system, other polynomials having the same structure, the minimum variance control law is a stable convergence point if $C(\lambda_i) > 0$, where λ_i are the zeros of $z^n B(z^{-1})$.

4. PRACTICAL CONSIDERATIONS

In any real application, there are practical issues that must be addressed to use the theory. We shall review some of these considerations here, with particular emphasis on the STR, since it has in fact been used more often in process control than other algorithms.

The first decision concerns model structure, including the delay. The latter is usually known, at least in cases where the sampling period is not too small; in any case, it may be overestimated but should under no circumstances be underestimated.

The choice of model order and structure depends on the identification method to be used. If the least squares method is used, we must contend with the possible lack of convergence due to the fact that $C(q^{-1})$ in Eq. 1 is not 1. Since the polynomial $C(q^{-1})$ may be approximated by $1/H(q^{-1})$ where $H(q^{-1})$ is another polynomial, Eq. 1 becomes

$$A(q^{-1})H(q^{-1})Y(t) = B(q^{-1})H(q^{-1})u(t-k) + e(t) \tag{7}$$

which is compatible with the least squares scheme, at the expense of higher order. If the recursive maximum likelihood method is used, the C polynomial is, of course, part of the model. The model may then have fewer parameters than that of Eq. 7, at the expense of greater complexity in the identification and the control calculations. To our knowledge, the structure of Eq. 7 has been used in all applications so far.

In process control applications, the structure must provide for integrating, or "reset" action; process outputs are regulated to some set-point value, and it is usually necessary for the output to equal the set-point. The Clarke and Gawthrop scheme [13] can be used with a performance index

$$J = E\{Y^2(t+k) + \lambda[u(t)-u(t-1)]^2\} \tag{8}$$

This leads to integration, but appears not to have been used in applications. In [16], Borrison and Syding used $\Delta u(t) = u(t) - u(t-1)$ as the input to their model. The self-tuning control law gives $\Delta u(t)$ as a function of $y(t)$ and past values of y and Δu: it is a "velocity algorithm", and hence introduces integration in a manner that is quite standard in process control.

It is usually necessary to include limits on the control, for theoretical and practical reasons. In process control, limits should also be set on the control increments. The setting of hard limits avoids the frequent STR effect of large initial control values.

The discount factor λ is usually set to some value ranging between .95 and .99. In [16], λ = .95 was used for startup, to speed up the initial transients of the parameter estimates, and increased to .99 in the steady-state. Even lower values have been used: Dumont and Bélanger [17] used λ = .70 to track changing parameters during a grade change on a kiln.

Several schemes have been used for STR algorithm startup. The simplest is to make all initial estimates zero and to use αI as the initial covariance matrix in the least squares algorithm:

This was done in [16], with $\alpha = 10$. The next method consists in using initial parameter estimates leading to a control law equivalent to that already in use. For example, the control law for a second-order plant with minimum variance control would be (see Eq. 3)

$$u(t) = -\frac{\alpha_1}{\beta_0} y(t) - \frac{\alpha_2}{\beta_0} y(t-1) + \frac{\beta_1}{\beta_0} u(t-1) + \frac{\beta_2}{\beta_2} u(t-2) \tag{9}$$

Suppose a PI controller was used, with

$$u(t) = -k_1 y(t) - k_2 y(t-1) + u(t-1) \tag{10}$$

Smooth transition from the PI controller is obtained by using the initial values

$$\hat{\alpha}_1(0) = \beta_0 k_1,\ \hat{\alpha}_2(0) = \beta_0 k_2,\ \hat{\beta}_1(0) = \beta_0,\ \hat{\beta}_2(0) = 0 \tag{11}$$

This startup method was used by Cegrell and Hedqvist [18].

Finally, the parameter estimates and the covariance matrix may be initialized by a trial period of straight identification. This was the course chosen by Sastry, Seborg and Wood [19]. If the identification stage takes place under fixed linear feedback control (as was the case in [19]), care must be taken to ensure identifiability of the parameters. Basically, the error-to-output transfer function must be in one-to-one correspondence with the parameters to be identified. A simple criterion for identifiability is given in [20].

The implementation of the STR on a minicomputer is relatively straightforward: Borisson and Syding [16] reported about 35 FORTRAN statements as the length of the program, with an execution time of 20 msec on a PDP-15 for 7 parameters. Some care must be taken to maintain the positive definiteness of the covariance matrix, especially if $\lambda = 1$: failure to do this will drive the algorithm into instability. Perhaps the best way to accomplish this is to use the square-root or UD factorization methods developed for on-line Kalman filtering [21],[22].

5. APPLICATIONS

In this section, we shall give an overview of the results of some applications of adaptive control to industrial processes.

The STR application to the control of moisture on a paper machine (Fig. 1) in [16] was partly a multivariable application, in that the controlled variable (moisture) is affected by two inputs (steam flow and thick stock flow). The other output (dry basis weight) depends only on one of the inputs (thick stock flow): the system can be said to have triangular structure. Moisture control is by steam flow, with thick stock flow used in the STR as a feed-forward variable. Six parameters were used, and good control was obtained in about 12 samples. The control in the steady state was not too different from that given by a well-tuned PI controller. That, in fact, is better than it sounds: paper machine dynamics vary considerably from grade to grade, and controllers are normally loosely tuned to avoid re-tuning at each grade change.

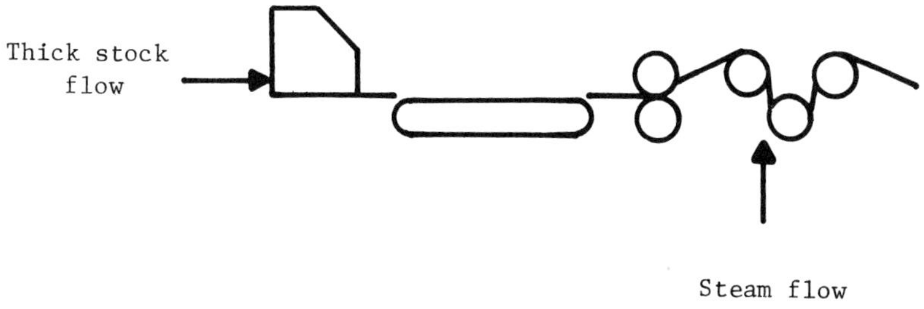

FIGURE 1. Paper machine.

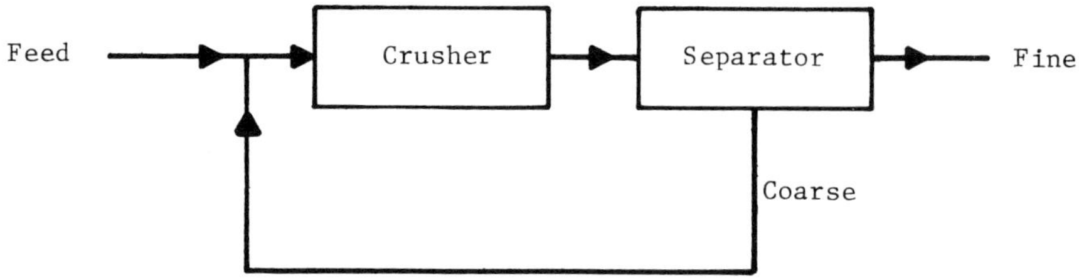

FIGURE 2. Ore crusher.

A second application was presented in [16], where the STR was applied to an ore crusher. The process is shown in Fig. 2. The basic problem is that the ability of the crusher to grind decreases as the feed is increased; more material comes back from the separator to increase the input flow further, and so on until the crusher "plugs up". The trick is to adjust the feed to keep the crusher power at a given level. The proper feed level varies with the grindability of the incoming material, which also affects the dynamics. A seven-parameter STR was used, with the feed as input and the power level as output. Satisfactory control was achieved in 20 samples. Appreciable economic gains were realized due to the ability to hold a power set point closer to the stability limit, i.e. 200 kw rather than 170 kw.

A third STR application [19] was concerned with a binary distillation column (Fig. 3). Steam flow was held fixed, and the top composition was regulated in the face of feed disturbances by manipulating the reflex ratio. Due to the nonlinear behaviour of the column, the best tuning of a PI controller is different for increases and decreases in feed. A four-parameter STR proved adept at handling both.

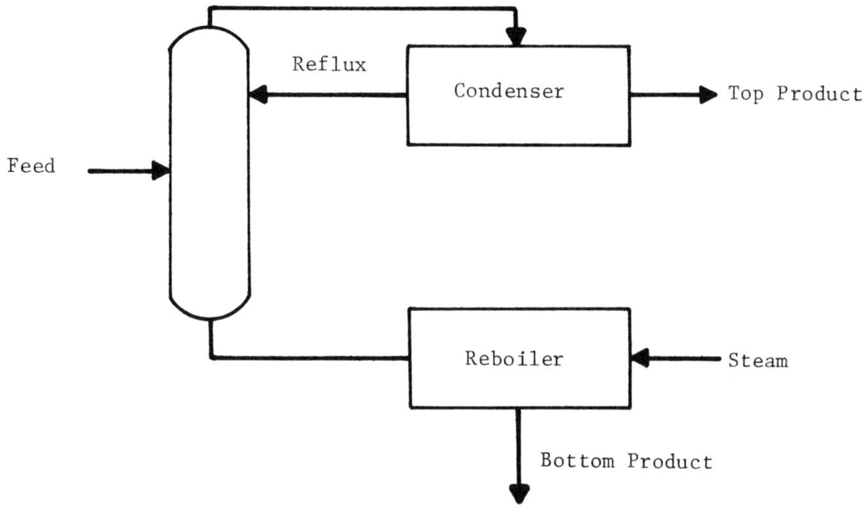

FIGURE 3. Distillation column.

In [17], the problem was to control a T_iO_2 kiln (Fig. 4), where the natural anatase form is converted into rutile. Product samples were taken every half hour and analyzed in the laboratory for rutile composition. The manipulated variable was the fuel. One novel element was the use of the STR algorithm to control a plant already controlled by a fixed regulator (Fig. 5). This possibility was raised by Wittenmark [12], and provides the STR with a backup controller. If the fixed controller contains integration, the STR does not need to address the problem of steady-state offset.

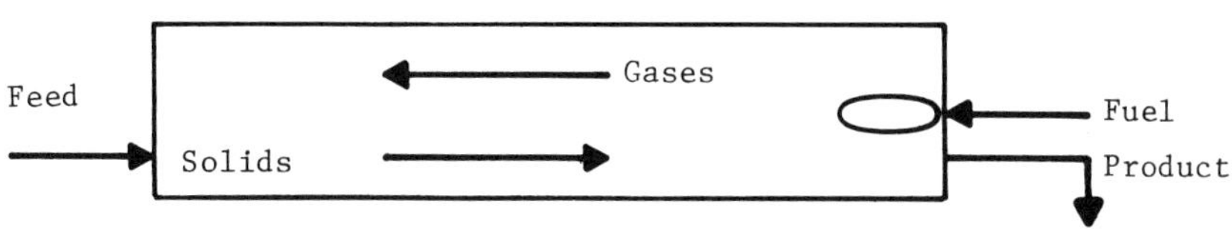

FIGURE 4. Kiln.

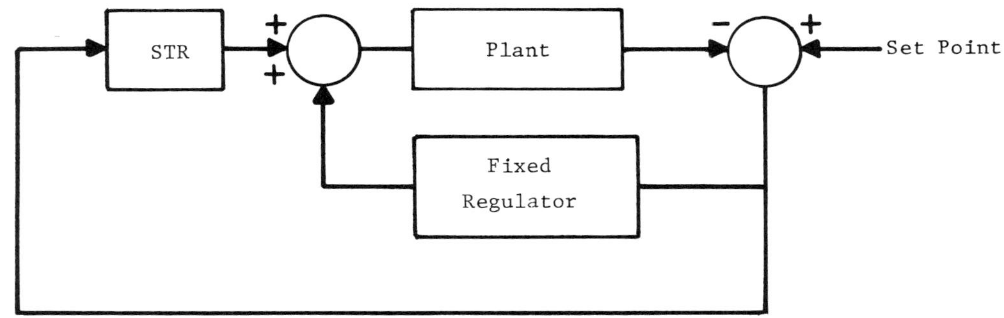

FIGURE 5. Combined STR + fixed regulator scheme.

Because of memory limitations in the process computer, only a two-parameter STR could be accommodated. The structure was chosen by analyzing the inner loop at nominal. Since the structure did not match that of the known plant model, the stability of the convergence point was analyzed by using Ljung's equation, for a range of values of plant gains and delays.

The fixed regulator had raised the in-spec percentage (percent of the time when the composition is within a desired range) from 68% to 75%: the STR raised it further to 85%. Tracking of the parameters was done during a grade change, with the discount factor $\lambda = .70$. The period of production of unusable product, normally about 10-12 hours, was cut down to 3 hours.

The model-reference scheme of [27] was applied to the same binary distillation column as in [19], with similar results: adaptive control successfully circumvented the asymmetric, nonlinear column behaviour. In addition, feedforward adaptive control was used here, with the feed acting as the disturbance variable; this proved quite successful at minimizing perturbation. Multivariable control was also used, with top and bottom compositions as outputs and reflex and steam as inputs. The model-reference system did reduce the loop interaction by using decoupled reference systems. We note that the column is rather quiet in the absence of feed perturbation: this means that the variables causing most of the variations can, in fact, be measured and included in the model.

6. DIFFICULTIES

Most of the difficulties standing in the way of more widespread applications of adaptive control are nontechnical.

As was stated in Section 2, it is not always easy to give clear-cut economics estimates of control improvement. Some of the benefits may be estimated from a guess in the reduction of the output variance, which may allow operation at a more economical steady-state level. For example, improved moisture control on a paper machine will allow the moisture set point to be increased without incurring a higher probability of excess moisture; this will enable the paper maker to sell the customer more water, which is cheaper than selling him fiber. In general, the only assessable benefits are those resulting from changes in the operating point: direct benefits can be ascribed to improved control if the steady-state can be changed.

There are other benefits due to control, but these may not be as direct. For example, better regulation of a paper machine may permit increased throughput and fewer paper breaks, but the improvements cannot be predicted.

The second, and perhaps more serious, difficulty is the feeling in much of the industry that what matters about control are the instrumentation, the process computer, the software and the operator interface. What academics know as control is put in a small box labelled "algorithms". That box is not considered to be too important: the feeling is that it will always be possible to tune a PID algorithm and obtain satisfactory control. This means that the personnel may include people knowledgeable in instrumentation, computer hardware and real-time software but often will not include an expert in control theory. This means that the universities have a specially heavy role to play, in education as well as in technical participation.

7. FUTURE RESEARCH

Several areas appear natural for further research. In self-tuning control, it is still not clear whether to use a model structure with or without the moving-average disturbance. There are still unanswered questions concerning non-minimum phase applications: the Clarke and Gawthrop scheme affords a simple solution, if one can guess the right control weight. Is it possible to adapt that weight on-line?

The self-tuning of multivariable regulators is another area that needs work. In a recent paper, Borisson [23] presented a STR using a minimum-variance control law. Other attempts (e.g. [24]) have been guided by ad-hoc reasoning. From the estimation or convergence points of view, there is nothing different about the multivariable case. The difficulty is to derive a simple control law, especially for the case where the controls do not have the same delay. We note that in [23], the controls did have the same delay. It is possible that an extension of the Clarke-Gawthrop work or a spectral factorization approach will lead to some kind of solution of this problem.

We may also suggest that the STR be used to tune simple controllers, e.g. PI or PID algorithms. A successful solution of this problem would be of great psychological value in the process industries. One approach could be to approximate the more complex controller obtained by the STR in the steady state by a simpler one [25]. As another approach, we may pose an inverse problem: for what class of plants will a STR leading to a given controller structure have stable convergence points?

For reasons given above, we do not think that the model reference approach will be as successful as the separation approach in process control. It may still be used as an identification method, to produce parameter estimates.

Finally, there are certain cases where a system can be modelled, with a few parameters not known. The wet end of a paper machine is an example of this situation: the unknown parameters would be the fiber retention coefficient of the wire and the effective volume (for mixing purposes) of the white water chest. In a recent paper by Ljung [26], modifications were introduced to the extended Kalman filter to guarantee convergence. It would seem natural to try to use this in an adaptive control loop.

8. CONCLUSIONS

It would appear that adaptive control is about ready for process control. For the next few years, most of the work will probably be done in collaboration with university personnel: There is still too much to be learned before adaptive control can be considered a routine industrial technique.

The advent of microprocessors should help the spread of adaptive control, since it is now possible to carry out experiments without interfering with the main process computer. The technique has come of age, and so have the tools to apply it.

REFERENCES

1. Astrom, K. S., Borisson, V., Ljung, L., and Wittenmark, B., "Theory and Applications of Self-Tuning Regulators," Automatica, Vol. 13, pp. 457-476, 1977.
2. Landau, I. D., "A Survey of Model Reference Adaptive Techniques - Theory and Applications," Automatica, Vol. 10, pp. 353-380, 1974.
3. Feuer, A., and Morse, A. S., "Adaptive Control of Single-Input, Single-Output Linear Systems," IEEE Trans. on Automatic Control, Vol. AC-23, pp. 557-569, 1978.
4. Narendra, K. S., and Valavani, L. S., "Stable Adaptive Controller Design - Direct Control," IEEE Trans. on Automatic Control, Vol. AC-23, pp. 570-582, 1978.

5. Landau, I. D., "Unbiased Recursive Identification Using Model Reference Adaptive Techniques," IEEE Trans. on Automatic Control, Vol. AC-21, pp. 194-202, 1976.

6. Ionescu, T., and Monopoli, R. V., "Discrete Model Reference Adaptive Control Systems with an Augmented Error Signal," Automatica, Vol. 13, pp. 507-517, Sept. 1977.

7. Lin, Y.-H., and Narendra, K. S., "A New Error Model for Discrete Systems and its Application to Adaptive Identification and Control," S & I S Report No. 7802, Yale University, Oct. 1978 (to appear in IEEE TAC).

8. Narendra, K. S., and Lin, Y.-H., "Stable Discrete Adaptive Control," S & I S Report No. 7901, Yale University, March 1979 (to appear in IEEE TAC).

9. Goodwin, C. C., Ramadge, P. J., and Caines, P. E., "Discrete Time Multivariable Adaptive Control," Technical Report, Harvard University, Nov. 1978 (to appear in IEEE TAC).

10. Martin-Sanchez, J. M., "A General Solution to the Adaptive Control of Linear Time-Variant Processes with Time Delays. Part II: Multivariable Control of a Binary Distillation Column," submitted manuscript, Dept. of Chemical Eng., U. of Alberta.

11. Peterka, V., "On the Steady State Minimum Variance Control Strategy," Kybernetica, Vol. 8, pp. 219-231, 1972.

12. Wittenmark, B., "Self-Tuning Regulators," Report TFRT-3054, Dept. of Automatic Control, Lund Inst. of Technology, 1973.

13. Clarke, D. W., and Gawthrop, P. J., "Self-Tuning Controller," Proc. IEEE, Vol. 122, pp. 929-934, 1975.

14. Astrom, K. J., and Wittenmark, B., "On Self-Tuning Regulators," Automatica, Vol. 9, pp. 185-200, 1973.

15. Ljung, L., "Analysis of Recursive Stochastic Algorithms," IEEE Trans. on Automatic Control, Vol. AC-22, pp. 551-575, 1977.

16. Borisson, V., and Syding, R., "Self-Tuning Control of an Ore Crusher," Automatica, Vol. 12, pp. 1-8, 1976.

17. Dumont, G. A., and Bélanger, P. R., "Self-Tuning Control of a Titanium Dioxide Kiln," IEEE Trans. on Automatic Control, Vol. AC-23, pp. 532-537, 1978.

18. Cegell, T., and Hedqvist, T., "Successful Adaptive Control of Paper Machines," Automatica, Vol. 11, pp. 53-60, 1975.

19. Sastry, V. A., Seborg, D. E., and Wood, R. K., "Self-Tuning Regulator Applied to a Binary Distillation Column," Automatica, Vol. 13, pp. 417-424, 1977.

20. Soderstrom, T., Ljung, L., and Gustavsson, I., "Identifiability Conditions for Linear Multivariable Systems Operating Under Feedback," IEEE Trans. on Automatic Control, Vol. AC-21, pp. 837-840, 1976.

21. Bierman, G. J., "Factorization Methods for Discrete Sequential Estimation," Academic Press, New York, 1977.

22. Thornton, C. L., and Bierman, G. J., "Filtering and Error Analysis via the UDU^T Covariance Factorization," IEEE Trans. on Automatic Control, Vol. AC-23, pp. 901-906, 1978.

23. Borisson, V., "Self-Tuning Regulator for a Class of Multivariable System," Aut. 15, pp. 209-216, 1979.

24. Keviczky, L., Hetthessy, J., Helga, M., and Kolostovi, J., "Self-Tuning Adaptive Control of Cement Raw Material Blending," Automatica, Vol. 14, pp. 525-532, 1978.

25. El-Attar, R. A., and Vidyasagar, M., "Order Reductions by l_1- and l_∞-norm Minimization," IEEE Trans. on Automatic Control, Vol. AC-23, pp. 731-734, 1978.

26. Ljung, L., "Asymptotic Behaviour of the Extended Kalman Filter as a Parameter Estimator for Linear Systems," IEEE Trans. on Automatic Control, Vol. AC-24, pp. 36-50, 1979.

27. Martin-Sanchez, J. M., "A General Solution to the Adaptive Control of Linear Time-Variant Processes with Time Delays. Part I: Theory," submitted manuscript, Dept. of Chemical Eng., U. of Alberta.

ADAPTIVE LEVEL CONTROL IN A GLASS FURNACE

R. Haber

J. Hetthéssy

L. Keviczky

I. Vajk

A. Kovács

Department of Automation
Technical University of Budapest, Hungary

Z. Csaszar

N. Czeiner

A. Feher

A. Turi

Research Group for Automation Control
Res. and Des. Inst. for Silicate Ind.
Budapest, Hungary

The paper deals with the self tuning control of the level of a glass furnace. First the furnace was analyzed and a control model was built up based on measured data. The most important parameter, the glass level was controlled adaptively. The prepara-

ISBN 0-12-514060-6

tions, as well as the hardware and software tools are reviewed. The control algorithm is written in detail. Finally, the experimental results are presented.

INTRODUCTION

The Department of Automation has been dealing with self-tuning control for a long time. We simulated the known algorithms and developed new variants of them. The Central Research and Design Institute for Silicate Industry entrusted us to build a control model of the second furnace of the Bellicd Glass Factory in Oroshaza (Hungary). Having finished the work there was a common interest to replace at least the most important traditional control loop of the furnace by an adaptive one. There was a need for the adaptive control because the fluctuation of the level disturbs the production very much, since the feeding is gravimetric and the gob-weight depends on the level. The paper introduces the glass furnace, the preparations of the adaptive control, the hardware and software tools, the applied algorithm and the experiments.

THE GLASS FURNACE

The glass furnace is a continuously operating tank furnace (Fig. 1). The raw material is fed into the top of the tank. The melted glass arrives at the working end due to its natural stream. The feeders are connected to the working tank. The feeders batch the glass in the form of gobs to the blow machine. The furnace is heated in turn from the right and the left side.

The glass level is measured by means of a pneumatic blower, while the feeding is taken place by shovels controlled by an excenter position. The quality of the glass depends on the gob

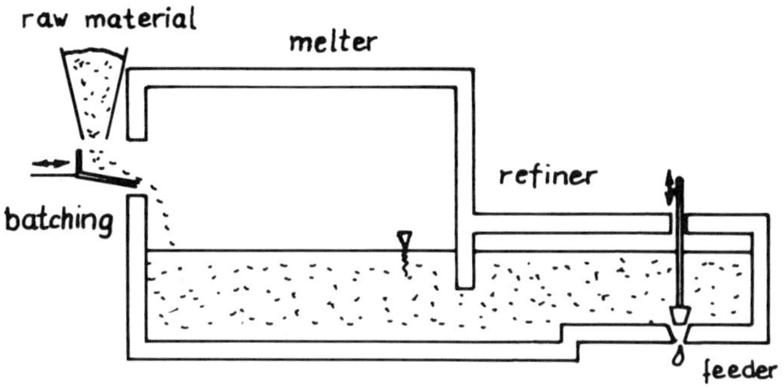

FIGURE 1. Glass furnace.

size, which is a function of the glass level because of the gravimetric feeding. The gob size depends on the viscosity too, which can be controlled through the temperature.

In order to ensure smooth production, the level, the temperature and the pressure in the furnace were controlled. The following traditional PI control loops were set up

- control of the level through feeding.
- control of the pressure through the flue gas valve.
- control of the temperature through the fuel gas quantity.

The furnace was heated by a mixture of air and natural gases of different caloric values. Thus the composition of the fuel gas was regulated, too. Fig. 2 shows the control loops applied to the furnace.

PRELIMINARY INVESTIGATIONS

Before designing the adaptive control system, preliminary investigations were made to determine which control signals have a basic influence on the level. Therefore a three-variable model of the furnace was set up [1]. The input signals of the model are:

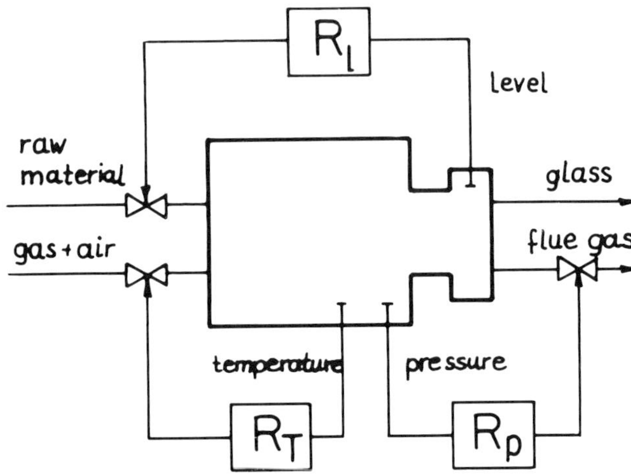

FIGURE 2. Traditional control loops.

1. batching
2. Quantity of the fuel gas
3. quantity of the flue gas

and the output signals are:

1. glass level
2. temperature
3. pressure.

In consequence with physical considerations, the following control loops were registered:

1. batching - glass level
2. quantity of the fuel gas - temperature
3. quantity of the flue gas - pressure.

Only linearized models about an operating point were determined.

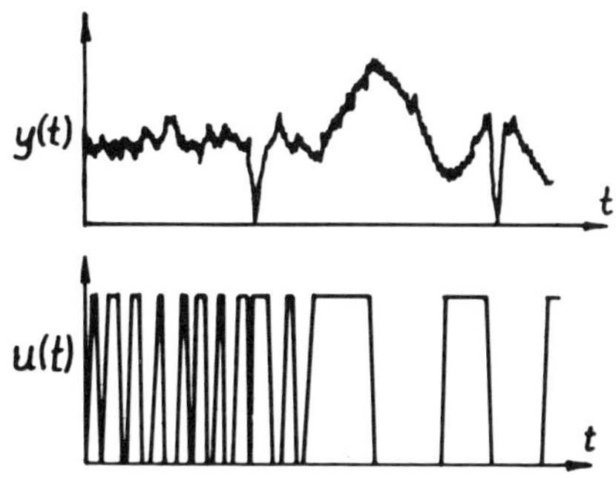

FIGURE 3. Batching and level records.

The records of batching (u(t)) and glass level (y(t)) are seen in Fig. 3. The batching varies between two values and the level decreases or increases almost inearly. The negative peaks arise from the change in firing. The batching/level model is free of delay time, is of the integrating type and has only a few memory terms. The identification procedure led to a third order difference equation which has no continuous equivalent [1]. First of all only the batching/level loop was set up to be replaced by an adaptive DDC (Direct Digital Control) one. The applied algorithm was a version of the "classical" minimum variance law for processes with integrator(s).

THE CONTROL ALGORITHM

Let us consider a linear SISO model described by the equation:

$$A(z^{-1})y(t) = B(z^{-1})u(t-d) + \lambda C(z^{-1})e(t) \tag{1}$$

where

$$
\begin{aligned}
A(z^{-1}) &= 1 + a_1 z^{-1} + \ldots + a_n z^{-n} \\
B(z^{-1}) &= b_o + b_1 z^{-1} + \ldots + b_m z^{-m} \\
C(z^{-1}) &= 1 + c_1 z^{-1} + \ldots + c_k z^{-k},
\end{aligned}
\tag{2}
$$

t is the discrete time, u(t) is the input signal, d the dead time and e(t) is white noise with variance λ.

The aim of the minimum variance control is to ensure the minimum of the functional

$$
V = \min_{u(t)} [y(t + d) - y_r]^2 \tag{3}
$$

where y_r is the reference value of the output signal. The optimal input signal can be computed as follows:

$$
u(t) = \frac{C(z^{-1})y_r - G(z^{-1})y(t)}{B(z^{-1})F(z^{-1})} \tag{4}
$$

Here

$$
C(z^{-1}) = A(z^{-1})F(z^{-1}) + z^{-d}G(z^{-1}) \tag{5}
$$

and

$$
\begin{aligned}
G(z^{-1}) &= g_o + g_1 z^{-1} + \ldots + g_{n-1} z^{1-n} \\
F(z^{-1}) &= 1 + f_1 z^{-1} + \ldots + f_{d-1} z^{1-d}.
\end{aligned}
\tag{6}
$$

By substituting Eqn. 5 to Eqn. 4 the control law can be expressed by the formula [2]:

$$
u(t) = \frac{A(z^{-1})}{B(z^{-1})} y_r - \frac{G(z^{-1})}{B(z^{-1})F(z^{-1})} [y(t) - y_r]. \tag{7}
$$

This form corresponds to the traditional control algorithms; the control signal is a weighted linear combination of the reference signal and the control error. In the case of an integrating plant and a constant reference value the first term of Eqn. 7 is zero, as for y_r = const.

$$y_r A(z^{-1})/B(z^{-1}) = y_r A(1)/B(1) = 0$$

and the control law is:

$$u(t) = \frac{G(z^{-1})}{B(z^{-1})F(z^{-1})} [y(t) - y_r]. \tag{8}$$

In the case of constant but unknown system parameters one can apply the self-tuning principle. The optimal input signal can be calculated by

$$u(t) = -\underline{\hat{p}}^T(t)\underline{x}(t)/\hat{q}_o \tag{9}$$

where $\underline{p}$ is the parameter of the regulator

$$\underline{p}^T = [q_1, q_2, \ldots, q_{m+d-1}, p_o, \ldots, p_{n-1}], \tag{10}$$

$\underline{x}$ is the vector defined as

$$\underline{x}^T(t) = [u(t-1), u(t-2), \ldots, u(t-m-d+1), r(t), \ldots, r(t-n+1)]; \tag{11}$$

and r(t) is the control error

$$r(t) = y(t) - y_r. \tag{12}$$

In Eqn. 10 $\hat{q}_o$ stands for a constant scale factor, which can be chosen in a wide region [3] in advance.

Using the relations of the recursive least squares estimation $\hat{\underline{p}}(t)$ can be derived from $\hat{\underline{p}}_{(t-1)}$ by the recursive relation;

$$\hat{\underline{p}}(t) = \hat{\underline{p}}(t-1) + \underline{\underline{R}}(t)[r(t) - \hat{q}_o u(t-d) - \hat{\underline{p}}^T(t)\underline{x}(t-d)]\underline{x}(t-d) \quad (13)$$

where

$$\underline{\underline{R}}(t) = \frac{1}{w^2}[\underline{\underline{R}}(t-1) - \frac{[\underline{\underline{R}}(t-1)\underline{x}(t-d)][\underline{\underline{R}}(t-1)\underline{x}(t-d)]^T}{w^2 + \underline{x}^T(t-d)\underline{\underline{R}}(t-1)\underline{x}(t-d)} \quad (14)$$

In Eqn. 14 w is the forgetting factor.

THE CONTROL TOOLS

For data logging, identification and control tasks a so-called Portable Process Computer Laboratory (PPCL) was used. It consists of a special trailer equipped with a small computer TPAi (∿PDP8) and CAMAC process peripherals. The work of the operators is simplified through an alphanumeric display and floppy discs [3].

The PPCL is supplied by the Process Control Program passage - MERCEDES. It contains three subsystems:

1. RTP - Real-Time Process Control Program System.
2. OLIP - Off-Line Identification Program System.
3. OCAD - Off-Line Computer Aided Design Program System.

MERCEDES is a user oriented interactive program having graphical/alphanumeric display/outputs [3].

The feeding of the raw material takes place by means of two feeders with pusher-type shovels. The rate of feeding depends on the stroke length of the shovels determined by the position of an excenter. The position of the excenter was divided into 4096 bits.

Level measurements were very noisy due to inaccuracies of the transducer which was used, a pneumatic pipe. In order to eliminate high frequency fluctuation in the control multiple sampling (8 samples between two control steps) was used. Samples between two control actions were filtered by the low-pass filter:

$$W_F(z^{-1}) = \frac{25 + 19z^{-1} + 14z^{-2} + 9z^{-3} + 7z^{-4} + 5z^{-5} + 3z^{-6} + z^{-7}}{83} \tag{15}$$

ADAPTIVE CONTROL OF THE FURNACE

Based on the identified model structure adaptive control experiments were executed. The best performance was achieved by the parameters

$$d = 1$$

$$m = 2$$

$$n = 3$$

$$T_{int} = 16 \text{ sec}$$

$$T_{samp} = 2 \text{ sec}$$

Here T_{int} denotes the time interval between two control actions. The initial value of the parameter vector - based on previous experiments - was:

$$\underline{p}_o^T = [-0.658, -0.042, 540.7, - 263{,}8, - 243{,}63].$$

The forgetting factor was

$$w = 0.99$$

and the weighting matrix was started from a unity matrix. After 210 steps (56 minutes) the parameter vector tuned to

$$\underline{p}^T = [0.67, 0.036, 525, -262, -244].$$

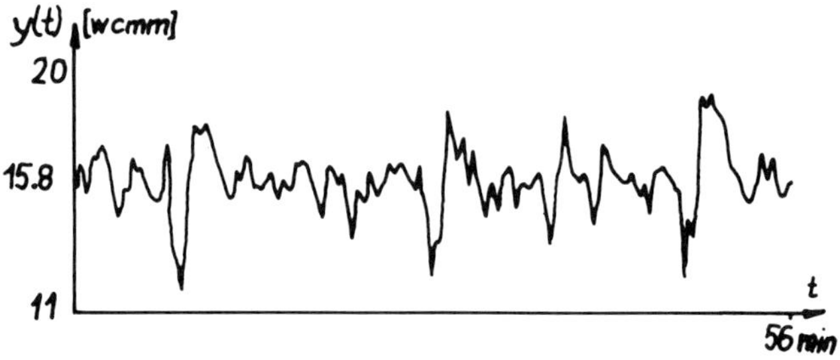

FIGURE 4. Traditional control.

Fig. 4 shows the traditional, Fig. 5 the adaptive control of the level. In Fig. 5 $y_F(t)$ stands for the filtered output:

$$y_F(t) = W_F(z^{-1})y(t).$$

(The level is scaled in Water Column MM as it is measured by a pneumatic pipe.)

It can be seen that the adaptive control ensured much smaller variance around the reference value y_r = 14.65 WCMM.

CONCLUSIONS

As is seen in Figs. 4 and 5, fluctuation of the glass level can be decreased by the adaptive control. In addition the operation of the furnace became smoother, the distribution of the mixture on the surface of the melt became more uniform and the fluctuation of the vault temperature decreased.

This control experiment had other advantages, as well. It was one of the first - maybe the first - industrial adaptive control application in Hungary. We hope that the common team of the Central

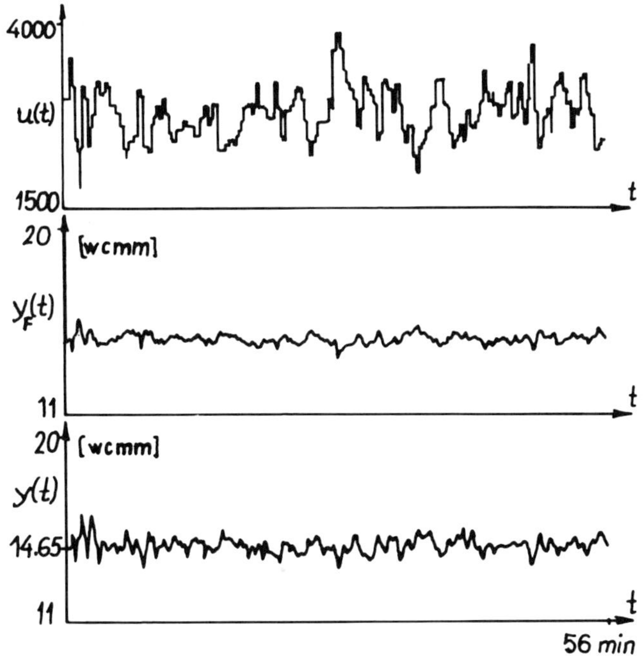

FIGURE 5. Adaptive control.

Research and Design Institute for Silicate Industry and the Department of Automation of the TU Budapest can solve further industrial problems applying up-to-date adaptive DDC algorithms. The Portable Process Computer Laboratory may be a good basis for it. In the future it may be that the control algorithm is realized by a microprocessor.

REFERENCES

1. Haber, R., and Co-workers, "Identification and Adaptive Control of a Glass Furnace by a Portable Process Computer Laboratory," Case Study, IFAC, Darmstadt, 1979.

2. Wittenmark, B., "A Self-Tuning Regulator," Report 7311, Division of Automatic Control, Lund Institute of Technology, 1973.
3. Feher, A., and Co-workers, "MERCEDES - Interactive Software Package for Identification and Control," IFAC, Prague, 1979.

EXPERIMENTAL EVALUATION OF HYPERSTABLE MODEL REFERENCE ADAPTIVE CONTROL[1]

R. M. Johnstone

S. L. Shah

D. G. Fisher

R. Wan

Department of Chemical Engineering
University of Alberta
Edmonton, Alberta, Canada

An experimental evaluation of two hyperstable model reference adaptive systems on a pilot plant is reported. The hyperstable single-input, single-output direct adaptive control techniques considered in this application rely only on plant input and output information. The performance of these two adaptive techniques is compared to conventional PID control of the same pilot plant. The pilot plant as used in this evaluation although quite simple has non-linear, time-varying characteristics. The adaptive control

[1]Financial assistance from the Natural Sciences and Engineering Research Council of Canada and Imperial Oil Enterprises Limited is gratefully acknowledged.

ISBN 0-12-514060-6

techniques are demonstrated to have robust and superior tracking and regulatory control properties compared to the classical PI control.

1. INTRODUCTION

The most challenging chemical process control problems involve tracking and regulation of complex, often poorly understood non-linear processes with time-varying parameters, significant pure time delays, and unknown external disturbances. These, and other characteristics led Foss [1] to suggest in his critique of process control theory, that most of the classical control theories are inadequate for the problems of chemical process control. Specifically, our own experience with control of process systems has shown that in many instances, state-space description of non-linear plants with time-delays is both difficult and inadequate. In many cases a dynamic model of the process is simply not available and hence many modern control techniques can not be applied directly. Therefore, direct or indirect adaptive control techniques that can control single-input, single-output, unknown, systems based on only input-output measurements offer a very attractive solution for a significant subset of process control problems.

In this paper we present results from an experimental application of two direct, hyperstable model reference adaptive control techniques to a pilot plant unit. Experimental evaluation of these two techniques is then compared to the conventional PID control of the same pilot plant. One of the direct adaptive control techniques as developed by Shah et. al. [2] is an extension to the input-output formulation of the MRAC technique first considered by Landau and Courtiol [3]. The second direct adaptive technique is based on a modification of the augmented error approach of Monopoli as reported by Suzuki and Takashima [4]. In a more recent paper this augmented error approach has been extended by Johnstone et. al. [5] to permit application to nonminimum phase systems. The concern for input and output boundedness in the stability of adaptive systems has been a subject of great interest. In this practical application and most others, bounds on input and output variables are a natural result of physical limits on the range of actual plant operations. Such limits are assumed by many approaches to stability analysis of adaptive systems. In summary, from a practical viewpoint the apriori assumption of input-output boundedness is entirely acceptable since in general both are bounded by virtue of physical constraints on the plant. Besides the problem of hard input or control constraints, in many practical implementations there arises a need for having incremental control bounds. In addition stability of the adaptive system needs to be assured in the presence of unmeasurable disturbances. These are points of great practical interest. In the present paper, the two MRAC techniques are experimentally evaluated in such an environment, i.e., in the presence of hard control constraints on inputs, noisy process measurements and/or unmeasurable disturbances.

2. DESCRIPTION OF THE PROCESS

The pilot plant unit used in this work consists of two 10 litre tanks in series interconnected by a variable resistance and a third linear tank to take up some of the flow out of tank 2 via a variable resistance. A simplified schematic of the process is shown in Figure 1. Tank 1 includes a tapered bottom section 15 cm long with the maximum and minimum diameter of 20 cm and 2.5 cm respectively. The liquid level in tank 1 is controlled by manipulating the flow inlet into tank 1. To best demonstrate the capability and performance of the adaptive and PID control algorithms, it was decided to control the level in tank 1 in both linear and non-linear

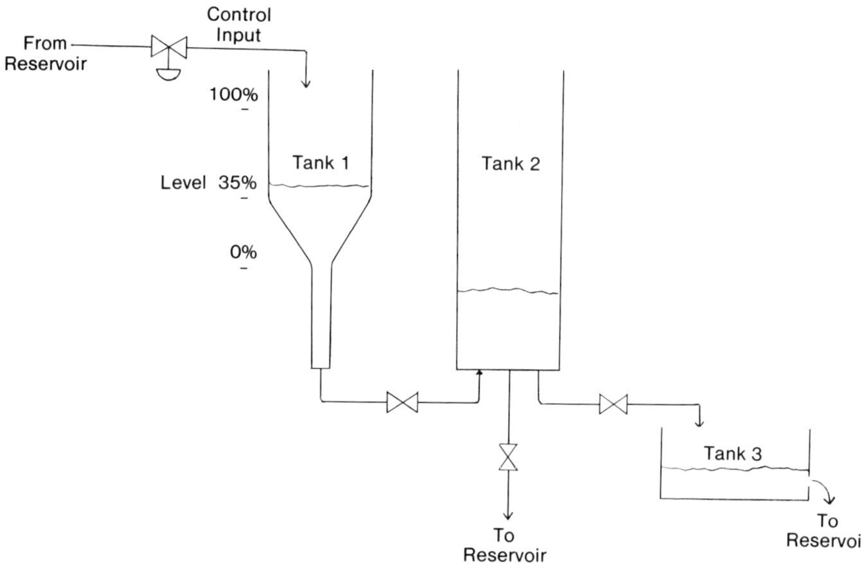

FIGURE 1. A simplified schematic of the pilot plant instrument trainer.

regions. The extent of the non-linearity and time-varying nature of this tank can best be appreciated by noting that the gain in the 15 cm non-linear region changes in the ratio of 64:1. This is indeed a significant amount of variation and makes the task of controlling level in the non-linear region non-trivial. This essentially represents a first-order process with a changing time-constant and gain. However, this is not the only non-linearity in this process. In addition, there are the usual non-linear control valve characteristics, changing line resistances and a variable hold-up in the second tank. This latter phenomenon contributed significantly to the non-linearities in the time-constant of the process. The level transducer for tank 1 uses a bubble system to simulate realistic industrial measurements. At high air flow rates, the bubbles in tank 1 and the inlet flow stream contribute considerable process noise and affect the signal to noise ratio of the level measurements in a non-linear fashion.

The process is equipped with commercial measurement transducers and final control elements of a type widely used in industry. It is interfaced to a distributed network of 3 HP1000 computers via an HP-MACS (measurement and control) system. A sampling time of 4 secs was used to control the unit.

The pilot plant unit used for this experimental work is simple, but is typical of many industrial units, and is similar to the unit selected by Taylor [6] to demonstrate the performance of their "adaptive" PID controller. Some similar non-linear liquid level control systems in industry include cylindrical horizontal storage tanks, "dished heads" on distillation columns, horizontal heat exchangers, etc. There appears to be considerable industrial interest in such non-linear problems that has prompted the Taylor Instrument Company to market a commercial "adaptive" controller [6]. This industrial controller requires some apriori knowledge of the non-linearity in a functional form. In contrast the adaptive control techniques require little or no apriori knowledge.

Work is presently under way to apply the same adaptive control techniques, plus some multivariable extensions, to the pilot plant evaporator used for the evaluation of a number of other modern control techniques, Fisher and Seborg [7].

3. CONVENTIONAL PROPORTIONAL PLUS INTEGRAL CONTROL

A conventional proportional plus integral DDC controller was designed for the unit under consideration. The choice of proportional and integral gains were made by using empirical tuning techniques. Values of the constants used are given in Appendix A. The PI controller was first tuned in the linear region. In this region the performance of the PI controller was excellent and at

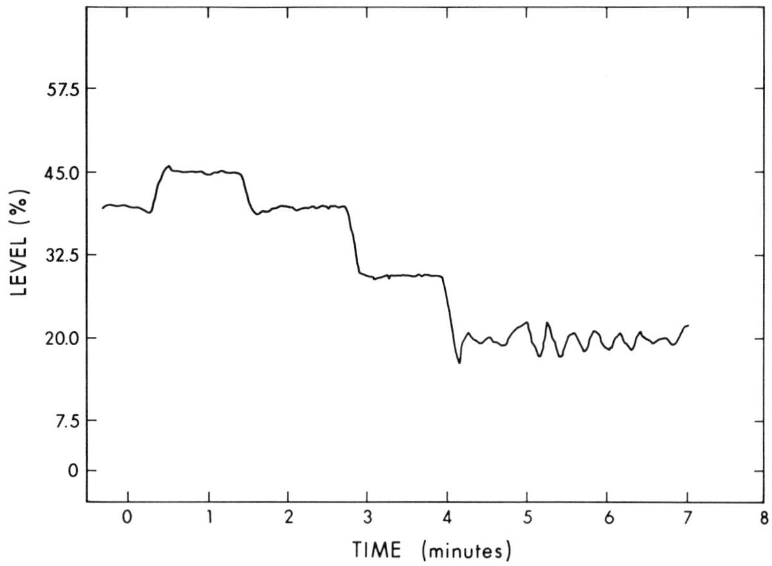

FIGURE 2. Level control of tank 1 using proportional plus integral DDC control, (cf. Appendix A for gain constants).

least as good as, if not better than, the adaptive system performance. Figures 2 and 3 show the output and control responses of the unit under PI control for a series of step changes in setpoint. Once operation is transferred to the non-linear region, however, PI control becomes very oscillatory and entirely unsatisfactory as evident in the region (b) of Figure 2 and in Figure 4. If the PI constants are retuned for the high gain portion of the non-linear region, control does improve as shown in Figure 4 (cf. region b). However, control performance once again deteriorates as soon as a new operating point in the non-linear region is sought. This is clearly evident in Figure 4. "Detuning" of the PI gains to give "acceptable" performance in the high process gain (low level) region results in a very sluggish system response in the low process gain

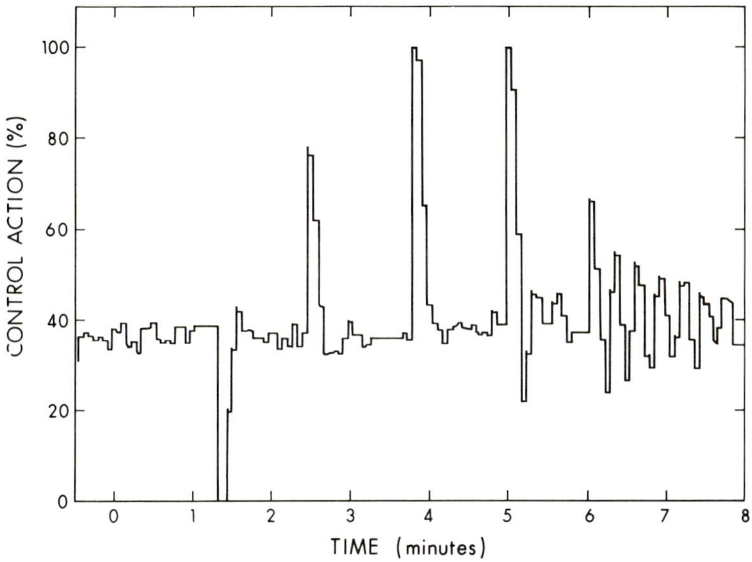

FIGURE 3. Control action or u_p(%) for PI control of level in tank 1 (cf. Fig. 2 and Appendix A. The valve is a reverse acting type (full shut) = 100%, u_{min} (full open = 0%^{max}).

region. The oscillatory performance of a simple PI controller on a simple process is useful in giving us an indication of the non-linear and time-varying nature of this system.

4. ADAPTIVE CONTROL APPLICATION 1

A deterministic direct adaptive scheme for single-input, single-output systems based on Popov's hyperstability criterion has been recently developed by Shah et. al. [2]. The method is an extension of the technique first proposed by Landau and Courtiol [3]. However, unlike their state-space scheme, the present technique relies only on plant input and output information. The adaptive mechanism

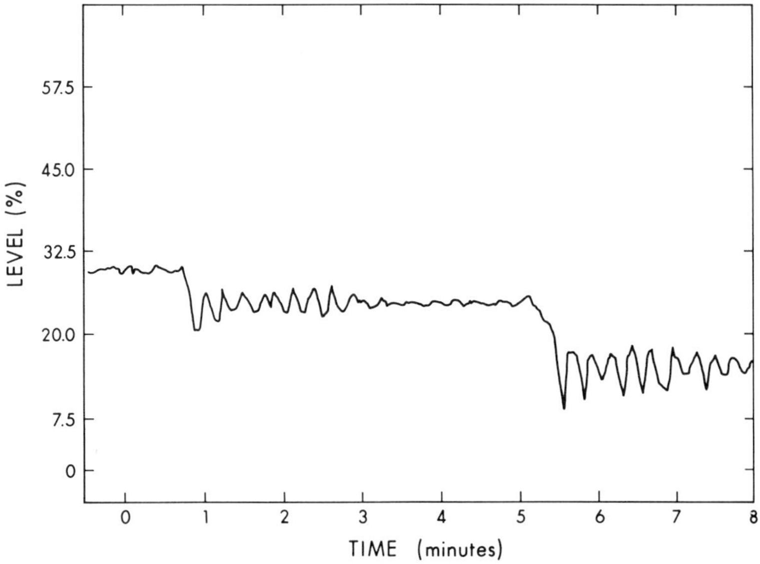

FIGURE 4. Level control of tank in the non-linear or high-gain region using PI control (cf. Figs. 2 and 3 and Appendix A).

generates values of the feedback, feedforward and setpoint controller matrices based only on measurements of plant output and input. The adaptive control law is

$$u_p(k) = K_p(k)y_p(k) + K_d(k)\xi_p(k) + K_m(k)u_m(k) \tag{1}$$

where u_p, y_p, ξ_p and u_m are the plant input, plant output, disturbance and reference inputs respectively. This control scheme is shown schematically in Figure 5. Stability of this system is based on a conjecture rather than a strict mathematical proof. The reader is referred to [2] for details of the theory and derivation of the control law.

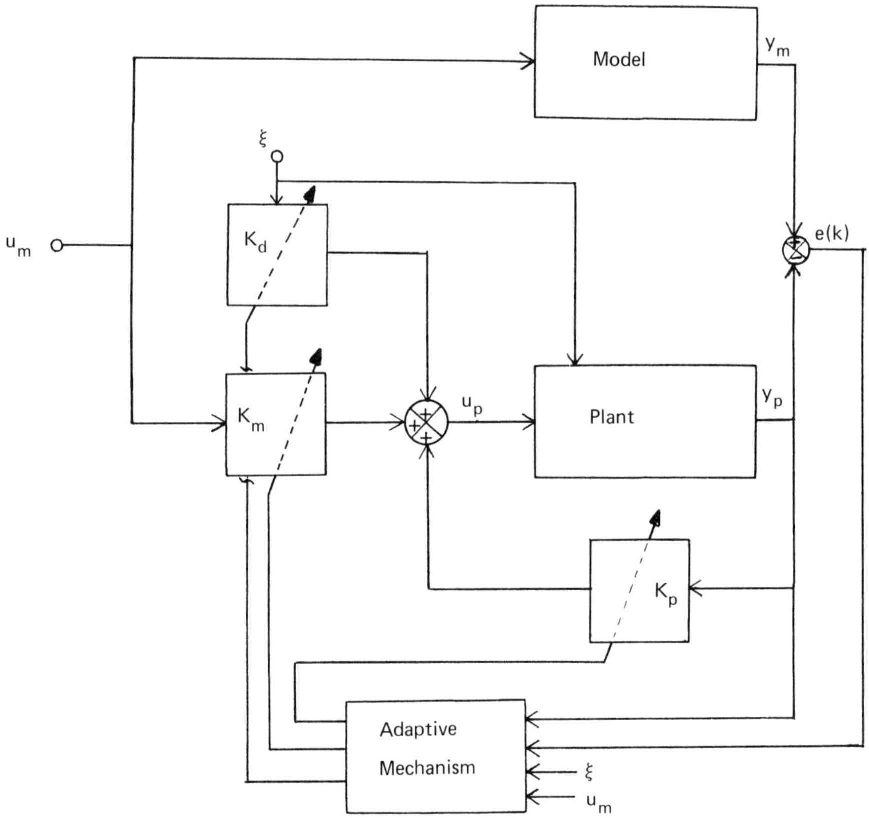

FIGURE 5. Configuration of the direct MRAC system 1.

Experimental Results

The objective of this series of runs was to evaluate the performance of the adaptive controller in the linear and non-linear regions under realistic plant conditions, i.e. in the presence of non-linearities, changing plant parameters and measurement noise conditions. A series of sustained step setpoint changes were made. The choice of initial parameters selected for the test run together with other details of the run is listed in Appendix B. The reference model was chosen to be unity, i.e. $y_m = u_m$. Such "instantaneous response" is, of course, not physically possible and results

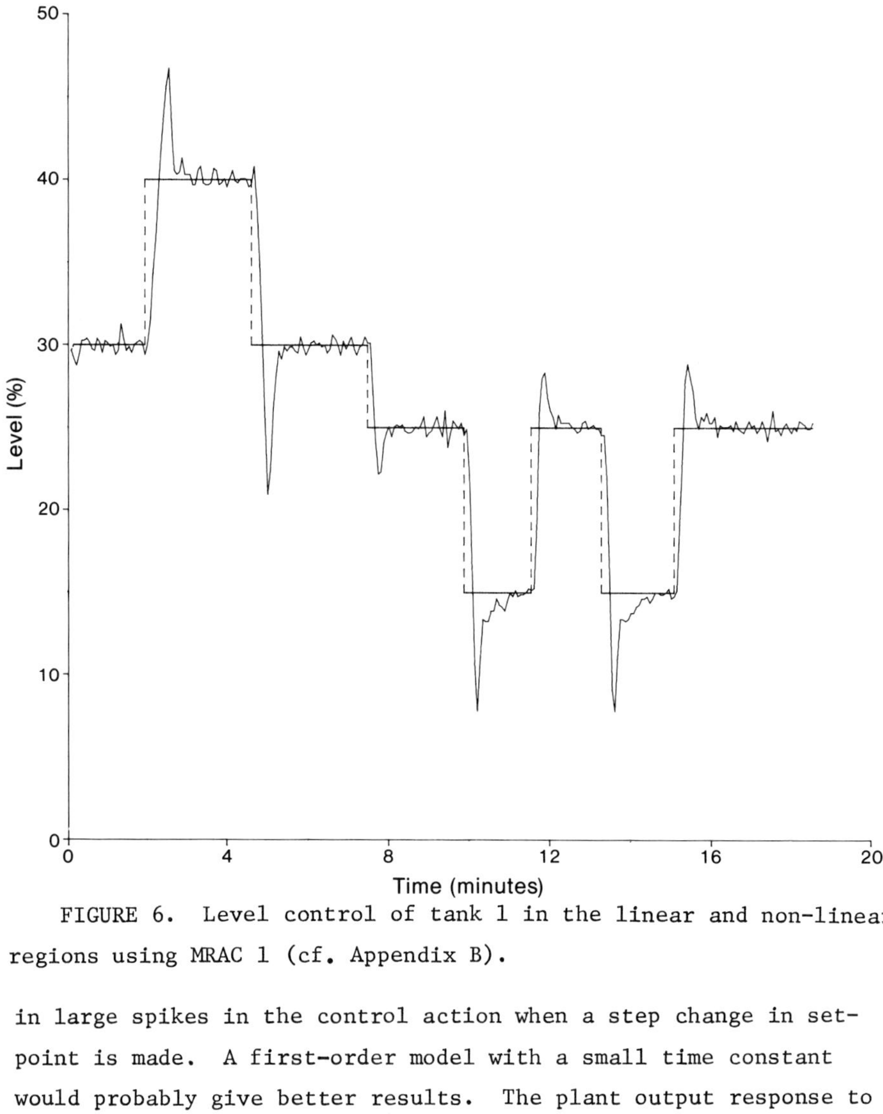

FIGURE 6. Level control of tank 1 in the linear and non-linear regions using MRAC 1 (cf. Appendix B).

in large spikes in the control action when a step change in setpoint is made. A first-order model with a small time constant would probably give better results. The plant output response to a series of sustained step setpoint changes is shown in Figure 6. The performance of the controller in the non-linear region is not very much different from its performance in the linear region. With the exception of a relatively large initial overshoot, the adaptive control system response is very satisfactory. The corresponding control input, $u_p(k)$ is shown in Figure 7. The control

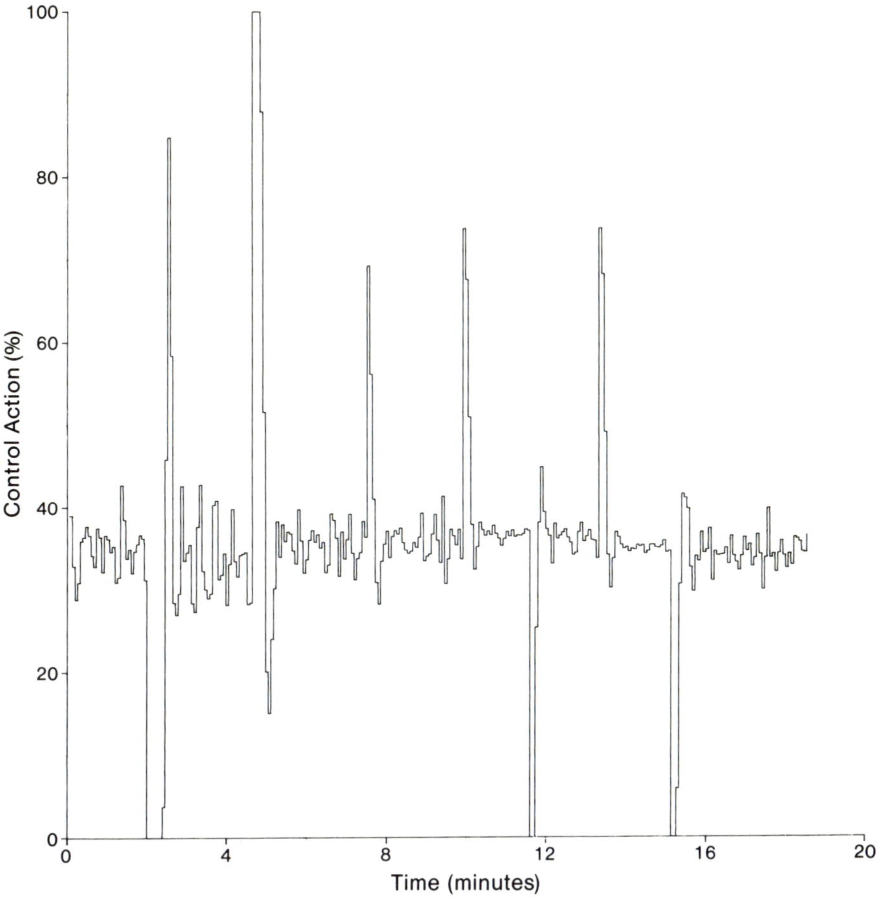

FIGURE 7. Control action or u_p(%) for level control of tank 1 under adaptive control (MRAC 1 cf. Appendix B).

excursions are not large and they only reach the maximum and minimum control limits (+100 and 0) a finite number of times and remain on these limits for only very short durations.

5. ADAPTIVE CONTROL APPLICATION 2

In a recent paper Suzuki and Takashima [4] proposed a particularly simple model reference adaptive control system. The scheme belongs to the augmented error subclass of adaptive schemes first

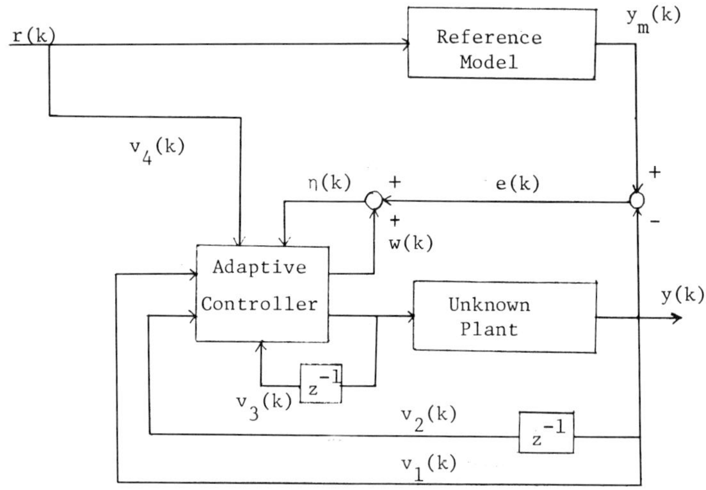

FIGURE 8. Configuration of Direct MRAC System 2 using an augmented error signal.

introduced by Monopoli [8]. The adaptive control law for the hyperstable MRAC scheme of Suzuki and Takashima [4] is described by

$$u(k) = \sum_{i=1}^{2n} g_i(k)v_i(k) \tag{2}$$

where the laws for the adjustable adaptive parameter, $g_i(k)$, are calculated by the following equation:

$$g_i(k) = g_i(k-1) + \gamma_i v_i(k-1)\eta(k) \qquad (i = 1,2,\ldots,2n) \tag{3}$$

The augmented error, $\eta(k)$, is obtained as a real root of a cubic polynomial as described in [4], and γ_i are user specified constant adaptive gains. The variables, $v_i(k)$ $(i = 1,2,\ldots,2n)$ for the present application $(n = 2)$ are defined in the overall adaptive system schematic shown in Figure 8. The augmented error adaptive scheme of Suzuki and Takashima has been modified to incorporate the following single-step objective functional [5].

$$J(k) = \{y_m(k)-y(k) + pu(k-1)\}^2 \triangleq \phi(k)^2 \qquad (4)$$

where y_m is the reference model output, y is the plant output and p is a constant polynomial in z^{-1}. The above modification permits application of the algorithm to non-minimum phase systems. By re-defining the error, $e(k) = y_m(k) - y(k)$, in the original algorithm [4] to a new term $e(k) = \phi(k)$ (cf. eqn. (4)), asymptotic stability of the modified system can be easily established to show that $\phi(k) \to o$ as $k \to \infty$, i.e. the objective function J is minimized. This modification allows excessive changes in the control inputs to be penalized which is a desirable feature in many practical implementations. The reader is referred to [4] and [5] for more details on the original algorithm and its extension.

Experimental Evaluation

As in Case I, a series of sustained setpoint changes were made to evaluate the performance of the adaptive controller. Appendix C contains details of the initial parameter values selected for this test run. The reference model was again chosen to be unity. Figures 9 and 10 show the plant output response and the corresponding control action. Both the responses are excellent in both the linear and the non-linear regions. As in Case 1, the control inputs reach the upper and lower limits only a finite number of times and only for short durations. The performance of this adaptive controller was also evaluated in the presence of a sustained (or step) unmeasurable disturbance. Satisfactory regulation with zero offset was achieved under these conditions as shown in Figure 11.

The experimental runs confirm the robust setpoint tracking and regulation properties of this adaptive controller observed previously in simulated studies (not reported here).

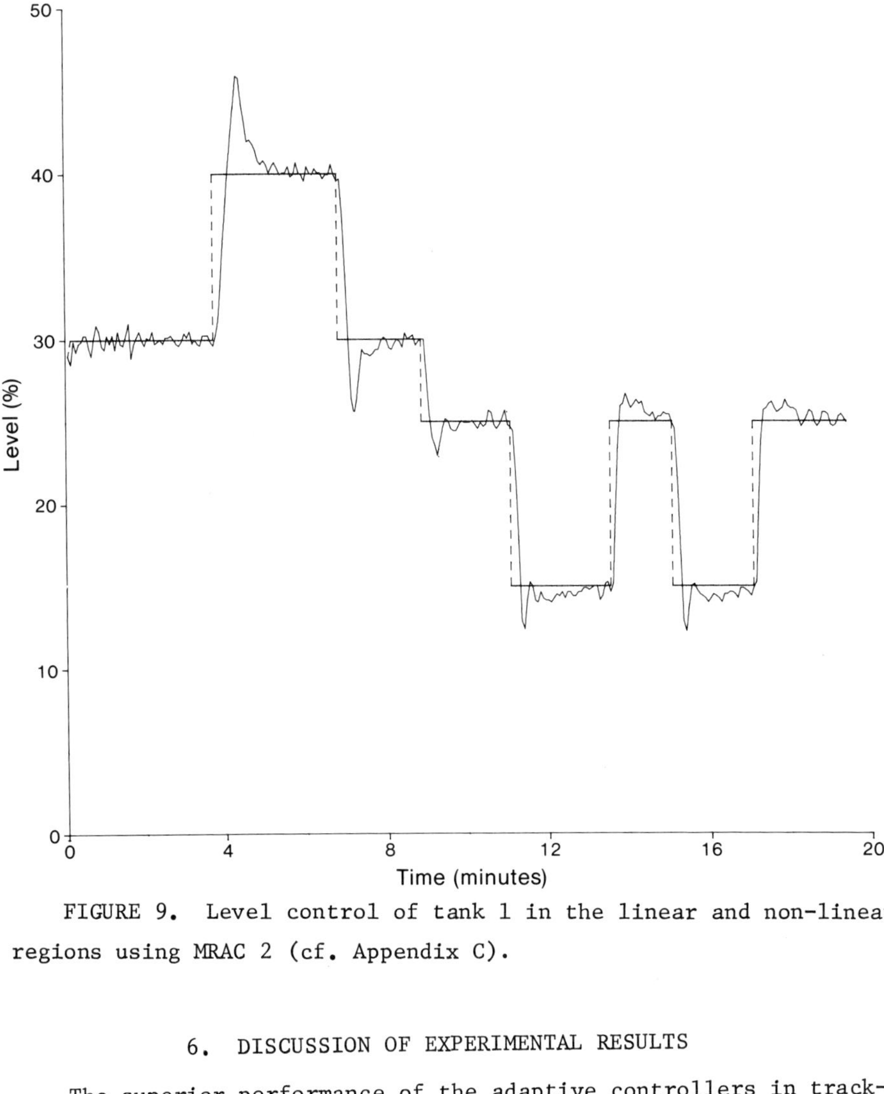

FIGURE 9. Level control of tank 1 in the linear and non-linear regions using MRAC 2 (cf. Appendix C).

6. DISCUSSION OF EXPERIMENTAL RESULTS

The superior performance of the adaptive controllers in tracking as well as regulation in the linear and non-linear regions of the plant in comparison to the PI controller is clearly evident from Figures 2-11. The oscillatory performance of the PI controller in the high-gain or non-linear region is expected because of the significant change in gain ratio (64:1) over a small level change in level.

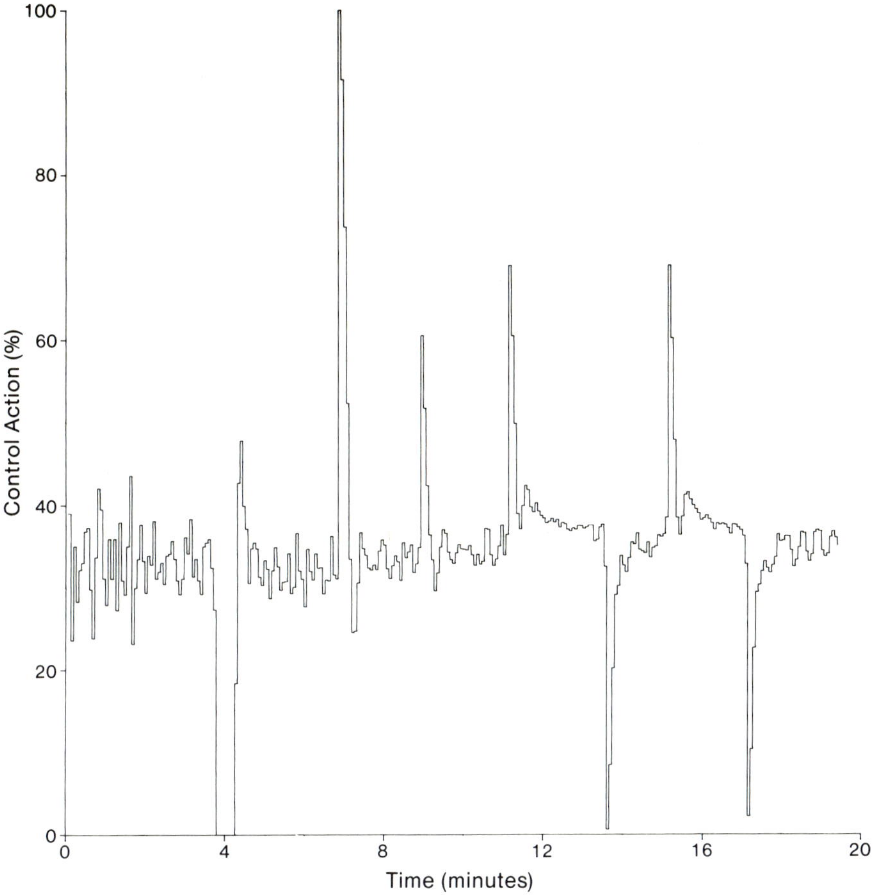

FIGURE 10. Control action or u_p(%) for level control of tank 1 under adaptive control (MRAC 2)(cf. Appendix C).

The two adaptive system performances are roughly similar. Both techniques are simple and easy to program. Each of the main adaptive control algorithms constitute only about 20 lines of Fortran code on the HP1000 computer and involve explicit calculation. No iterative or converging calculations are involved. The robust properties of both the algorithms are evident from the runs with measurement noise and/or unmeasured disturbances. The excellent tracking performance of the algorithm in the non-linear region where the gain or time-constant of the first order system changes significantly are clearly evident from the experimental results.

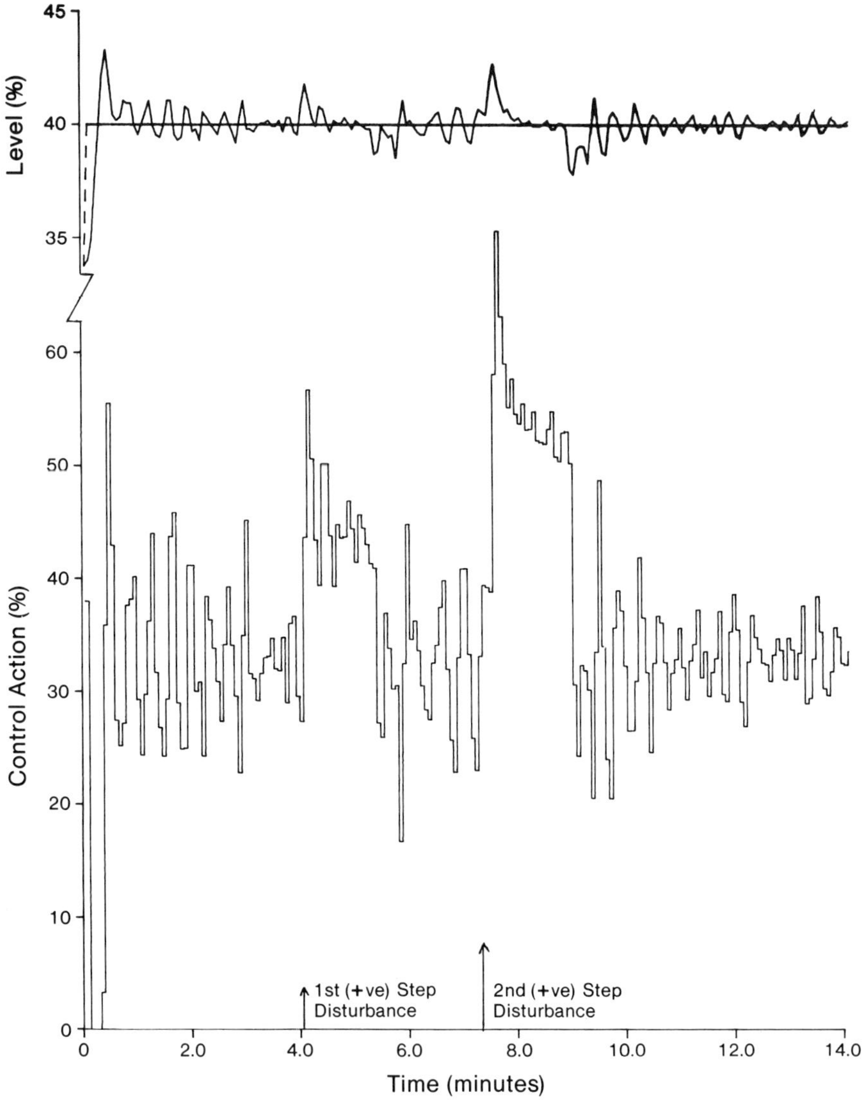

FIGURE 11. Level control of tank 1 using MRAC 2 in the presence of unmeasurable step disturbances (as indicated by arrows on time axis).

7. CONCLUSIONS

1) This experimental evaluation of two SISO MRAC techniques which require only I/O information from the process is one of the very few experimental applications reported to date in the literature. Both applications were successful and gave better performance than classical PI control.

2) The robust nature of the adaptive controller was illustrated by the ability of the controller to handle a non-linear/time-varying process with measurement noise and unmeasurable disturbances. The significant non-linear/time-varying nature of the process is best appreciated by knowing that the gain in the non-linear region changes in the ratio of 64:1 over a 15 cm level change.

APPENDIX A

Run 2 (cf. Figure 2)

proportional gain = -7.5

integral gain = -0.15

Run 4 (cf. Figure 4)

proportional gain = -7.5

integral gain = -0.15

retuned at ∿ 3 min to new gains:

proportional gain = -7

integral gain = -0.14

APPENDIX B

Run 1 (cf. Figures 6-7)

proportional adaptive gains = 0.003 (r_p, q_p, s_p)

integral adaptive gains = 0.001 (r_I, q_I, s_I)

initial controller parameters = 0

APPENDIX C

Run 2 (cf. Figures 9-11)

adaptive gains γ_i = 0.006 (i = 1,2,3,4)

initial controller parameters = 0

REFERENCES

1. Foss, A. S., "Critique of Chemical Process Control Theory," AIChE Journal, Vol. 19, pp. 209-214, 1973.
2. Shah, S. L., Fisher, D. G., and Karim, M. N., "Hyperstable Adaptive Control - A Direct Input-Output Approach without Explicit Model Identification," Proc. of 1979 JACC, Denver, Colorado, pp. 481-486, June 1979.
3. Landau, I. D., and Courtiol, B., "Design of Multivariable Adaptive Model Following Control Systems," Automatica, 10, pp. 483-494, 1974.
4. Suzuki, T., and Takashima, S., "A Hyperstable Scheme for Discrete Model Reference Adaptive Control System," Int. J. Control, Vol. 28, No. 2, pp. 245-252, 1978.
5. Johnstone, R. M., Shah, S. L., and Fisher, D. G., "An Extension of Hyperstable Adaptive Control to Non-Minimum Phase Systems," accepted for publication in Int. J. of Control (in press) 1980.
6. Andreiev, N., "A Process Controller (Taylor's Micro-Scan 1300) that Adapts to Signal and Process Conditions," Control Engineering, Dec. 1977.
7. Fisher, D. G., and Seborg, D. E., "Multivariable Computer Control - A Case Study," American Elsevier, New York, 1976.
8. Monopoli, R. V., "Model Reference Adaptive Control with an Augmented Error Signal," IEEE Trans. on Auto. Control, AC-19, pp. 474-484, 1974.

DISCRETE ADAPTIVE CONTROL OF A RADIO TELESCOPE[1]

S. I. Haque[2]

Intermetrics, Incorporated
Cambridge, Massachusetts

R. V. Monopoli

Department of Electrical and Computer Engineering
University of Massachusetts
Amherst, Massachusetts

A discrete model reference adaptive control algorithm is applied to the control of a large Radio Telescope to reduce the effects of friction and unknown parameters. Results are presented for field tests conducted with the adaptive controller in

[1]This research was partially supported by the National Science Foundation under Grants NSF ENG 74-03399, and NSF ENG 79-06641.

[2]Formerly with ECE Department, University of Massachusetts, Amherst.

ISBN 0-12-514060-6

operation in the azimuth loop only with elevation held constant. Initial testing demonstrates an improvement in tracking accuracy by a factor of 2:1.

1. INTRODUCTION

This paper describes the design of a Discrete Model Reference Adaptive Control System (DMRACS) used to compensate for the effects of unknown parameters in the stabilization and tracking loops of a Radio Telescope. The Radio Telescope is located in the vicinity of the Quabbin Reservoir near Amherst, Massachusetts, and is part of a larger facility known as the Five College Radio Astronomy Observatory (FCRAO). The FCRAO is a cooperative project between four small private colleges (Amherst, Hampshire, Mount Holyoke, and Smith) and the state-supported University of Massachusetts. The Radio Telescope has a 45 foot paraboloidal dish antenna in a radome. It is used for observations at millimeter wave lengths. The antenna rides on an altazimuth mounting that is computer controlled. Sighting on a celestial object can be done to about two seconds of arc, and tracking can be maintained to within a few seconds of arc. The beam width at 1 mm., the shortest wavelength at which observations are planned, is 24 seconds of arc.

The overall goal of this effort was to improve the telescope's pointing and tracking accuracy by compensating for inertial variations, static and coulomb friction, and other anomalies in the telescope's mount. The Discrete Model Reference Adaptive Control technique used for this work was developed in [1] and is a discrete version of the results presented in [2]. For a more detailed description of Model Reference Adaptive Control techniques, the interested reader is referred to references [3], [4].

In the discussion that follows, section 2 presents a mathematical development of the Adaptive Control scheme; section 3 gives a description of the telescope control system, and a model for it; section 4 presents the results of simulations and certain considerations for a practical application of this type; section 5 gives the results of the field tests and section 6 presents the conclusions along with some ideas for further improvement in the results already obtained.

2. MATHEMATICAL DEVELOPMENT

This section presents an application of the theory developed by Ionescu and Monopoli in [1]. Figure 1 gives a block diagram of

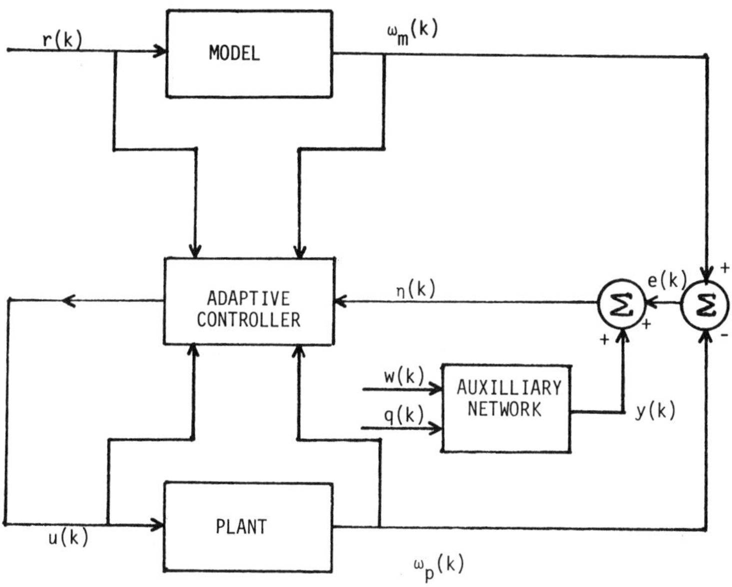

FIGURE 1. System configuration for discrete adaptive controller.

the adaptive control system. Plant refers to the system to be controlled, and Model is the reference model which specifies the desired behavior of the plant. An auxiliary network is required to generate an additional control signal (y in the figure). An augmented error, η, is then used to generate the signals needed for the synthesis of the plant input u. Use of the augmented error circumvents the problem of needing anticipatory values of signals in generating the control input u. The objective is to generate a control sequence u(k) such that the error $e(k) = \omega_m(k) - \omega_p(k)$ between model and plant outputs goes to zero in spite of imperfect knowledge of plant parameters.

The theory is now presented for a first-order system described by the differential equation

$$\dot{\omega}_p(t) + a\,\omega_p(t) = b\,u(t) + f(\omega_p(t)) \tag{1}$$

where a, b, are system parameters, u is the control signal, and $f(\omega_p(t))$ is a non-linear term describing the static and coulomb friction in the system. Figure 2 gives a plot of the friction characteristics. Equation (1) is an approximate model for the telescope when ω_p is considered to be the angular velocity in azimuth, u(t) the applied torque, and elevation is held constant.

An approximate difference equation for the system, derived from (1), may be written as

$$\omega_p(k+1) + a_s\omega_p(k) = b_s u(k) + f_s(\omega_p(k)) \tag{2}$$

where $a_s = -e^{-aT}$, $b_s = \frac{b}{a}(1-e^{-aT})$, T is the sampling period, and $f_s(\omega_p(k))$ is the discretized form of the non-linearity defined in (1).

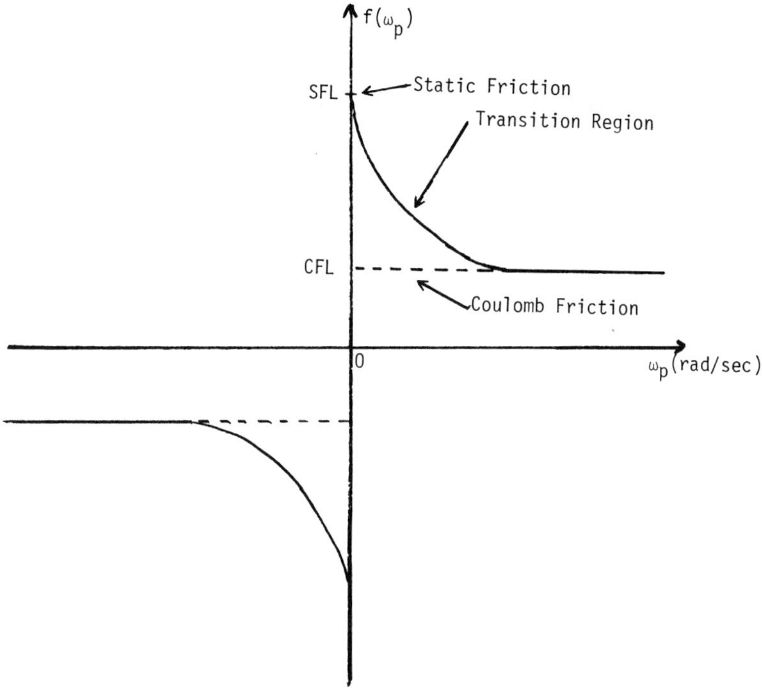

FIGURE 2. Coulomb and static friction.

The exact difference equation corresponding to (1) is impossible to write in terms of $f_s(\omega_p(k))$. The difficulty lies in the fact that the static and coulomb friction forces may change sign between sampling instants. The approach taken here is to assume that $\omega_p(t)$ does not change sign between sampling instants too often. This is a reasonable assumption to make for this particular application because the telescope is used primarily for star-tracking and therefore is not subject to command inputs forcing the antenna velocity to change sign. If $\omega_p(t)$ does not change sign, (2) is exact.

The desired behavior of the plant is the output of a stable model which satisfies the equation

$$\omega_m(k+1) + a_d\omega_m(k) = K_o r(k) \tag{3}$$

where K_o and a_d relate to nominal values of plant parameters.

The design problem is, then, the development of a parameter adaptive control system for (1) which will cause the error

$$e(k) = \omega_m(k) - \omega_p(k) \tag{4}$$

between model and plant outputs to go to zero as $k \to \infty$.

In order to avoid anticipatory values of the error e, we need to define an augmented error signal η, as explained in [1].

$$\eta(k) = e(k) + y(k) \tag{5}$$

where y(k) is the output of an auxiliary network defined as

$$(1+a_d z^{-1})y(k+1) = Z_w(0)(w(k)+q(k+1)) \tag{6}$$

For the first-order case, the Z_w filter reduces to a simple constant, called c_0 in this paper. Signals w(k) and q(k) are auxiliary inputs to be determined along with the control input u(k) as part of the design.

Subtracting (2) from (3), and adding the result to (6), an expression for $\eta(k)$ is obtained as

$$(1+a_d z^{-1})\eta(k+1) = K_o r(k) - b_s u(k) + (a_s-a_d)\omega_p(k)$$

$$- f_s(\omega_p(k)) + c_0(w(k) + q(k+1)) \tag{7}$$

In order to introduce adaptive gains to attempt to minimize the right-hand-side of equation (7), the control input u(k) is defined as

$$u(k) = k^0(k-1)\phi^0(k) + k^1(k-1)\phi^1(k) + k^3(k-1)\phi^3(k) \tag{8}$$

where

$$\phi^0(k) = \frac{1}{c_0} r(k) \tag{9}$$

$$\phi^1(k) = \frac{1}{c_0} f_s(\omega_p(k)) \tag{10}$$

and

$$\phi^3(k) = \frac{1}{c_0} \omega_p(k) \tag{11}$$

and k^i, i = 0,1,3 are adjustable parameters ("gains") to be defined on the basis of Lyapunov's direct method. A term for i = 2 is not included in order to keep the notation here consistent with that in [1].

If we define the additional signals

$$\phi^4(k) = k^0(k)\phi^0(k) + k^1(k)\phi^1(k) + k^3(k)\phi^3(k) - u(k) \tag{12}$$

$$w(k) = k^4(k)\phi^4(k) \tag{13}$$

$$q(k) = -K_q\eta(k) \sum_{\substack{i=0 \\ i\neq 2}}^{4} (\phi^i(k-1))^2 \tag{14}$$

and the adjustable gains k^i , i = 0,1,3,4 as

$$k^i(k) - k^i(k-1) = \frac{c_0}{b_s\lambda_i} \phi^i(k-1)\eta(k) \qquad i = 0,1,3 \tag{15}$$

$$k^4(k) - k^4(k-1) = -\frac{1}{\lambda_4} \phi^4(k-1)\eta(k) \tag{16}$$

then, by the theory presented in [1] and [5], $\lim_{k\to\infty} \eta(k) = 0$, $\lim_{k\to\infty} y(k) = 0$, and therefore $\lim_{k\to\infty} e(k) = 0$, provided that the constants c_0, K_q, λ_{imin} satisfy the following inequalities

$$c_0 > 0 \tag{17}$$

$$K_q > 0 \tag{18}$$

$$\lambda_{imin} > \frac{1}{2K_q} \tag{19}$$

Equations (8) through (19) completely define the adaptive control algorithm for this problem.

3. THE PLANT

The Radio Telescope located at Quabbin is used for star-tracking purposes. There are two principal modes of operation, namely, a program controlled mode, and a slew, or search mode. In the program controlled mode, the telescope responds to instantaneous predicted data commands which are based on computer generated information. Since tracking velocities are in general very low, effects of friction are especially dominant. The slew mode can be used as a manual or computer velocity control. The telescope is used in this mode to acquire celestial objects inside its field of view (or "capture" region, as it is often termed). The mount's response to a rate step input is important for this purpose. (By rate step input is meant a step input to the velocity or tachometer loop (see Figure 4). If computer generated, this "step" has a finite rise time; for slewing to a position removed from the present position of the antenna, the tachometer loop is fed with an error signal which is built up from almost zero for tracking to $\pm$ 10 volts at a rate of approximately 0.4 volts every 50 msec. (which is the sampling rate.) If the mount does not drive smoothly, it becomes difficult to "lock" on to a particular celestial object. The telescope can be slewed at two rates. The faster slewing mode is utilized when the antenna has to be moved to a position removed more than one degree from its existing position. When the error in position is less than one degree, the telescope is slewed at a slower rate.

The adaptive controller was designed for the telescope operating in the program controlled mode. The same concepts may be extended to design a controller for the slewing mode of operation. This is actually a special case of the larger control system configuration.

The configuration in Figure 3 represents a simplified control system diagram for control of one axis of the telescope mount as it exists at present, without the adaptive controller. The axis under study is the azimuth axis (or east-west axis). The load represents the antenna, the angular velocity of which is controlled by generating rate commands for the tachometer loop through the error in position $e_1(k)$. The total non-linear frictional torque acting on the system is a combination of two frictional functions, namely static and coulomb friction. These are typically as shown in Figure 2. The preamplifier is a lag network designed to boost the low frequency forward loop gain of the tachometer loop. The transfer function constants of the preamplifier, power amplifier, and tachometer are known. The telescope has, in addition to the various elements shown in Figure 3, a computer controlled feed-forward network which is used when the antenna has to be slewed to a new position removed more than one degree from its present position. This network is not shown in the block diagram as it was not intended to operate the adaptive controller in either of the two slewing modes for the initial testing. Also not shown is a saturation circuit that is used in order to circumvent the

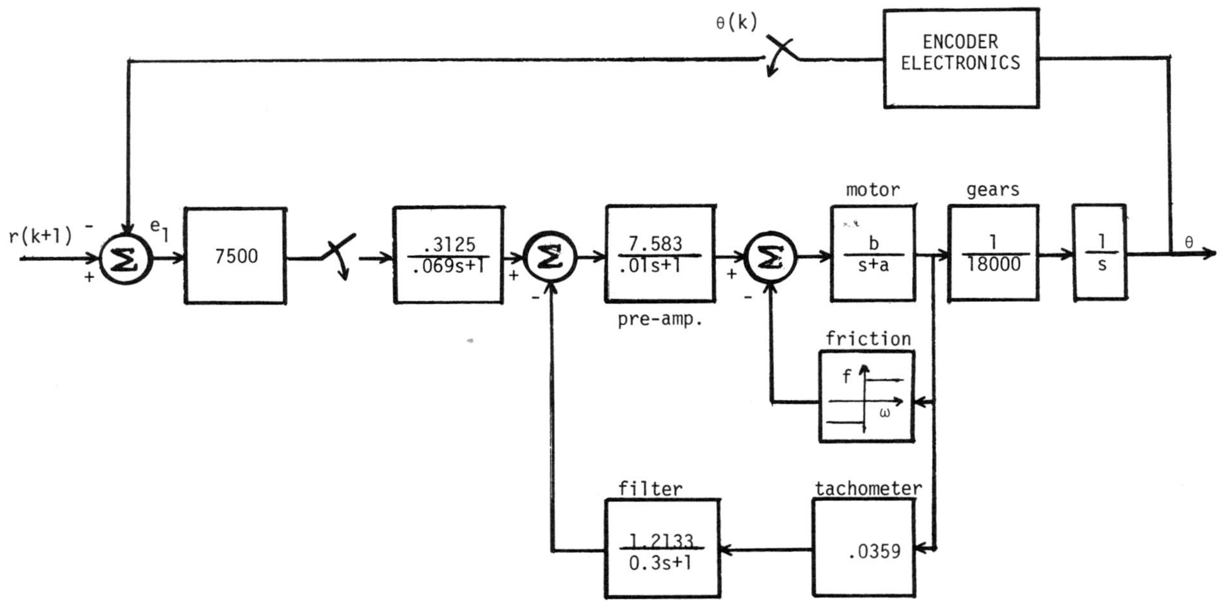

FIGURE 3. Telescope control system.

problem of preamplifier saturation for large step commands. When the error in position, $e_1(k)$, reaches a certain threshold, the voltage at the input of the compensation circuit is frozen and the closed loop (position) integration is removed until $e_1(k)$ settles down.

Some initial data on the telescope antenna was available enabling us to obtain an estimate of the motor, load and tachometer transfer function parameters. This data is as follows:

Azimuth Inertia (essentially identical to elevation)
I_L = 95,000 ft-lb-sec^2.
Gear Train Ratio = 18,000:1
Gear Train Inertia = 10,000 lb-ft-sec^2 (at output)
Output Element Friction = 34,000 ft-lbs/(rad/sec)(at output)
Drive train and motor friction = 4.7x10^{-4} ft-lb/rad-sec.
(at motor shaft)
Tachometer constant 0.0359 volts/(rad/sec)

The above data leads to the following transfer function for the power amplifier, motor, load and tachometer:

$$\frac{\omega_p(s)}{V(s)} = \frac{3.61}{s + 1.732} \tag{20}$$

where V(s) is the Laplace Transform of the output of the preamplifier in Figure 3. The closed loop transfer function from u to ω_p,

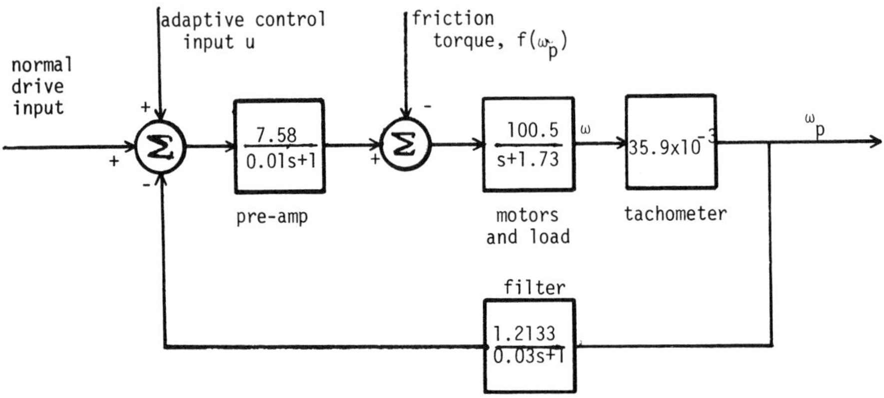

FIGURE 4. Tachometer loop - "plant" as seen by adaptive controller.

for the tachometer loop in Figure 4, derived using (20) and the known transfer functions for the electronic circuits, is

$$G(s) = \frac{2.7363 \times 10^{3}(s+33.3)}{(s+112.6)(s^{2}+22.46s+1033.3)} \tag{21}$$

The validity of this model was tested by comparing the actual system response to the response of a simulation using (21). The results are shown in Figure 5. Comparison was only made once the telescope had reached the slow slewing mode ($t > 11$ seconds). These results show the model to be quite accurate.

A frequency response study made of (21) notes that the complex conjugate pole pair and the system zero have corner frequencies

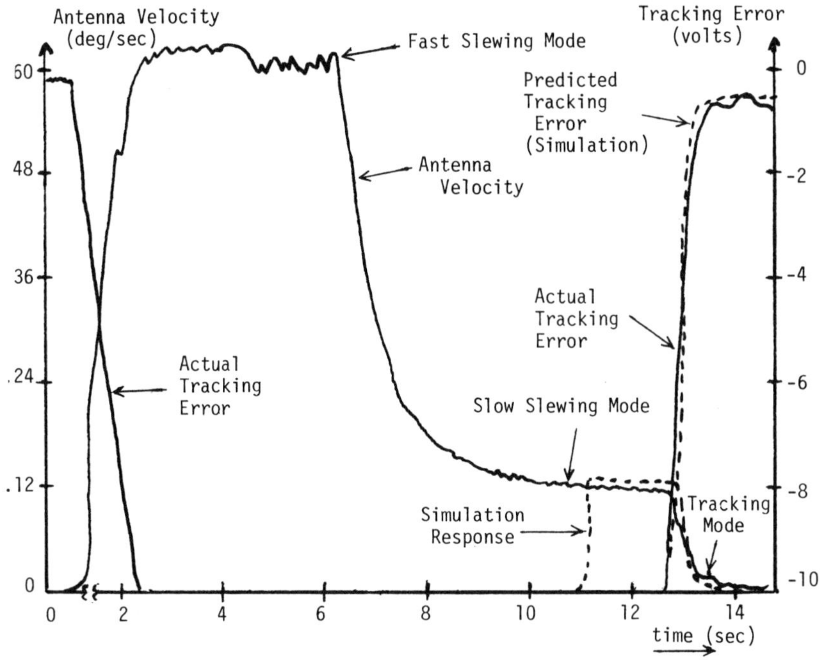

FIGURE 5. Actual system response vs. simulation response.

at ω = 32.145 rad/sec and ω = 33.3 rad/sec, respectively. Thus the combined log magnitude plot of these would yield an asymptote of approximately -20 db/dec above the corner frequency of ω = 33.3 rad/sec. In addition, a 3rd pole at s = −112.6 contributes an additional -20 db/dec beyond the corner frequency of ω = -112.6 rad/sec. Figure 6 gives the Bode plot. Because of this analysis, it was decided to model the system as a first order system for the range of frequencies of interest. This means that the dynamics of the control amplifier (Figure 3) have been modelled as a constant, known gain. The transfer function of the approximate first order model of the system was chosen as

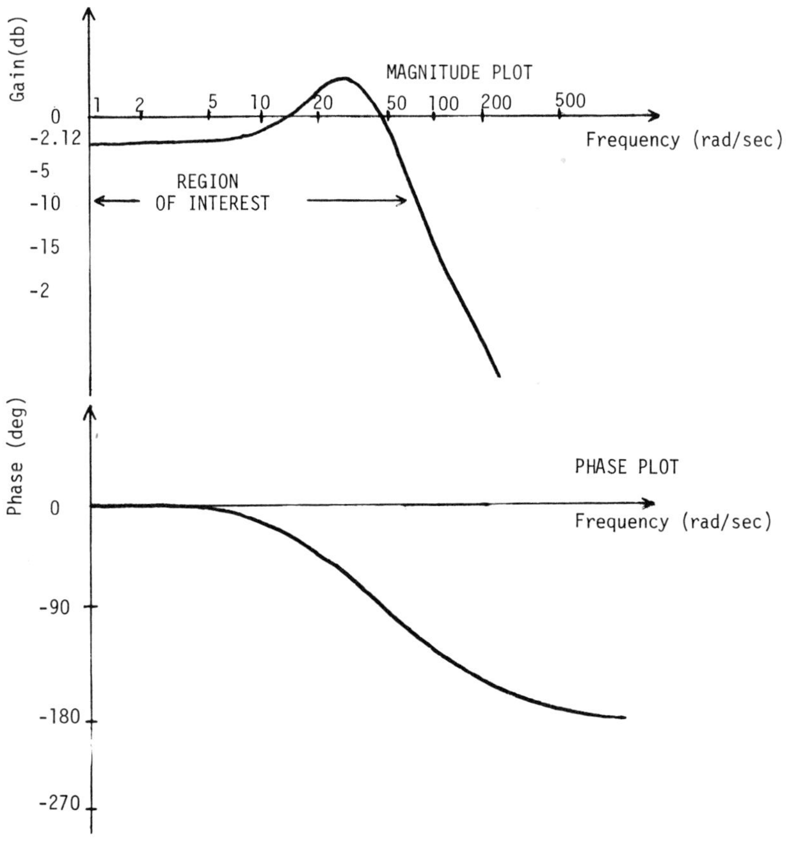

FIGURE 6. Bode plot for tachometer loop.

$$G_{approx}(s) = \frac{0.7832}{0.03115s+1} \tag{22}$$

which has a break frequency close to the first break frequency of (21) and the same d.c. gain. It is this transfer function, (22) which relates to (2) and the development of the adaptive control algorithm in section 2. The following reasons serve to justify the modelling assumption:

(i) it is much simpler to obtain convergence of the adaptive error signal for a first-order system than it is for a higher order system because there are fewer adjustable gains.

(ii) it was essential to keep the software for the implementation of the adaptive controller to an absolute minimum.

(iii) the frequency content of signals normally encountered in operation do not require perfect model matching, i.e. convergence of parameter errors to zero, to achieve good tracking of the model by the plant.

4. SIMULATIONS

A simulation study was made on the complete model of the real system (see Figure 3) in order to

(a) test the validity of the modelling approximation made in the previous section,

(b) determine a set of constants for the controller software essential for good convergence, and acceptable transients during convergence,

(c) make a measurement noise study on the algorithm, and to study the effects of possible drift in any of the adjustable gains and in the control signal itself. A previous study [6], made on the same algorithm had confirmed its feasibility under noisy conditions for a linear plant with the implicit assumption that the results could be carried over to non-linear plants.

Simulations were made using the Advanced Continuous Simulation Language (ACSL). The language has certain features which can be exploited to advantage in a hybrid analog-digital simulation. Thus the simulation model gave us a close approximation to the real system. In the simulation model, coulomb friction was modelled as a constant torque opposed to motion, provided that the servo is in motion. If the servo is not in motion, static friction is assumed to be equal to, and opposite to the applied torque, up to a maximum friction level.

Simulations were made for the two cases:

(a) with 10% uniformly distributed measurement noise, and

(b) with no measurement noise.

Table 1 summarizes the results of the simulations, confirming the validity of the assumptions, and also the ability of the model reference scheme to improve the tracking accuracy of the telescope system. In this table, comparisons are made between the system without adaptive gains and three cases with adaptive gains. The three cases are (a) adaptive gains initialized so that parameter errors are zero (impossible to do on the real system); (b) adaptive gains initialized at -50% of "correct" values in (a); (c) adaptive gains initialized at 150% of "correct" values in (a). The comparisons are made with and without measurement noise. Criterion used are the sum squared error, max error, and steady state tracking accuracy. It may be seen that

(1) measurement noise does not cause a significant deterioration of performance; and

(2) the steady state tracking error is reduced by almost a factor of 2 in the worst case.

Previous studies [7-11] have shown that it is not worthwhile to include an adaptive gain for static friction. This result was confirmed in this study. Therefore, a gain k^1 was included for coloumb friction, but none was included for static friction. Figure 5 illustrates the close agreement between the transient responses of the simulation model and the actual system.

TABLE 1. Simulation results.

DESCRIPTION OF TEST	ADAPTIVE CONTROLLER PERFORMANCE				SYSTEM PERFORMANCE
	WITH MEASUREMENT NOISE		WITHOUT MEASUREMENT NOISE		
	sum square error	max error	sum square error	max error	STEADY-STATE TRACKING ERROR
	$(volts)^2$	(volts)	$(volts)^2$	(volts)	(degrees)
No Adaptive Gains	4.23 (increasing)	1.86×10^{-1}	3.79	1.47×10^{-1}	0.0059
4 Adaptive Gains with "correct" initial conditions	2.36×10^{-4}	-1.11×10^{-2}	2.24×10^{-4}	-1.04×10^{-2}	0.0014
4 Adaptive Gains −50% of correct i.c.'s	6.65×10^{-3}	-2.84×10^{-2}	4.907×10^{-3}	-2.52×10^{-2}	0.0030
4 Adaptive Gains +150% of correct i.c.'s	5.21×10^{-3}	-1.94×10^{-2}	4.43×10^{-3}	-1.82×10^{-2}	0.0009

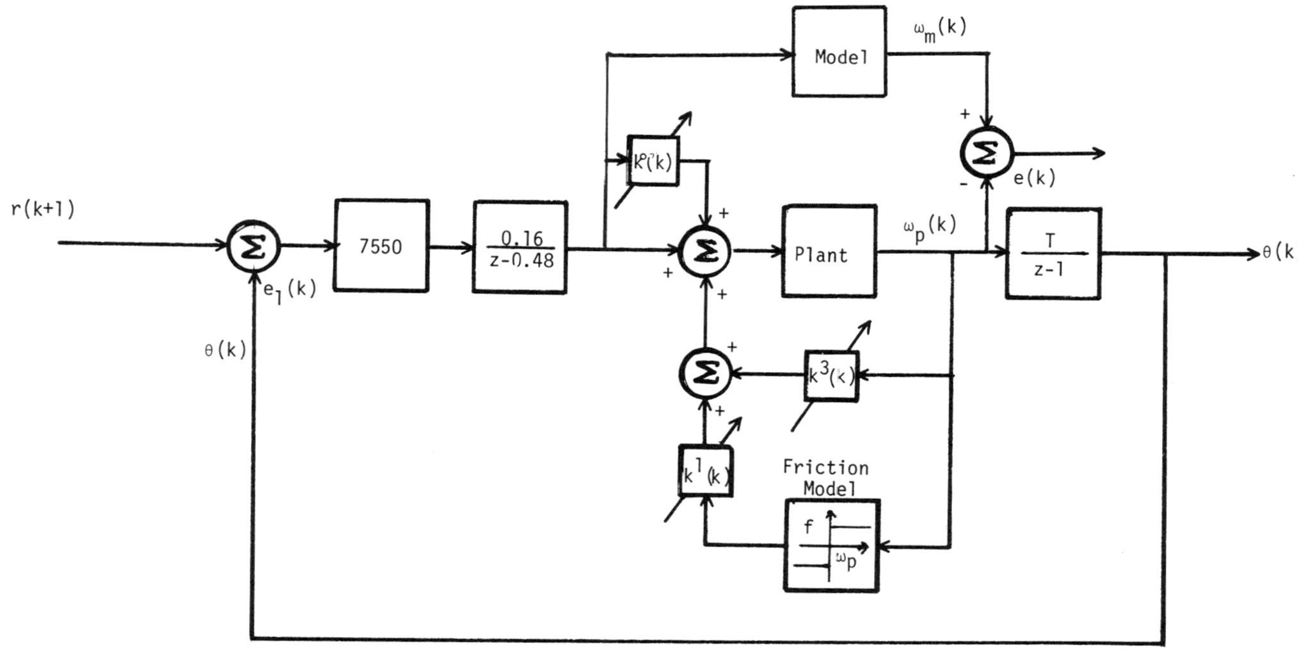

FIGURE 7. System configuration with adaptive gains.

5. FIELD TESTS

A discrete version of the adaptive control system configuration used in field tests is shown in Figure 7. The model (eqn. (3)), adaptive gain algorithm (eqns. (15),(16)), and the control signal (eqn. (8)) were all "implemented" in software on a MODCOMP IV minicomputer. This is a 16-bit machine, which is also used to perform other functions associated with antenna operation. Therefore, time and memory allowed for adaptive control were limited. MODCOMP IV has a total memory size of 96K 16-bit words, and has a 500 μsecond memory access time. A typical calculation in floating point multiplication, done in hardware, is 5 μseconds. The computer puts out a command signal for the antenna position every 50 mseconds, which is also the sampling frequency. The command signal is essentially a sine wave of very low frequency, the frequency being determined by the position in the sky of the object being tracked. In the time scale at which our testing was done, the command input could be approximated as a ramp function.

Before implementing the adaptive control system, it was decided to go through a series of tests on the non-adaptive system in order to determine the limits between which the adaptive controller should operate. These tests are described below:

(i) As seen from Figure 7, k^0 modifies the loop gain of the closed loop control system. Too large a value of k^0 might send the system into its unstable region. It was decided, therefore, to operate the system with just this one adaptive gain to determine bounds for stability. An upper bound of 0.4 and a lower bound of -1.0 were found. Therefore, during adaptation, the algorithm for this gain had to be modified so that it could not vary outside these limits.

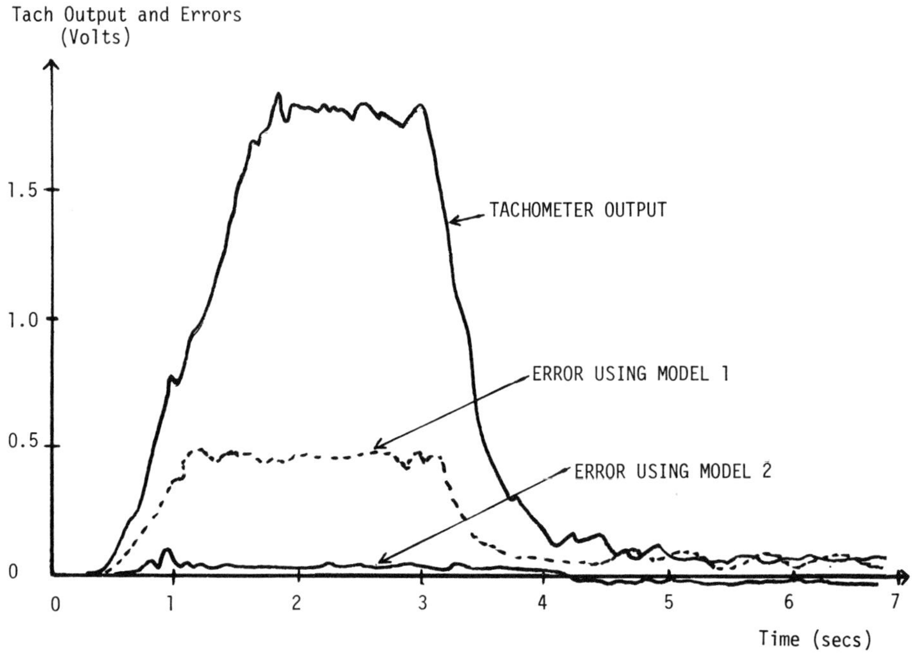

FIGURE 8. Results of off-line model adjustment.

(ii) The system was run with only the adaptive gain k^1, which is intended to compensate for friction. By trying different values for this gain, it was found that $k^1 = 0.24$ reduced the tracking error to almost zero. This procedure is helpful in determining an initial value for this gain.

(iii) A procedure for "tuning" the model to the plant (off-line) was used in order that k^0 would not have to vary outside the bounds mentioned in (i) above. Two error responses, one for one set of model parameters (model 1) and one for a second set (model 2) are shown in Figure 8, where it is seen that the error using model 2 is much less than that using model 1. Therefore, model 2 was chosen to be used in the adaptive scheme. This off-line,

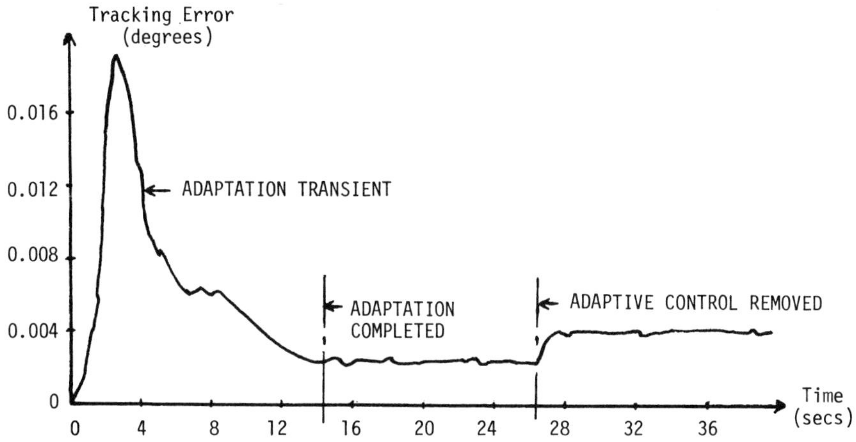

FIGURE 9. Tracking error with and without adaptive control.

crude identification scheme was very useful in limiting the transients in the system during the adaptation phase of operation. Values of model 2 parameters K_o and a_d are listed below.

Figure 9 illustrates the adaptive control system behavior when the following design values were used:

$LA = c_0(b_s\lambda_i)^{-1} = 10.0$ for $i = 0,1,3$; $LB = \lambda_4^{-1} = 10.0$; $c_0 = 1.1$; $a_d = -.1$; and $K_o = 0.5$.

From Figure 9 it can be seen that the error in tracking, after the initial transient, is reduced to almost one-half of the error in tracking without adaptive compensation. These error values are given in Table 2 for two different positions of the telescope antenna.

TABLE 2. Tracking accuracy of radio telescope (elevation at +100°).

POSITION IN AZIMUTH (degrees)	TRACKING ERROR	
	WITHOUT ADAPTIVE CONTROL (degrees)	WITH ADAPTIVE CONTROL (degrees)
250°	0.0043	0.0022
290°	0.0033	0.0016

It is also instructive to study the behavior of the control signal u during the adaptation phase. Figure 10 gives the behavior of u and the adaptive error signal for one run. Table 3 gives the initial and steady state values of the gains k^0, k^1, k^3, k^4 for different positions of the antenna. It can be seen that different steady state values of adaptive gains are required for different telescope positions. This indicates that system parameters are indeed changing as a function of azimuth position. These changes would be even more pronounced if elevation were also changing.

6. REMARKS AND CONCLUSIONS

Results from this study show that the discrete adaptive controller does improve the tracking accuracy of the radio telescope considerably. The maximum improvement which might be achieved could not be determined from this study due to limits on time available for field testing imposed by the heavy demand for the

TABLE 3. Steady state values of the adaptive gains (elevation at +100 degrees).

POSITION IN AZIMUTH (degrees)	ADAPTIVE GAINS		ADAPTIVE GAINS		ADAPTATION ERROR e of Fig. 7 (volts)
	k^0 initial= 0.0	k^1 initial=0.2	k^3 initial=0.0	k^4 initial=0.45	
250	0.3737	0.07341	-0.0309	-0.487	-0.1065×10^{-4}
255	0.3692	0.03137	0.4656	-0.508	-0.1839×10^{-4}
260	0.3715	0.02154	0.4470	-0.511	0.3378×10^{-4}
265	0.3646	0.02419	0.4292	-0.512	0.1320×10^{-3}
290	0.3953	0.1146	-0.0428	-0.476	-0.3018×10^{-4}
295	0.3995	0.1056	-0.0383	-0.476	-0.3018×10^{-4}

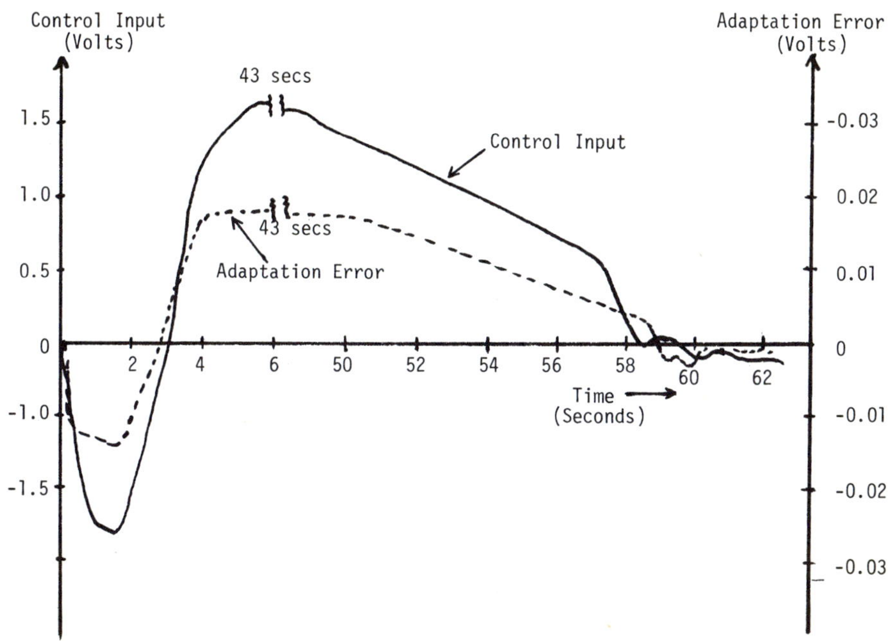

FIGURE 10. Control input and adaptation error for one run.

telescope for radio astronomy purposes. Several possibilities exist for improving on the results reported here. First, better position encoders could help improve performance. Current encoders are only accurate to within ±0.6 arc-seconds. Second, the concept of feedforward compensation could be combined with the adaptive controller as was successfully done in [7] and [11]. These possibilities should be investigated before finalizing the design of the adaptive controller.

REFERENCES

1. Ionescu, T. and Monopoli, R. V., "Discrete Model Reference Adaptive Control Systems with an Augmented Error Signal," Automatica, Vol. 13, pp. 507-517, Sept. 1977.
2. Monopoli, R. V., "Model Reference Adaptive Control with an Augmented Error Signal," IEEE-TAC, Vol. AC-19, pp. 474-484, October 1974.
3. Landau, I. D., "A Survey of MRA Techniques - Theory and Applications," Automatica, Vol. 10, pp. 353-379, July 1974.
4. Egardt, Bo, "Stability of Model Reference Adaptive and Self-Tuning Regulators," Department of A. C., Lund Inst. of Technology, Report No. LUTFD2/(TFRT-1017)/1-163/(1978).
5. Narendra, K. S. and Lin, Y., "Stable Discrete Adaptive Control," S & I S Report No. 7901, Yale University, March 1979.
6. Monopoli, R. V. and Troiani, M., "Discrete MRAS with Measurement Noise," Proceedings of the 7th Triennial World Conference of IFAC, Helsinki, pp. 1989-1994, June 1978.
7. Gilbart, J. W. and Winston, G. C., "Adaptive Compensation for an Optical Tracking Telescope," Automatica, Vol. 10, pp. 125-131, March 1974.
8. Reed, R. M., "Digital Implementation of a Lyapunov Designed Model Reference Adaptive Controller," Masters Thesis, University of Massachusetts, Amherst, May 1977.

9. Gutmann, R. L. and Monopoli, R. V., "Adaptive Control for a Laser Tracing System," Proceedings of the 21st Midwest Symposium on Circuits and Systems, ISU, Ames, Iowa, August 1978.
10. Sciabarassi, "A Real Time Lyapunov Designed Adaptive Controller," Masters Thesis, University of Massachusetts, Amherst, May 1975.
11. Gilbart, J. W., Winston, G. C., and Monopoli, R. V., "Combination Open Loop, Closed Loop, and Adaptive Compensation: Its Application to an Optical Tracking Telescope," Proceedings of the 1972 Int. Conf. Cybernetics and Society, Washington, D. C., October 1972.

ADAPTIVE CONTROL OF A LASER POINTING AND TRACKING SYSTEM: A FEASIBILITY STUDY[1]

R. L. Gutmann

Boeing Company
Seattle, Washington

R. V. Monopoli

University of Massachusetts
Amherst, Massachusetts

R. Van Allen
USAF, Kirtland A. F. B.
Albuquerque, New Mexico

A complete simulation study for an adaptive control technique applied to a laser pointing and tracking system is presented. It is shown that the adaptive gains very effectively reduce the unde-

[1]This work was supported by Applied Technology Associates under Air Force Contract F29601-77-C-0008.

ISBN 0-12-514060-6

sirable effects of static friction, coloumb friction, and unknown system parameters. Furthermore, the control is effective even though the sixth order system is modeled as a second order one for design purposes, and even with realistic levels of measurement noise present.

1. INTRODUCTION

In recent months an adaptive control technique has been developed to reduce many of the effects of static friction, coloumb friction, and unknown system parameters in a precision gimballed pointing and tracking system. The technique, known as Lyapunov-based model reference adaptive control, has been demonstrated through simulation to produce excellent results, even in the presence of a modest noise environment.

In this paper the Lyapunov-based model reference control algorithm for a gimballed laser pointing and tracking system is derived in detail; however, this should be considered as a generic example of the application of this technique. The intent is to provide the interested reader with a case history of the utilization of an adaptive control theory to a practical problem, thereby facilitating the application of the theory to new situations. The results of the simulation study presented here indicate the feasibility of applying this theory to real world problems.

Several excellent references exist on model-reference adaptive control theory. (Refer, for example, to [1],[10-12],[14], or [15].) The Lyapunov-based designs have an advantage over other adaptive controllers in that they guarantee overall system stability for the given assumptions on the plant (e.g., system order and form of the nonlinearities).

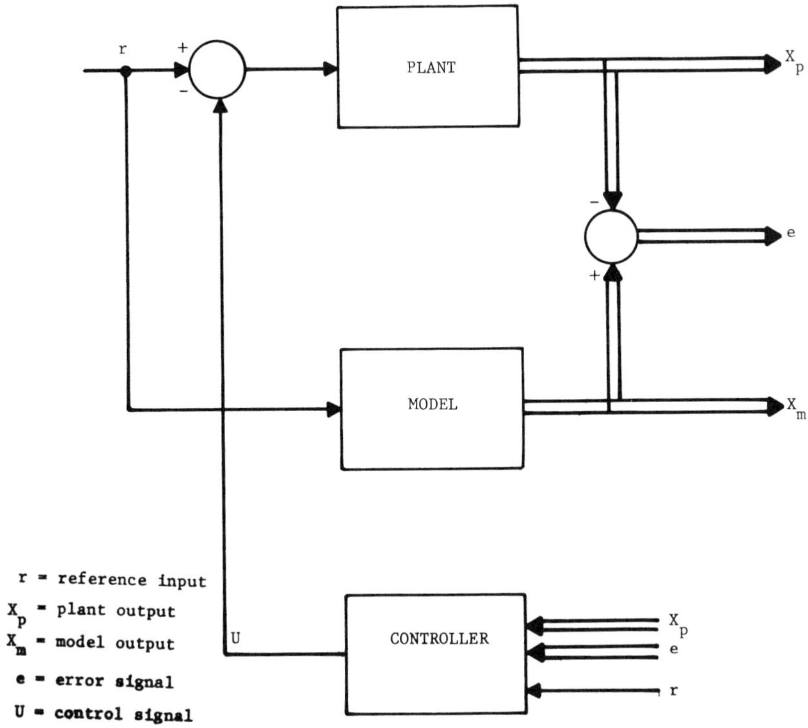

FIGURE 1. Model reference adaptive control system.

A block diagram of a functional model reference adaptive control system is shown in Figure 1. With respect to this figure, the plant is the system to be controlled. It is assumed that not all of the plant parameters are known. A model of the same order (number of state variables) as the plant is chosen by the system designer. The input to the model is the same as the reference input to the system. An adaptive controller is chosen so that the difference between the plant output (plant state variables) and the model output (state variables) goes to zero. The net closed loop system has the same input-output properties as the model. That is, the plant now behaves like the model.

The difference between the model outputs and plant outputs is known as the error signal. The intent of the control scheme is to drive the error signal to zero. Differential equations for the behavior of the error signal may be written (refer to Section 4), and the stability properties of these differential equations may be determined by the system designer. That is, it is possible to design a controller such that the system of differential equations describing the error signal is asymptotically stable (i.e., the error signal tends to zero). Lyapunov's direct method is utilized to show the asymptotic stability of this system (refer to [11] or [14]).

In the past, these techniques have been applied to an electrically activated NASA telescope as described in [2] and [3]. The undesirable effects of friction and unknown parameters were successfully reduced in this application. Similar studies have been applied to a simple d.c. electric motor and reported in [16] and [17]. Both of these previous results are indicative of the fact that one should expect similar improvements in accuracy when adaptive control techniques are applied to the gimballed pointing and tracking system described in this paper.

In the remainder of this paper the theoretical adaptive controller is derived, computer simulation test results are presented, practical implementation problems (such as noisy measurements and unavailable signals) are discussed, potential solutions are presented, and simulation results incorporating these solutions are given. A summary of this work is provided in Section 7. Earlier and related work on the adaptive control of this system is reported in [4-7].

2. THE PLANT

A simplified schematic of the system under consideration is depicted in Figure 2. The relative target position is resolved into azimuth and elevation loop commands, which pass through a series of compensation and stabilization networks and become inputs to the hydraulic actuator units that translate the commands into gimbal drive signals. This system utilizes four gimbals: an inner gimbal and an outer gimbal for each channel (azimuth and elevation). In order to provide a problem and subsequently a computer simulation of manageable dimensions, the subsection of the elevation loop shown in Figure 3 was chosen for study (the "Inner Gimbal" in Figure 2). It is felt that the effects of unknown parameters and friction are most serious in the inner gimbal structure. It should also be noted that the azimuth inner gimbal assembly is functionally equivalent, so that the controller developed for the elevation channel is expected to operate equally well in the azimuth channel.

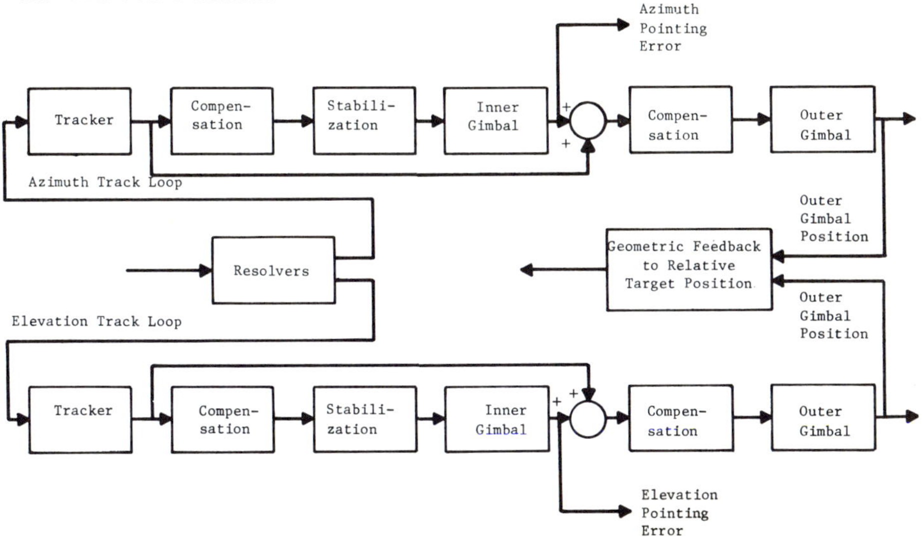

FIGURE 2. Simplified characterization of the pointing and tracking loops.

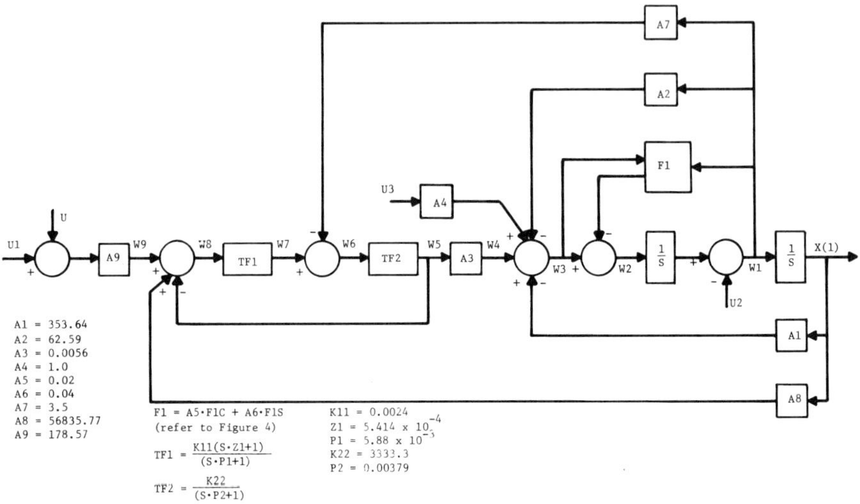

FIGURE 3. The elevation pressure loop and inner gimbal assembly.

Ideally, only the inner gimbal structure would be included in the system to be investigated, however it is necessary to inject control signals into the system, and the nearest access point to the gimbal is the pressure loop electronic input. Signals generated by the adaptive controller will be added to the channel at this point (input to the pressure loop). The effects of the pressure loop dynamics must be considered in the design of a controller (refer to Section 3 for a discussion of this aspect). With respect to Figure 3, U1 represents the commanded inner gimbal torque from the elevation stabilization loop (refer to Figure 2) and A9 represents electronic gains. The transfer functions TF1 and TF2 represent the hydraulic pressure loop, A3 is a pressure to torque conversion, and the two integrators represent the inner gimbal structure. Of the other two inputs, U2 represents outer gimbal motion coupled to inner gimbal dynamics and U3 represents linear acceleration disturbances due to gimbal imbalances. The gain A1 represents the gimbal spring term and is assumed in practice to be unknown. It is also assumed that A2 (gimbal damping) and A4 (acceleration disturbance moment arm), as well as constants associated with F1 (friction) are unknown in practice (the friction models are discussed below). The gain A7 represents hydraulic flow coupling between gimbal motion and the pressure unit; and A8 represents a spring cancellation term (since the function of this term is included in the adaptive controller, it will be set to zero when the controller is installed). The system output X(1) represents inner-outer gimbal angle which is proportional to gimbal tracking error. The input signal labeled U in Figure 3 is the control signal generated by the adaptive controller discussed in this paper. It is present only when the controller is present.

The compensation circuits, stability loops and system gains built into the system of Figure 2 were designed under the assumption that the inner gimbal assemblies (Figure 3) were _linear_ systems with _known_ parameters. As suggested in Figure 3, these assemblies contain unknown parameters (A1, A2, and A4, among others) and nonlinearities (friction). Collectively, the assumptions made about the inner gimbal assemblies by the system designers are known as the _design model_ (for the inner gimbals). The discrepancies between the design model and reality can lead to potentially significant pointing errors. In addition, the system designer is usually aware that his design model is not perfect and therefore designs compensators to handle a wide class of potential inner gimbal assemblies. If the exact inner gimbal model were at hand, these compensators could be simplified and optimized. Adaptive control techniques are applied to force the real plant of Figure 3 to behave like (have the same input-output properties as) the assumed design model (this design model then becomes the "Model" of Figure 1, and is discussed in detail below in Section 3). This allows the designer to simplify and optimize the compensators since they must now control a much narrower class of gimbal

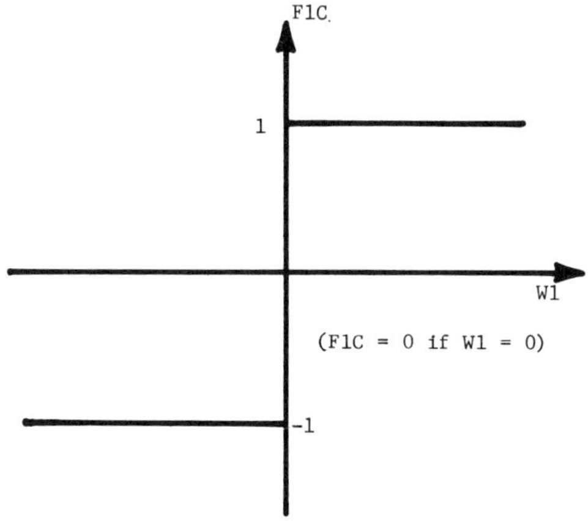

FIGURE 4a. Coloumb friction model.

assemblies. Furthermore some of the nonlinear effects of friction are removed by the adaptive controller (no previous effort was made to compensate directly for these nonlinear effects). Hence the net effect of the adaptive controller is to allow more accurate compensators to be built and to remove some of the system nonlinearities.

The friction model used represents static plus coloumb friction. Figures 4a and 4b depict these models. Coloumb friction force is assumed to be at a constant level and opposed to motion provided

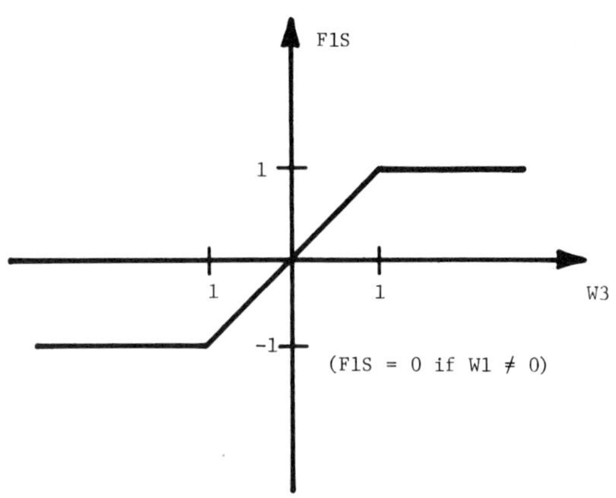

FIGURE 4b. Static friction model.

that the gimbal is in motion. If the gimbal is not in motion static friction force is assumed to be equal to, and opposed to, applied torque (up to the maximum static friction level, known as the "breakaway point"). The maximum friction level is assumed to be somewhat higher than coloumb friction level. However in practice both friction levels are assumed to be unknown.

3. THE MODEL

In general, the choice of model parameters is the prerogative of the system designer. In this instance the parameters were chosen so that the adaptive control "Model" (Figure 1) is the same as the design model discussed in Section 2. The model is linear and, of course, all of the coefficients are known. Therefore driving the error between plant and model to zero is equivalent to removing the effects of friction and uncertainties in parameters.

One of the primary concerns in designing this adaptive controller and the associated model is the effect of the hydraulic pressure loop dynamics. If the effects of the pressure loop are highly dependent on the frequency content of the input signals (over the range of input frequencies of concern) then a dynamic model of the pressure loop must be included as part of the overall model. On the other hand, if the frequency response of the pressure loop is relatively flat over the input range and if the delay (phase shift) is not a significant factor in compensation, then the pressure loop may be modeled as an (unknown) constant gain.

In order to appreciate these effects, the system of Figure 3 is redrawn in Figure 5. In Figure 5 the effects of flow coupling (A3•T2 in the figure) are separated from the effects of the remainder of the pressure loop (A9•T1•A3). The gains A9 and A3 are reciprocal so that A9•A3 = 1.0. The transfer function T1 is given by

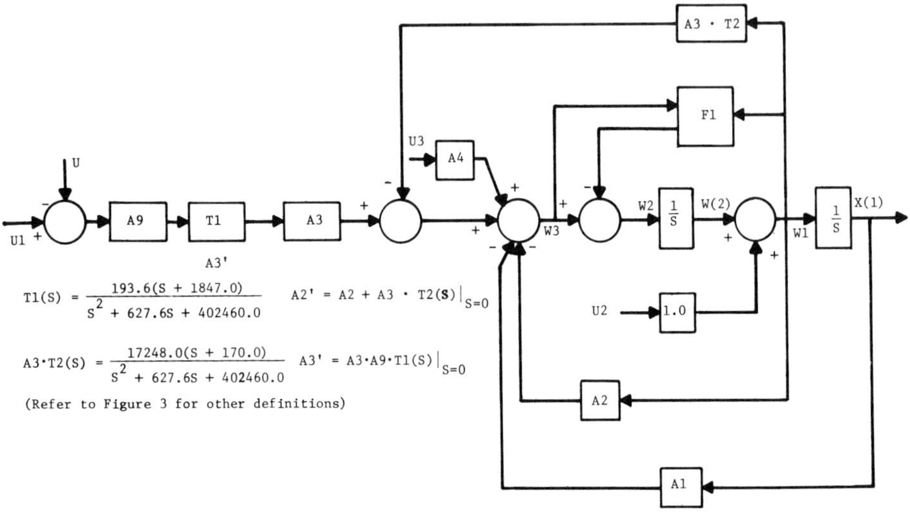

FIGURE 5. Simplified and reduced version of Figure 3.

$$T1(S) = \frac{193.6\ (S + 1847.0)}{S^2 + 627.6S + 402460.0} \tag{3-1}$$

This has a cutoff of approximately 100 Hz (the natural frequency of the inner gimbal assembly is approximately 1 Hz). It is assumed that the cutoff frequency is high enough to consider $|T1(j\omega)|$ flat over the region of interest. Therefore the adaptive compensation signals, which are introduced into the system as electronic signals into the pressure loop, can be thought of as being inputs to the inner gimbal loop applied directly through a constant gain. In order to determine the adequacy of this assumption, the effects of pressure loop delay on friction compensation were determined by simulations. With respect to Figure 3, the friction signal F1 is subtracted from W3. When F1 (modified by the pressure loop gain) is added to W9 (the signal access point) the net result is the elimination of practically all of the degradation effects due to friction. This demonstrates that the delay due to T1 is not significant in friction compensation. Note that this experiment employs perfect knowledge of friction torques, a situation never achieved in practice.

Hence we are able to choose the latter option of dealing with the pressure loop: treat it as an unknown, constant gain. This gain is labeled A2' (refer to Figure 5) and is defined by

$$A3' = A3 \cdot A9 \cdot T1(S)\Big|_{S=0}$$

where T1(S) is defined in Eq. (3-1).

Furthermore, with respect to Figure 5, the flow coupling effects are included in the transfer function A3•T2. This is given by

$$A3 \cdot T2 = \frac{17248.0\ (S + 170.0)}{S^2 + 627.6S + 402460.0} \tag{3-2}$$

which has a gain of approximately 7.3 at $\omega = 0$, and remains flat over the frequency range of interest. Hence A3•T2, in practice, acts as an additional (unknown) gain term operating in parallel with A2 (refer to Figure 5). That is, the combined effects of flow coupling and A2 are given by an unknown system gain, A2', where

$$A2' = A2 + A3 \cdot T2\Big|_{S=0} \tag{3-3}$$

Hence, even though we theoretically compensate for A2 with the adaptive controller, we will actually be compensating for A2'.

4. THE ADAPTIVE CONTROLLER

The function of the adaptive controller is to provide a control signal, U, that is used to drive the error between the model and the plant to zero. The control signal is a linear combination of various system signals whose coefficients are time-variable gains. These gains are adjusted by the adaptive controller in such a fashion as to force the error signal to zero. The algorithm discussed below embodies six adaptive gains. Adaptive gains may be easily added or deleted from this controller by following a procedure parallel to the one presented here. An explicit summary of the controller derived here is given at the end of this section.

From Figure 5, the differential equation describing the plant is given by

$$\dot{X}(1) = W1 = X(2) - U2$$

$$\dot{X}(2) = W2 = W3 - F1 \qquad (4\text{-}1)$$

$$= A3'(U1-U) + A4 \cdot U3 - A1 \cdot X(1) - A2' \cdot W1$$

$$- A5 \cdot F1C - A6 \cdot F1S$$

In order to simplify our discussion we make the assumptions:

1. that the functions F1C and F1S are of known form;
2. that A2' and A3' may be treated as constants as discussed in the previous section; and
3. that U1, U2 and U3 are measurable signals.

These assumptions are discussed further at the conclusion of this section.

On the basis of the discussion in Section 3, a second order model is chosen of the form

$$\dot{XM}(1) = XM(2) - U2$$

$$\dot{XM}(2) = -B1 \cdot XM(1) - B2 \cdot (XM(2) - U2) + B3 \cdot U1 \qquad (4\text{-}2)$$

where B1, B2 and B3 are known constants (chosen by the designer).

The error signal is defined as

$$E \triangleq (e_1, e_2)^T \qquad (4\text{-}3)$$

where

$$e_1 \triangleq XM(1) - X(1)$$

$$e_2 \triangleq (XM(2) - U2) - (X(2) - U2) = XM(2) - X(2)$$

We may write differential equations describing e_1 and e_2:

$$\dot{e}_1 = \dot{X}M(1) - \dot{X}(1) = (XM(2) - U2) - (X(2) - U2) = e_2$$

$$\dot{e}_2 = \dot{X}M(2) - \dot{X}(2) = -B1\cdot XM(1) - B2\cdot(XM(2) - U2) + (B3 - A3')\cdot U1 - A4\cdot U3 + A1\cdot X(1) + A2'\cdot W1 + A5\cdot F1C + A6\cdot F1S + A3'\cdot U$$

We add and subtract B1·X(1) and B2·W1 from this equation giving

$$\begin{aligned}\dot{e}_1 &= e_2 \\ \dot{e}_2 &= -B1(XM(1) - X(1)) - B1\cdot X(1) \\ &\quad -B2[(XM(2) - U2) - W1] - B2\cdot W1 + (B3 - A3\ )U1 \\ &\quad -A4\cdot U3 + A1\cdot X(1) + A2'\cdot W1 + A5\cdot F1C + AG\cdot F1S + A3'\cdot U \\ &= -B1e_1 - B2e_2 + (A1 - B1)\cdot X(1) + (A2' - B2)W1 \\ &\quad + (B3 - A3')U1 - A4\cdot U3 + A5\cdot F1C + A6\cdot F1S + A3'U\end{aligned} \tag{4-4}$$

If we choose

$$U = K1(t)\cdot X(1) + K2(t)\cdot W1 + K3(t)\cdot U1 + K4(t)\cdot U3 + K5(t)\cdot F1C + K6(t)\cdot F1S$$

(assumptions (1) and (2) allow us to construct such a signal, U), then we obtain

$$\begin{aligned}\dot{e}_1 &= e_2\\ \dot{e}_2 &= -B1e_1 - B2e_2 + (A1 - B1 + A3'K1)X(1)\\ &\quad + (A2' - B2 + A3'K2)\cdot W1 + (B3 - A3' + A3'K3)U1\\ &\quad + (A3'K4 - A4)\cdot U3 + (A3'K5 + A5)F1C\\ &\quad + (A3'K6 + A6)F1S \end{aligned} \tag{4-5}$$

If we define parameter errors:

$$\begin{aligned}\delta_1(t) &= A1 - B1 + A3'K1(t)\\ \delta_2(t) &= A2' - B2 + A3'K2(t)\\ \delta_3(t) &= B3 - A3 + A3'K3(t)\\ \delta_4(t) &= A3'K4(t) - A4\\ \delta_5(t) &= A3'K5(t) + A5\\ \delta_6(t) &= A3'K6(t) + A6\end{aligned} \tag{4-6}$$

then

$$\begin{aligned}\dot{e}_1 &= e_2\\ \dot{e}_2 &= -B1\cdot e_1 - B_2e_2 + \delta_1\cdot X(1) + \delta_2\cdot W1 + \delta_3\cdot U1\\ &\quad + \delta_4\cdot U3 + \delta_5\cdot F1C + \delta_6\cdot F1S\end{aligned}$$

This may be written in vector form as

$$\dot{E} = AE + \underline{b}w \tag{4-7}$$

where

$$A = \begin{bmatrix} 0 & 1\\ -a2 & -a1\end{bmatrix}; \quad \underline{b} = \begin{bmatrix} 0\\ 1\end{bmatrix}; \text{ and}$$

$$w = \delta_1\cdot X(1) + \delta_2\cdot W1 + \delta_3\cdot U1 + \delta_4\cdot U3 + \delta_5\cdot F1C + \delta_6\cdot F1S$$

Classic Lyapunov stability theory is used at this point to develop differential equations for δ_i, $i = 1,2,\ldots,6$ such that Eq. (4-7) is asymptotically stable (i.e., so that $E \to \underline{0}$ as $t \to \infty$). If we pick a Lyapunov function (V-function) of the form

$$V(E,\delta_i) = E^T PE + \sum_{i=1}^{6} \lambda_i \delta_i^2 \tag{4-8}$$

where P is a positive definite matrix described below, and $\lambda_i > 0$, $i = 1,2,\ldots,6$, its time derivative is

$$\dot{V} = \dot{E}^T PE + E^T P\dot{E} + \sum_{i=1}^{6} 2\lambda_i \delta_i \dot{\delta}_i \tag{4-9}$$

which can be written as

$$\dot{V} = E^T QE + 2E^T P\underline{b}w + 2 \sum_{i=1}^{6} \lambda_i \delta_i \dot{\delta}_i \tag{4-10}$$

where $P = [p_{ij}]$ satisfies the Lyapunov matrix equation $A^T P + PA = -Q$ and Q is an arbitrary positive definite matrix. We now write

$$\begin{aligned}\dot{V} = -E^T QE &+ 2(p_{12}e_1 + p_{22}e_2)(\delta_1 X(1)) + \delta_2 W1 \\ &+ \delta_3 U1 + \delta_4 U3 + \delta_5 F1C + \delta_6 F1S) \\ &+ 2 \sum_{i=1}^{6} \lambda_i \delta_i \dot{\delta}_i .\end{aligned} \tag{4-11}$$

It can now be seen that if

$$\begin{aligned}
\dot{\delta}_1 &= A3'\dot{K}1 = -(1/\lambda_1)(p_{12}e_1 + p_{22}e_2)X(1) \\
\dot{\delta}_2 &= A3'\dot{K}2 = -(1/\lambda_2)(p_{12}e_1 + p_{22}e_2)W1 \\
\dot{\delta}_3 &= A3'\dot{K}3 = -(1/\lambda_3)(p_{12}e_1 + p_{22}e_2)U1 \\
\dot{\delta}_4 &= A3'\dot{K}4 = -(1/\lambda_4)(p_{12}e_1 + p_{22}e_2)U3 \\
\dot{\delta}_5 &= A3'\dot{K}5 = -(1/\lambda_5)(p_{12}e_1 + p_{22}e_2)F1C \\
\dot{\delta}_6 &= A3'\dot{K}6 = -(1/\lambda_6)(p_{12}e_1 + p_{22}e_2)F1S
\end{aligned} \tag{4-12}$$

then

$$\dot{V} = -E^T QE. \tag{4-13}$$

Since $\dot{V}$ is negative semidefinite in all the coordinates and V is positive definite in all coordinates, by Lyapunov's direct stability theorems [8], the system is stable. Furthermore since $\dot{V}$ is negative definite with respect to E (termed <u>partial asymptotic stability</u> in [8]), we can guarantee a system that is stable in E and δ_1 through δ_6, and that $E(t) \to \underline{0}$ as $t \to \infty$.

Note that Eq. (4-12) implies that we must choose the ith gain by

$$\dot{K}i = -(1/A3'\lambda_i)(p_{12}e_1 + p_{22}e_2)S_i$$

where S_i is a system signal (X(1),W1, and so forth). Even though A3' is unknown, the λ_i is arbitrary so that we may define an arbitrary positive constant γ_{i2} by

$$\gamma_{i2} \triangleq (1/A3'\lambda_i)$$

where we use the fact that we know A3' to be positive (by gross inspection of the physical system). Henceforth by choosing

$$\dot{K}i = -\gamma_{i2}(p_{12}e_1 + p_{22}e_2)S_i$$

where $\gamma_{i2} > 0$, we satisfy Eq. (4-12) for some λ_i, $i = 1,2,\ldots,6$.

Although we have not shown it, Equation (4-12) is also asymptotically stable if the inputs (U1,U2, and U3) are sufficiently rich in frequency content (refer to [13] for a discussion of the proof). This means that generally we will also have $\delta_i(t) \to 0$, $i = 1,2,\ldots,6$. From the definition of the δ_i(Equation (4-6)) we see that this implies that

$$
\begin{aligned}
K1(t) &\rightarrow (B1 - A1)/A3' \\
K2(t) &\rightarrow (B2 - A2)/A3' \\
K3(t) &\rightarrow (A3 - B3)/A3' \\
K4(t) &\rightarrow A4/A3' \\
K5(t) &\rightarrow -A5/A3' \\
K6(t) &\rightarrow -A6/A3'
\end{aligned}
\tag{4-14}
$$

as $t \rightarrow \infty$.

Since the gain differential equations given above have the form:

$$\dot{K}i = -\gamma_{i2}(p_{12}e_1 + p_{33}e_2)S_i, \quad i = 1,2,\ldots,6. \tag{4-15}$$

where S_i is a system signal, the resulting gain is given by

$$Ki = Ki_o - \int_o^t \gamma_{i2}(p_{12}e_1 + p_{22}e_2)S_i(t)dt \tag{4-16}$$

where Ki_o is the initial value of the ith gain, and S_i is a system signal. It has been found in practice ([11],[16]) that if Ki has a "proportional part" in addition to the "integral part" given above, then the error terms (e_1,e_2) converge to zero more quickly. That is, the equation for the ith gain is generally given by

$$
\begin{aligned}
Ki = Ki_o &- \gamma_{i1}(p_{12}e_1 + p_{22}e_2)S_i(t) \\
&- \int_o^t \gamma_{i2}(p_{12}e_1 + p_{22}e_2)S_i(t)dt
\end{aligned}
\tag{4-17}
$$

where the system designer is free to choose the positive constants, γ_{ij}, $i = 1,2,\ldots,6$ and $j = 1,2$.

We must account for the three assumptions made earlier. All three assumptions are true only in approximation. The effects of the difference between the assumptions made and reality are addressed using computer simulations which are discussed in detail in the following section. First, the functions F1C and F1S cannot be measured nor directly generated (since the signal W3 is not measurable). Several attempts were made to replicate these functions (with, for example,

$$F2' = \begin{cases} \text{SIGN(W1)} & \text{if } W1 \neq 0 \\ \text{SIGN(E)} & \text{if } W1 = 0 \end{cases}$$

and

$$F3' = \begin{cases} 0 & \text{if } W1 \neq 0 \\ \text{SIGN(E)} & \text{if } W1 = 0 \end{cases}$$

used to compensate for F1C and F1S respectively); however it was found that performance equal to or better than any of the combinations of two functions could be obtained with a single function, F2, given by

$$F2 = \begin{cases} \text{SIGN(W1)} & \text{if } W1 \neq 0 \\ \text{SIGN(XM(2) - U2)} & \text{if } W1 = 0 \end{cases} \quad (4\text{-}18)$$

Using F2 to compensate for friction has the added benefit of reducing the number of adjustable gains from six to five.

The justification for the second assumption (treating A2' and A3' as constants) was presented in the previous section and has been supported by computer simulations as shown in the following section. For the third assumption note that the signal U1 is electronic and may be used directly. The other two (U2 and U3) must be obtained from transducers. Since at this time the exact nature of the transducers to be used is unknown, it is assumed for this paper that U2 and U3 are explicitly available.

The net result of these modifications is that the model equations become

$$\dot{XM}(1) = XM(2) - U2$$
$$\dot{XM}(2) = -B1 \cdot XM(1) - B2(XM(2) - U2) + B3 \cdot U1 \quad (4\text{-}19)$$

The controller becomes

$$U = K1 \cdot X(1) + K2 \cdot W1 + K3 \cdot U1 + K4 \cdot U3 + K5 \cdot F2 \tag{4-20}$$

and the equations for the gains become

$$\begin{aligned}
K1 &= K1_o - \gamma_{11} \cdot h \cdot X(1) - \int_o^t \gamma_{12} \cdot h \cdot X(1)dt \\
K2 &= K2_o - \gamma_{21} \cdot h \cdot W1 - \int_o^t \gamma_{22} \cdot h \cdot W1dt \\
K3 &= K3_o - \gamma_{31} \cdot h \cdot U1 - \int_o^t \gamma_{32} \cdot h \cdot U1dt \\
K4 &= K4_o - \gamma_{41} \cdot h \cdot U3 - \int_o^t \gamma_{42} \cdot h \cdot U3dt \\
K5 &= K5_o - \gamma_{51} \cdot h \cdot F2 - \int_o^t \gamma_{52} \cdot h \cdot F2dt
\end{aligned} \tag{4-21}$$

where

$K1_o, K2_o, \ldots, K5_o$ are the initial values of the respective gains; $h = (p_{12}e_1 + p_{22}e_2)$; and $\gamma_{11}, \gamma_{12}, \ldots, \gamma_{15}, \gamma_{21}, \ldots, \gamma_{25}$ are positive constants chosen by the system designer.

Recall that

$$\begin{aligned}
e_1 &= XM(1) - X(1) \\
e_2 &= (XM(2) - U2) - W1
\end{aligned} \tag{4-22}$$

and that

$$F2 = \begin{cases} \text{SIGN (W1)} & \text{if } W1 \neq 0 \\ \text{SIGN (XM2 - U2)} & \text{if } W1 = 0 \end{cases} \tag{4-23}$$

Equations (4-19) through (4-23) must be implemented in real time in order to utilize the adaptive controller. Either digital or analog implementation may be used; however, the implementation issue is not discussed here as it is considered tangential. Equations (4-19) through (4-23) represent a summary of the adaptive control scheme developed in this paper.

5. SIMULATIONS OF THE BASIC ALGORITHM

In this section we represent some simulation results representative of the algorithm just described. In these simulations, the parameters A1, A2' (refer to Figure 5), A3', and A5 are assumed to be unknown. Only one system input, U1, is used for these simulations (hence A4 does not enter the system so K4 $\equiv$ 0, and X(2) = W1). We also assumed that (spring cancellation term) A8 has been removed as discussed in Section 2. Referring to Figures 3, 4a and 4b, the friction levels used for these simulations are A5 = 0.02 and A6 = 0.04 (of course, these levels are assumed to be unknown in practice and for the purposes of the following simulations). The friction levels (A5 and A6) were chosen on the basis of preliminary experimental data. The values chosen appear to be quite conservative. The improvements exhibited below would be even more dramatic with higher friction levels.

There appears to be no scientific method for choosing the constants γ_{ij}, i = 1,2,...,5, j = 1 or 2, of Equation (4-21) at this time. The larger these constants are, the faster the gains converge (hence the faster $E \rightarrow \underline{0}$). However, large constants cause large transients in controller signals at start-up. These transients may or may not be a problem in the actual system; they are a problem in computer simulations of the system in that they are difficult to numerically integrate. In addition, the sensitivity of the system to measurement noise is directly proportional to certain of these constants. Hence the ad hoc optimization scheme used to set these constants was to increase their value (speed up convergence) as much as possible, while maintaining numerical integration stability. Research is being conducted by various parties into finding a more suitable algorithm for setting the constants.

In the following plots, we show the system input, U1 (refer to Figure 2 or Figure 4), the system output, X(1), and the error signal, e_1, (difference between model output and plant output). The test input signal (U1) used in these cases is a 0.2 Hz, sine wave, which produces approximately 0.5 mrad. output sine wave.

The first input-output pair (Figure 6a) represents the system with friction, but with no adaptive compensation. The most significant feature here is the nonlinear appearance of the output signal (flat tops on the sine wave and zero output for the initial 0.5 sec. of simulation).

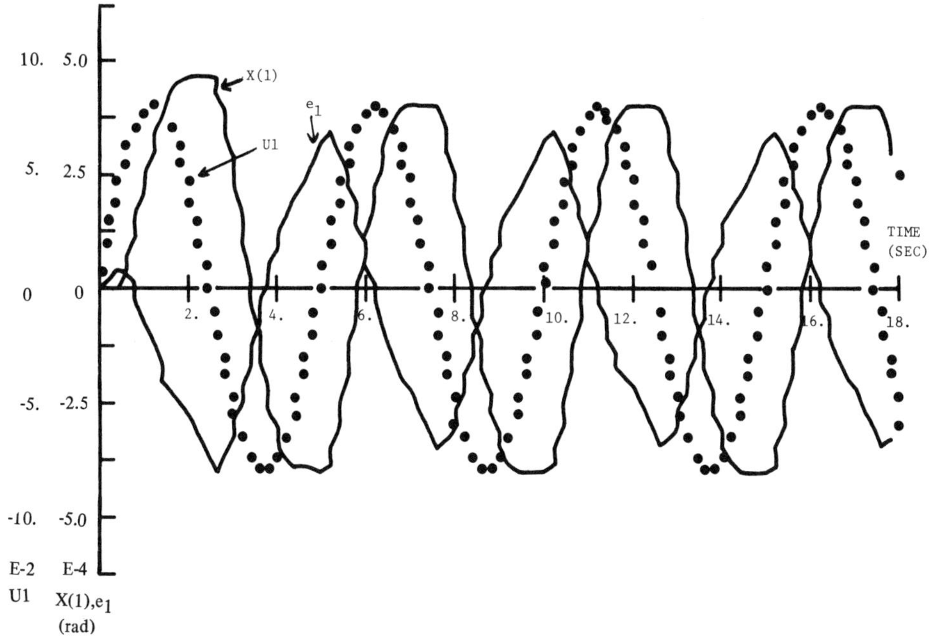

FIGURE 6a. No adaptive gains.

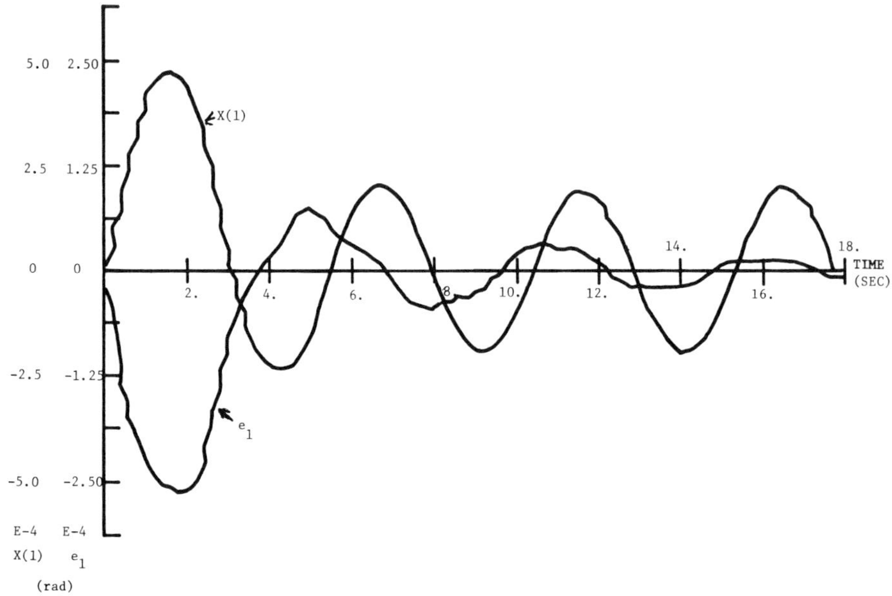

FIGURE 6b. Four adaptive gains, no noise.

Figure 6b depicts the behavior of the tracking system when the adaptive controller is used. This controller is the one discussed in the previous section with K4 deleted and only one friction gain, K5. This figure is important for several reasons. First it demonstrates some of the advantages of the adaptive controller. The output is missing the nonlinear characteristics of Figure 6a. Also, we are assured that the problems associated with the unknown parameters are being mitigated since the error signal is converging to zero. Secondly, the assumptions made explicitly in previous sections are found not to cause major problems. Particularly:

1. The pressure loop may be considered a constant gain for some frequencies. This implies that we may ignore the pressure loop in the Model (as we have done) and that we may add the

control signal, U, at the summing junction indicated in Figure 3 without worrying about the dynamic delays associated with the pressure loops.

2. One friction compensation function, F2, is adequate for this application.

These initial simulation results are promising; therefore a more detailed analysis is performed to investigate some problems associated with the practical application of this algorithm.

6. PRACTICAL CONSIDERATIONS

In this section we address two problems commonly met in the application of adaptive controllers:

i) not all of the signals necessary to implement the controller are available; and

ii) the signals that are available are corrupted by additive noise.

Research into these problems may be found in [9],[11], or [13]. These problems are considered together here since one set of solutions is found that alleviates both problems simultaneously. In our case the signal W1 = X(2) - U2 is unavailable. However since X(1) = W1, we generate W1 by differentiating, then smoothing X(1). Furthermore X(1) is expected to be somewhat noisy. This tends to complicate the generation of W1 since the differentiation operator is very sensitive to noise. The derivative of X(1) was taken and smoothed using the combined first order transfer function.

$$\hat{W1} = X(1) \cdot \frac{\alpha S}{S + \alpha} \tag{6-1}$$

The signal X(1) is smoothed in the same filter that computes W1. The transfer function for this smoothing is given by

$$\hat{X}(1) = X(1) \cdot \frac{\alpha}{S + \alpha} \tag{6-2}$$

The error signals $\hat{e}_1$ and $\hat{e}_2$ are formed in a similar manner:

$$\hat{e}_1 = (XM(1) - X(1)) \cdot \frac{\alpha}{S + \alpha} \tag{6-3}$$

and

$$\hat{e}_2 = (XM(1) - X(1)) \cdot \frac{\alpha S}{S + \alpha} \tag{6-4}$$

We now use $\hat{e}_1$ and $\hat{e}_2$ in place of e_1 and e_2 in Eq. (4-21). Figure 7 demonstrates the effects of noise distributed uniformly over $\pm 1.0 \times 10^{-5}$ added to X(1). For this simulation the value of α in Eq. (6-1)-(6-4) was set to 60.0 (this constitutes a filter with a broad passband hence little filtering for these signals). This simulation demonstrates that noise has the effect of slowing or disrupting the convergence of E to $\underline{0}$.

Tightening the bandwidths of the filters described above improves the convergence of E to $\underline{0}$ in the presence of noise, however additional modifications were found to further improve convergence. First, in obtaining the results shown in Figure 7, it was noted that the gain K2 was quite noisy due to the noise present in the signals used to generate K2. Hence, a first order filter was added to smooth K2. This filter and the filters of Eq. (6-1)-(6-4) are shown in Figure 8.

A second modification was found to further improve convergence. This modification involves the addition of another adjustable gain. In order to explain the function of this gain, the reader is re-

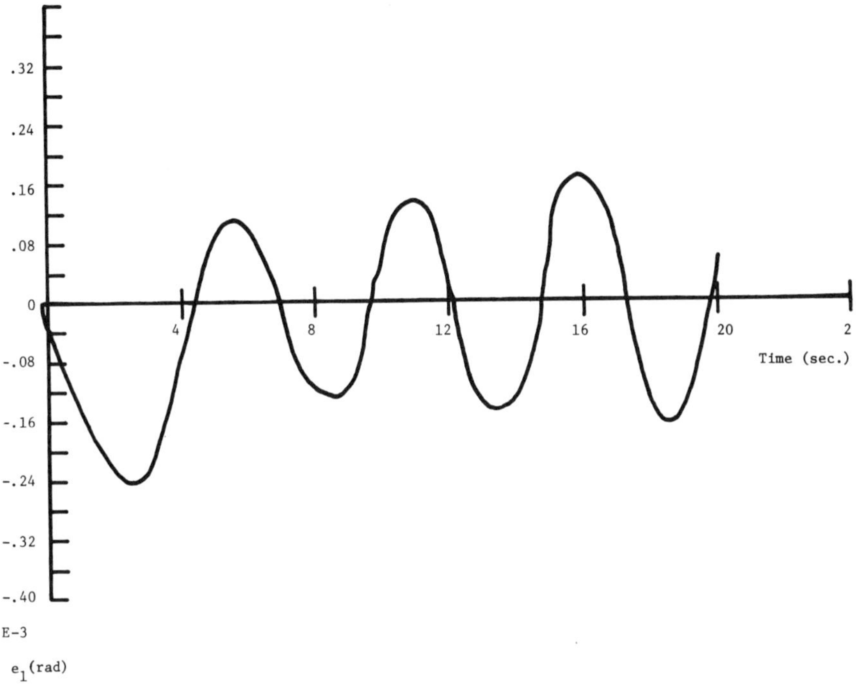

FIGURE 7. 5% Noise, 4 gains, broad band filters.

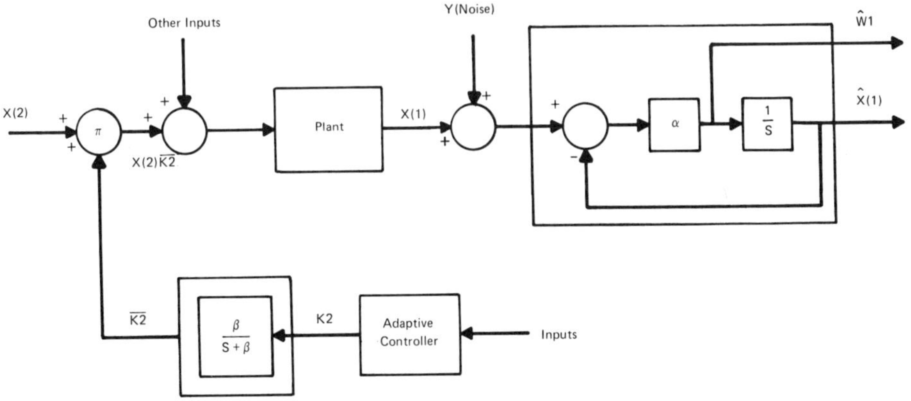

FIGURE 8. Location and nature of noise filters.

ferred to the development of the algorithm in Section 4. The noise associated with the various system signals tends to introduce positive terms that make $\dot{V}$ (Eq. 4-13) less negative. This is directly relateable to slower convergence. Hence if $\dot{V}$ can be made more negative without changing any of the functions of the terms already established, better (faster) convergence can generally be established. A pseudo gain (K7) is used to do this. The gain K7 can be thought of as compensating for a fictitious bias in the input (U1).

The differential equation for K7 would then be given by

$$\dot{K}7 = -(1/\lambda_7)(p_{12}e_1 + p_{22}e_2)\cdot 1.0 \tag{6-5}$$

Following the pattern established in Section 4 we assume that K7 is composed of a proportional part plus an integral part. That is

$$K7 = K7_o - \gamma_{71}(p_{12}e_1 + p_{22}e_2) + \int_o^t \gamma_{72}(p_{12}e_1 + p_{22}e_2)dt \tag{6-6}$$

(S = 1.0 in equation 4-17.) It was found through simulation that the integral part added little to the final result. The equation used for K7 is therefore given by

$$K7 = K7_o - \gamma_{71}(p_{12}e_1 + p_{22}e_2) \tag{6-7}$$

where

$$K7_o = 0$$

(note $\hat{e}_1, \hat{e}_2$ are used in place of e_1, e_2 in practice).

The control signal, U, of Eq. (4-20) is modified to incorporate K7:

$$U = K1 \cdot X(1) + K2 \cdot W1 + K3 \cdot U1 + K5 \cdot F2 + K7 \quad (6\text{-}8)$$

(note that K4 has been dropped since U3 = 0).

Figure 9 depicts the convergence of X(1) to XM(1) using the filters of Figure 8 with $\alpha = 20$, $\beta = 10$ and using the pseudo-gain of Eq. (6-7). The peak to peak error in Figure 9 is 6×10^{-5} as compared to 3.5×10^{-4} in Figure 7. Integral squared error for the first 20 seconds is 2.6×10^{-8} for Figure 9 as compared to 3×10^{-7} for Figure 7. This represents significant improvement. Using our filtering and pseudo-gains, when the noise level is doubled (2.0×10^{-5}) the peak to peak error becomes 8×10^{-5} and the integral squared error becomes 2.8×10^{-8}, not too different from Figure 9.

7. CONCLUSION

The adaptive controller derived in this paper eliminates the uncertainties of unknown parameters and many of the effects of friction in simulations of a pointing and tracking system. The

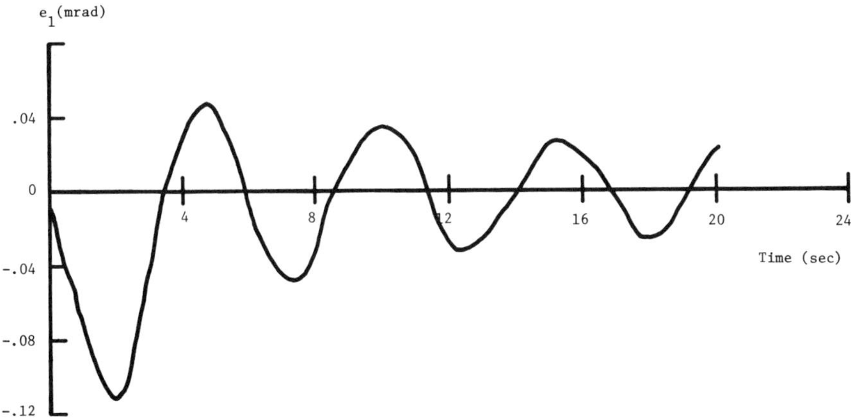

FIGURE 9. 5% Noise, 5 gains, all filters.

adaptive controller was modified so that it operates well with limited signal information and with noisy signals. More elaborate simulation will be conducted following a more accurate characterization of system signals. A trial controller will then be tested on the system hardware.

As a result of adding this controller, the remaining portions of the control system can be optimized to drive a known linear system.

REFERENCES

1. Feuer, A., and Morse, A. S., "Adaptive Control of Single Input Single Output Linear Systems," IEEE Trans. Auto. Control, Vol. AC-23, August 1978.
2. Gilbart, J. W., Monopoli, R. V., and Winston, G. C., "Combination Open Loop, Closed Loop, and Adaptive Compensation: Its Application to an Optical Tracking System," Proc. 1972 Int. Conf. Cybernetics and Society, Washington, D. C., Oct. 1972.
3. Gilbart, J. W., and Winston, G. C., "Adaptive Compensation for an Optical Telescope," Automatica, April 1974.
4. Gutmann, R. L., "Adaptive Identification and Control of a Gimballed Laser Pointing and Tracking System," Report No. 41, Vol. II, Research Reports, 1978, USAF-ASEE Summer Faculty Research Program, Conducted by Auburn University, Auburn, Alabama, September 1978.
5. Gutmann, R. L., and Monopoli, R. V., "Adaptive Control of a Laser Tracking System: A Feasibility Study," Applied Technology Associates, Publication Number ATA-R-5, June 1978.
6. Gutmann, R. L., and Monopoli, R. V., "A Supplement to 'Adaptive Control of a Laser Tracking System: A Feasibility Study'," Applied Technology Associates, September 1978.

7. Gutmann, R. L., and Monopoli, R. V., "Adaptive Control of a Laser Tracking System," Proceedings of the 21st Midwest Symposium on Circuits and Systems, August 14-15, 1978, Ames, Iowa.
8. Hahn, W., "Stability of Motion," Springer-Verlag, New York, 1967.
9. Kessler, F. W., Jr., "Design of Adaptive Control for Systems with Measurement Noise," M. Sc. Thesis, University of Massachusetts, Amherst, May 1977.
10. Landau, I. D., "A Survey of Model Reference Adaptive Techniques, Theory and Applications," Automatica, July 1974.
11. Monopoli, R. V., "Model Reference Adaptive Control with an Augmented Error Signal," IEEE Trans. Auto. Control, Vol. AC-19, 1974.
12. Monopoli, R. V., "On Model Reference Adaptive Control Systems," Proc. 1977 Allerton Conf. on Circuit and System Theory, October 1977.
13. Monopoli, R. V., and Kessler, F. M., Jr., "Measurement Noise in Model Reference Adaptive Systems," Proc. 1977 Joint Auto. Control Conf., San Francisco, Ca.
14. Narendra, K. S., and Kudva, P., "Stable Adaptive Schemes for System Identification and Control--Parts I and II," IEEE Trans. on Syst., Man, and Cybernetics, Vol. SMC-4, No. 6, November 1974.
15. Narendra, K. S., and Valavani, L. S., "Stable Adaptive Control Design--Direct Control," IEEE Trans. Auto. Control, Vol. AC-23, August 1978.
16. Reed, R. M., "Digital Implementation of a Lyapunov Designed Model Reference Adaptive Controller," M. S. Thesis, University of Massachusetts, Amherst, 1977.
17. Sciabarrasi, J. E., and Monopoli, R. V., "A Real Time, Liapunov Designed, Adaptive Controller," NASA Grant NGL-22-010-018 Report No. 75-2, 1975.

CLASSICAL CONTROL INTERPRETATION AND DESIGN OF MICROCOMPUTER ADAPTIVE CONTROLLERS[1]

C. P. Neuman
R. L. Morris[2]

Carnegie Institute of Technology
Carnegie-Mellon University
Pittsburgh, Pennsylvania

Technological advances in large-scale integration (LSI) and integrated circuit (IC) design and fabrication are increasing the capabilities of microprocessors and pertinent interfaces while reducing their cost. These advances offer the control engineer both low-cost computing power and the prospect of implementing control theoretic methods which have been developed over the past thirty years.

This paper focuses on the classical control interpretation of three classes of model reference adaptive controllers (MRAC) for process control applications:

---

[1]This work has been supported in part by the National Science Foundation under Grant ENG 76-81346.

[2]Present address: Control Systems Department, Scientific Research Laboratory, Ford Motor Company, Dearborn, Mi.

ISBN 0-12-514060-6

- o Identification error method
- o Output error method
- o Input error method.

These MRAC are compared in a common framework from the viewpoint of microcomputer implementation. Techniques of classical control engineering are applied to examine the adaptive pole shifting and zero cancellation characteristics of each design. It is demonstrated that the choice of either a full-parallel (FP) or a series-parallel (SP) reference model in the adaptive controller dramatically affects the performance of the closed-loop MRAC. The distinguishing features of gradient and least-squares adjustment mechanisms are summarized in the context of practical considerations imposed by the microcomputer.

The MRAC are evaluated by simulating the analog process on an EAI TR-20 analog computer and testing the control algorithm on a PDP-11/40 minicomputer and then on a MDS-800 Microcomputer Development System designed around an Intel 8080A microprocessor. Guidelines for the design and application of these microcomputer MRAC are developed in this unified theoretical and experimental framework. These digital controllers can effectively utilize the computational power and intelligence of the microcomputer.

I. INTRODUCTION

Advanced technological processes, such as high-performance aircraft and robots, must be controlled accurately over a wide range of operating conditions. This environment establishes the characteristics of the feedback system which are required to control the process within the design specifications. Design of such controllers is a formidable engineering problem and fixed controllers may compromise the performance of the process over the entire operating range.

Adaptive controllers can alleviate these design problems. Since adaptive controllers are self-tuning, the performance of the process remains at the specified level in spite of changes in, or lack of knowledge of, the process characteristics. Through advances in large-scale integration (LSI) technology, microcomputers are inexpensive and powerful tools for implementing adaptive controllers. Adaptive control algorithms that are effective in the face of practical constraints remain to be developed. While designs of microcomputer adaptive controllers emerge from a spectrum of approaches, relationships between these designs are lacking in the literature. Little attention is paid to practical constraints that arise in microcomputer implementation. Algorithms are exemplified through large-scale digital simulation, without consideration of practical constraints.

The practical constraints may be imposed by the process, microcomputer, sensors or actuators. Process characteristics dictate the speed of the adjustment mechanism and the quality of the control signal. A fast process requires an adjustment mechanism with similar dynamic characteristics. Processes with higher-order dynamics, such as robot arms, cannot tolerate noisy control signals. The microcomputer limits the magnitude of the control signal and sets the minimum cycle time for the algorithm. Thus, algorithm complexity is a key consideration. Errors in the available signals can degrade the performance of algorithms that rely on accurate information.

The goal of this paper is to highlight discrete model reference adaptive controller (MRAC) designs in a common framework and provide guidelines for their microcomputer implementation. The analytic approaches provide tools for developing practical and robust microcomputer adaptive controllers. The salient literature is referenced and new results are stated. Their complete development will appear elsewhere.

This paper is organized as follows. In Section II, the underlying concepts of identification [1-2], output [3-5], and input [6-7] error MRAC designs are reviewed from the viewpoint of microcomputer implementation. Then, in Section III, these designs are compared in a common framework. Classical control engineering techniques are applied in Section IV to interpret the adaptive pole shifting and zero cancellation action of each design upon convergence. In Section V, the distinguishing characteristics of the adjustment mechanisms are summarized in the light of practical considerations. Finally, the design implications of this unified approach to microcomputer MRAC development are indicated in Section VI.

II. MICROCOMPUTER MRAC

The model reference adaptive controllers (MRAC) applicable for the microcomputer control of linear (single-input---single-output) processes are composed of the three functional blocks:

① Process

② Reference Model

③ Adaptive Controller (a digital feedback controller with adjustable gains).

Block ① is the analog process under microcomputer control and blocks ② and ③ are algorithms stored in the read only memory (ROM) of the microcomputer. The parameters of the process under control are unknown and the reference model output x(j) prescribes the dynamic response of the closed-loop control system. Microcomputer implementation of a MRAC is illustrated in Figure 1.

The analog reference signal r(t) and the process output y(t) are fed to the microcomputer through analog-to-digital (A/D) converters. The control signal u(j) is fed to the process through

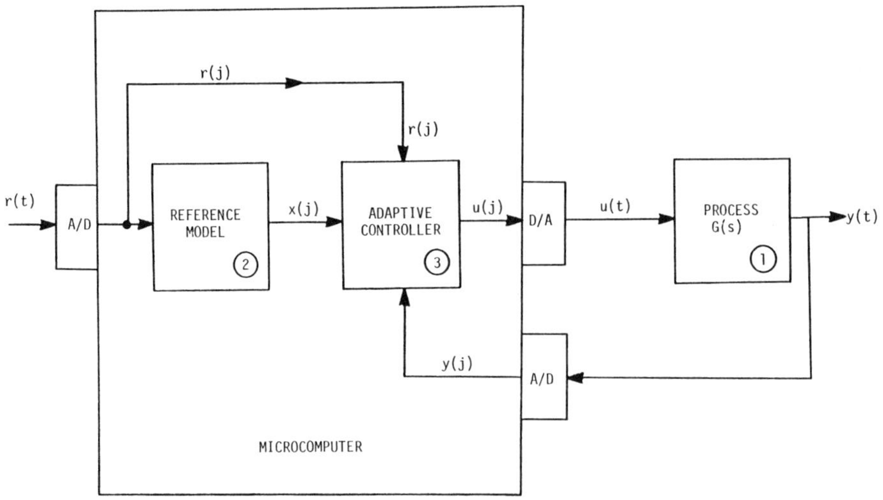

FIGURE 1. Microcomputer model reference adaptive controller (MRAC).

the digital-to-analog (D/A) converter. The D/A and A/D are the interfaces between the digital controller and the process. Each of the functional blocks is described in turn.

2.1 Process Modeling for Microcomputer Control

Through the application of the pertinent physical laws or experimentation, the (single-input---single-output) physical process is modeled (around operating points) by the (lumped parameter) analog transfer function

$$G(s) = \frac{Y(s)}{U(s)} = \frac{f_{n_1} s^{n_1} + f_{n_1-1} s^{n_1-1} + \ldots. + f_1 s + f_o}{g_n s^n + g_{n-1} s^{n-1} + \ldots. + g_1 s + g_o} . \quad (1)$$

In (1), n is the order of the process and $n_1 < n$. While both n and n_1 are specified by the engineer, the process parameters $\{f_i\}$ and $\{g_\ell\}$ are unknown. Since the analog process is under microcomputer control, the input is updated at each sampling instant and maintained piecewise constant by the digital-to-analog converter (D/A). The D/A is the interface between the digital controller and the process. At each sampling instant, the digital-to-analog converter transmits a piecewise constant input to the process. To facilitate control system analysis and design, the D/A is modeled by the equivalent cascade of an ideal impulse sampler and zero-order-hold (ZOH). In turn, the cascade of the ideal impulse sampler, ZOH, and analog process is completely equivalent (at the sampling times) to the discrete-time model characterized by a digital transfer function. Accordingly, the digital transfer function $W(z^{-1})$ of the cascade of the D/A converter and analog process can be obtained from the analog transfer function G(s) through the formula [8]

$$W(z^{-1}) = [1 - z^{-1}] \, Z \left\{ [L^{-1} \{\frac{G(s)}{s}\}] \Big|_{t \leftarrow kT} \right\},$$

where T is the constant sampling period. In signal processing, $W(z^{-1})$ is the digital transfer function of the step-invariant filter. The step-response of the step-invariant filter [9] matches the step-response of G(s) at the sampling times. The FORTRAN IV program DIGIT [10] has been implemented to generate the digital transfer function $W(z^{-1})$ of the step-invariant digital filter directly from the analog transfer function G(s). The salient outputs of the program are the digital transfer functions expressed as ratios of polynomials, their partial-fraction expansions, loci of poles and zeros as a function of the sampling period, and frequency responses.

The digital transfer function[3] of the discrete-time model is therefore [10-11]

$$W(z^{-1}) \triangleq W(z^{-1};T)$$

$$= \frac{B_p(z^{-1})}{A_p(z^{-1})}$$

$$= \frac{b_1(T)z^{-1}+b_2(T)z^{-2}+\ldots\ldots+b_{n-1}(T)z^{-(n-1)} + b_n(T)z^{-n}}{1-a_1(T)z^{-1}-a_2(T)z^{-2}-\ldots\ldots-a_{n-1}(T)z^{-(n-1)}-a_n(T)z^{-n}}$$

$$= \frac{b_1 z^{-1} + B(z^{-1})}{1 - A(z^{-1})} \quad . \tag{2}$$

The output of the discrete-time model matches the output of the analog process at the sampling instants.

If

$$s = \lambda_\ell \quad \text{for} \quad \ell = 1,2,\ldots,n$$

are the poles of the analog process G(s), then

$$z = \exp\,[\lambda_\ell T] \quad \text{for} \quad \ell = 1,2,\ldots,n$$

---

[3] The z-transform $U(z^{-1})$ of the sampled time-signal u(kT) is defined according to [8]

$$U(z^{-1}) = U(z^{-1};T) \triangleq \sum_{k=0}^{\infty} u(kT)z^{-k} = Z[u(kT)].$$

Multiplication of $U(z^{-1};T)$ by z^{-i} corresponds to delaying u(kT) by iT seconds, where T is the constant sampling period. That is,

$$z^{-i}U(z^{-1};T) = Z\,\{u[(k-i)T]\} \ .$$

are the corresponding poles of $W(z^{-1})$. In contrast to the specification of the denominator $A_p(z^{-1})$ of $W(z^{-1})$, conversion of the analog process dynamics into the numerator $B_p(z^{-1})$ of the digital transfer function is not straightforward. Even an analog process without numerator dynamics leads to a digital transfer function with numerator dynamics. The numerator coefficients $b_i(T)$ of $W(z^{-1})$ are nonlinear functions of all of the process parameters and the sampling period T. For processes without pure delays, the leading coefficient $b_1(T) \neq 0$ regardless of the order n_1 of the numerator of the process transfer function G(s). In fact, the digital transfer function $W(z^{-1})$ of the cascade of a D/A and an analog process with n poles and n_1 zeros ($n_1 \leq n$) has n poles and

$$(n-1) \text{ zeros if } n_1 < n$$

and

$$n \text{ zeros if } n_1 = n.$$

The process can therefore be modeled by the input-output difference equation

$$y(j) = \left[\sum_{i=1}^{n} b_i q^{-i}\right] u(j) + \left[\sum_{i=1}^{n} a_i q^{-i}\right] y(j) \tag{3a}$$

$$= b_1 q^{-1} u(j) + B(q^{-1})u(j) + A(q^{-1})y(j), \tag{3b}$$

where q^{-1} is the backward-shift or delay operator[4] [12]. In (3), $u(j) \triangleq u(jT)$ and $y(j) \triangleq y(jT)$ are, respectively, the process input

---

[4]The q^{-1} operator is defined according to

$$\{q^{-1}\}y(j) = y(j-1)$$

and

$$\{q^{-N}\}y(j) = y(j-N)$$

where N is a positive integer. Furthermore,

$$\{q^{0}\}y(j) \triangleq y(j).$$

and output at the j-th sampling instant and T is the constant sampling period. The unknown coefficients $\{a_i\}$ and $\{b_i\}$ are nonlinear functions of the unknown process parameters $\{f_i\}$ and $\{g_\ell\}$ and the sampling period T. From the authors' experience, estimating b_1 is a key problem in MRAC implementation. This problem is addressed in Section III.

2.2 Reference Model

To define a <u>realizable</u> output for the process, the reference model is specified by the cascade of a D/A converter and an analog process with optimum response characteristics [13]. The sampled reference signal r(j) drives the reference model. The output of the full-parallel (FP) reference model x(j) can thereby be represented by the m-th order input-output difference equation

$$x(j) = \left[\sum_{i=1}^{m} b_i^o q^{-i}\right] r(j) + \left[\sum_{i=1}^{m} a_i^o q^{-i}\right] x(j) \tag{4a}$$

$$= b_1^o q^{-1} r(j) + B^o(q^{-1}) r(j) + A^o(q^{-1}) x(j) \tag{4b}$$

$$= B_m(q^{-1}) r(j) + [1-A_m(q^{-1})] x(j) \ . \tag{4c}$$

The process (3) and the reference model (4) are characterized by similar difference equations; the superscript zero in (4) is introduced to signify the fixed coefficients of the reference model. In practice [14], the reference model is often selected to respond as a second-order process with a tunable undamped natural frequency ω_n (seconds^{-1}) and dimensionless damping ratio ξ.

2.3 Controller

To implement adaptive controllers on a microcomputer for processes with dominant time-constants on the order of one second, algorithm complexity is a key consideration. This eliminates adaptive state-variable controllers because an adaptive state observer/parameter estimator would be required for single-input/single-output processes, and such controllers complicate the algorithm beyond the capabilities of presently available microcomputers. Discrete-time, asymptotically stable, adaptive control algorithms are sought for microcomputer implementation

Within this framework, the block ③ controllers which are applicable to microcomputer MRAC can be designed from three approaches:

- o Identification error method
- o Output error method
- o Input error method.

The identification error method [1,14 and 15] utilizes a series-parallel (SP) identifier to estimate the process coefficients. The identification error

$$e_{ID}(j) = y(j) - t(j) \quad , \tag{5}$$

which is the difference between the process output y(j) and the identifier output t(j), drives the adjustment mechanism for the identifier coefficients. These coefficients are then utilized to update the controller gains. Although Landau [16] has developed the full-parallel (FP) identifier, it has not, to the best of the authors' knowledge, been applied to the closed-loop adaptive control designs reviewed in this paper.

In the output error method [3-5], the difference

$$e_o(j) = x(j) - y(j) \tag{6}$$

between the reference model output x(j) and the process output y(j) drives the adjustment mechanism which directly updates the controller gains.

In the input error method [6-7], a one-step-ahead optimal controller utilizes information available at the j-th sampling instant to calculate what the optimal control signal $u^*(j-1)$ should have been at the previous sampling instant. This control signal is compared with the control signal u(j-1) that was actually produced by the adaptive controller at the (j-1)th sampling instant to form the input error

$$e_I(j) = u^*(j-1) - u(j-1) \; . \tag{7}$$

The input error drives the adjustment mechanism which directly updates the controller gains.

For each method, with the exception of the self-tuning regulator [15], closed-loop stability and reference model tracking are guaranteed by demonstrating that the appropriate error signal is asymptotically stable. The stability of the self-tuning regulator is proved from the stochastic point of view. Theoretical considerations alone do not lead to guidelines for selecting which design to apply. Practical considerations do detect the differences.

As described in Section III, the resulting controllers are intimately related. MRAC designs incorporate such enhancements as: the state variable filters and auxiliary signals of Ionescu and Monopoli [3] and Narendra and Lin [5], the SP reference model of Baradello [14], the control signal penalty of Johnson and Tse [6], and the auxiliary fixed controller of Neuman and Morris [7]. As illustrated in Section IV, each of these designs results in adaptive pole shifting and zero cancellation. When the enhancements are removed, each implements the identical controller following the convergence of the adjustment mechanism.

III. A COMMON FRAMEWORK

Two features tie the aforementioned approaches into a common framework:

(i) These MRAC can be written in the equation error [17] format for adjustment mechanism design, and

(ii) The form of the resulting controller is identical for each design.

If the control signal is sufficiently rich [18], the adaptive mechanism drives the controller gains to their true values. The unenhanced designs are reviewed to relate these MRAC.

3.1 Identification Error Method

This approach utilizes an SP identifier having the input-output dynamics of the process model (3). The identifier output is calculated according to

$$t(j) = \left[\sum_{i=1}^{n} \hat{b}_i(j-1)q^{-i}\right] u(j) + \left[\sum_{i=1}^{n} \hat{a}_i(j-1)q^{-i}\right] y(j), \tag{8}$$

where $\{\hat{b}_i(j-1)\}$ and $\{\hat{a}_i(j-1)\}$ estimate the unknown process coefficients at the (j-1)-th sampling instant. The equation error form of the identification error (5) is

$$e_{ID}(j) = \left[\sum_{i=1}^{n} [b_i - \hat{b}_i(j-1)]q^{-i}\right] u(j)$$

$$+ \left[\sum_{i=1}^{n} [a_i - \hat{a}_i(j-1)]q^{-i}\right] y(j) . \tag{9}$$

Adjustment mechanisms are designed in [19] to guarantee the asymptotic stability of the identification error.

The controller is constructed by solving (8) for u(j-1) in terms of t(j). Then the identifier output

$$t(j) = x^{R}(j-1) \tag{10}$$

is forced to follow the model output $x^{R}(j)$ with one computational cycle delay. The resulting controller is

$$u(j) = \frac{1}{\hat{b}_1(j)} \left\{ x^{R}(j) - \left[\sum_{i=2}^{n} \hat{b}_i(j) q^{-(i-1)} \right] u(j) - \left[\sum_{i=1}^{n} \hat{a}_i(j) q^{-(i-1)} \right] y(j) \right\} . \tag{11}$$

When

$$x^{R}(j) = x(j+1) \tag{12}$$

which is physically realizable since x(j) is the output of the FP reference model (4), the full-parallel output error (6) becomes

$$\begin{aligned} e_o(j) &= x(j) - y(j) \\ &= x^{R}(j-1) - y(j) \\ &= t(j) - y(j) \\ &= - e_{ID}(j) . \end{aligned} \tag{13}$$

If $e_{ID}(j)$ is guaranteed to be asymptotically stable, then the output error $e_o(j)$ is also asymptotically stable, and the process output (3) tracks the reference model (4).

When the adjustment mechanism converges, the control signal (11) is calculated according to

$$u(j) = \frac{1}{b_1}\left\{x^R(j) - \left[\sum_{i=2}^{n} b_i q^{-(i-1)}\right] u(j) - \left[\sum_{i=1}^{n} a_i q^{-(i-1)}\right] y(j)\right\}. \tag{14}$$

3.2 The Output Error Method

In this approach the controller gains $\{\hat{K}(j),\ \hat{K}_i^u(j)$, and $\hat{K}_i^y(j)\}$ are adjusted directly. Upon substituting (3), (4), and (12) into (6), the output error becomes

$$e_o(j) = x^R(j-1) - \left[\sum_{i=2}^{n} b_i q^{-i}\right] u(j) - \left[\sum_{i=1}^{n} a_i q^{-i}\right] y(j) - b_1 u(j-1). \tag{15}$$

When the control u(j) is calculated according to

$$u(j) = \hat{K}(j) x^R(j) + \left[\sum_{i=2}^{n} \hat{K}_i^u(j) q^{-(i-1)}\right] u(j) + \left[\sum_{i=1}^{n} \hat{K}_i^y(j) q^{-(i-1)}\right] y(j), \tag{16}$$

the output error (15) becomes

$$e_o(j) = b_1 \left\{ [K-\hat{K}(j-1)x^R(j-1) + \left[\sum_{i=2}^{n} [K_i^u - \hat{K}_i^u(j-1)]q^{-i} \right] u(j) \right.$$
$$\left. + \left[\sum_{i=1}^{n} [K_i^y - \hat{K}_i^y(j-1)]q^{-i} \right] y(j) \right\}, \tag{17}$$

which is in the equation error form and the starting point for the design of asymptotically stable adjustment mechanisms. In (17), $K = \frac{1}{b_1}$, $K_i^u = \frac{-b_i}{b_1}$ for i = 2,3,..,n and $K_i^y = \frac{-a_i}{b_1}$ for i = 1,2,..,n. When the adjustment mechanism converges, the controller (16) becomes (14) and the output error $e_o(j)$ is asymptotically stable.

3.3 The Input Error Method

In the input error method, the quadratic cost function

$$J[u(j)] = \frac{1}{2} [x^R(j) - y(j+1)]^2 + \frac{Q^2}{2} [u(j)]^2 \tag{18}$$

is defined to penalize the <u>predicted</u> output error $e_o(j+1)$ at the (j+1)-th sampling instant and the applied control signal u(j) at the j-th sampling instant. By penalizing the control signal, the luxury of having zero-output error is sacrificed for the reality of applying a reasonable control signal. The control which minimizes (18) is

$$u^*(j) = \frac{\pi(b_1,Q)}{b_1} \left\{ x^R(j) - \left[\sum_{i=2}^{n} b_i q^{-(i-1)} \right] u(j) \right.$$
$$\left. - \left[\sum_{i=1}^{n} a_i q^{-(i-1)} \right] y(j) \right\}, \tag{19}$$

where

$$\pi(b_1,Q) \triangleq [1 + (\frac{Q}{b_1})^2]^{-1} . \tag{20}$$

As in the output error method, the control signal u(j) is estimated according to (16). The input error (7) becomes

$$e_I(j) = \pi(\hat{b}_1,Q)\left\{\frac{1}{\hat{b}_1}\left[x^R(j-1)-y(j)\right]\right\} - \left[1-\pi(\hat{b}_1,Q)\right] u(j-1) \quad (21)$$

where

$$\pi(\hat{b}_1,Q) \triangleq [1 + (\frac{Q}{\hat{b}_1})^2]^{-1} . \quad (22)$$

In turn, (21) is equivalent to

$$\begin{aligned} e_I(j) = \sigma(b_1,\hat{b}_1,Q) \Bigg\{ & [K^* - \hat{K}(j-1)]x^R(j-1) \\ & + \left[\sum_{i=2}^{n} [K_i^{*u} - \hat{K}_i^u(j-1)]q^{-i}\right] u(j) \\ & + \left[\sum_{i=1}^{n} [K_i^{*y} - \hat{K}_i^y(j-1)]q^{-i}\right] y(j) \Bigg\}, \end{aligned} \quad (23)$$

which is again in the equation error form. In (23),

$$\sigma(b_1,\hat{b}_1,Q) = \frac{\pi(\hat{b}_1,Q)}{\Omega(b_1,\hat{b}_1,Q)} \quad (24)$$

$$\Omega(b_1,\hat{b}_1,Q) = [(\frac{b_1}{\hat{b}_1}) + (\frac{Q}{\hat{b}_1})^2]^{-1} \quad (25)$$

$$K^* = \frac{\Omega(b_1,\hat{b}_1,Q)}{\hat{b}_1}$$

$$K_i^{*u} = -\frac{\Omega(b_1,\hat{b}_1,Q)}{\hat{b}_1} b_i \quad \text{for } i = 2,3,\ldots,n$$

and

$$K_i^{*y} = -\frac{\Omega(b_1,\hat{b}_1,Q)}{\hat{b}_1} a_i \quad \text{for } i = 1,2,\ldots,n.$$

When an asymptotically stable adjustment mechanism drives $e_I(j)$ to zero, the control signal converges to

$$u(j) = \frac{\Omega(b_1, \hat{b}_1, Q)}{\hat{b}_1} \left\{ x^R(j) - \left[\sum_{i=2}^{n} b_i q^{-(i-1)} \right] u(j) - \left[\sum_{i=1}^{n} a_i q^{-(i-1)} \right] y(j) \right\} . \qquad (26)$$

When the control penalty $Q = 0$, equation (26) reduces to (14). In this framework, each of the unenhanced MRAC converges to the identical controller (14). Furthermore, each can employ the same equation error adjustment mechanism to guarantee asymptotic stability.

When the control signal saturates, by exceeding the dynamic range of the D/A interface, only the identification error (9) remains valid. The output and input error equations, (17) and (23) respectively, are violated during control signal saturation. Their adjustment mechanisms must then be turned-off until the control signal again falls within the dynamic range of the D/A. The output and input error methods lose important information during control signal saturation, while the identification error method is unaffected.

The maximum adjustment at each computational cycle is determined by relaxation gains which are chosen to satisfy stability criteria. In both the output and input error methods, these gains depend upon an estimate of the leading coefficient b_1 in the numerator of the process transfer function (2). If there is significant error in this estimate of b_1, then either the closed-loop system becomes unstable or the adjustment at each computational cycle becomes negligible [19]. The relaxation gains for the identification error method do not require an estimate of b_1.

IV. CLASSICAL CONTROL INTERPRETATIONS OF ADAPTIVE CONTROLLERS

Enhanced MRAC are designed in the literature [3,5–7 and 14]. The enhancements prescribe the controller, detail the actions of the feedback filters, and influence the performance of the closed-loop MRAC. Knowledge of these effects can enable microcomputer adaptive controllers to become a practical reality.

The control penalty [6] and the SP reference model [14] in the adaptive controller block ③ of Figure 1 are the two enhancements considered in this section. Comparison of (11) and (26) indicates that a control penalty, heretofore associated with the input error method alone, can be incorporated into the identification error method. This is accomplished by multiplying (11) by a positive constant π less than unity. Inclusion of a control penalty in the output error method, however, would destroy the equivalence of (6) and (17).

The SP reference model

$$x^R(j-1) = B_m(q^{-1})r(j) + [1-A_m(q^{-1})]y(j) \tag{27}$$

is realized by replacing the recursive component of (4) by non-recursive filtering of the process output y(j). The SP reference signal (27) is then substituted directly into the identification (11), output (16), and input (16) error controllers.

When the adjustment mechanisms converge and the adjustable gains become constant parameters, the controllers can be analyzed and interpreted in the context of classical control engineering. Since each controller can be implemented according to (16), the results presented in this section are independent of the adjustment mechanism. When the gains converge, this controller can be rewritten in terms of z-transform transfer functions as follows:

$$U(z^{-1}) = \frac{\hat{K}}{1-\hat{K}^u(z^{-1})} X^R(z^{-1}) + \frac{\hat{K}^y(z^{-1})}{1-\hat{K}^u(z^{-1})} Y(z^{-1}) \tag{28}$$

where

$$\hat{K}^u(z^{-1}) = \sum_{i=2}^{n} \hat{K}_i^u z^{-(i-1)} , \tag{29a}$$

$$\hat{K}^y(z^{-1}) = \sum_{i=1}^{n} \hat{K}_i^y z^{-(i-1)} , \tag{29b}$$

and the $\{\hat{K}_i^u\}$ and $\{\hat{K}_i^y\}$ are constant gains. The reference model signal $X^R(z^{-1})$ in the adaptive controller block ③ can be generated by either the FP reference model (4) and (12) or the SP reference model (27). The impact of FP and SP reference models is examined in Sections 4.1 and 4.2, respectively.

4.1 Root-Locus Interpretation of a FP Reference Model Adaptive Controller

From (4) and (12), the reference model signal

$$X^R(z^{-1}) = \frac{B_m(z^{-1})}{z^{-1}A_m(z^{-1})} R(z^{-1}) . \tag{30}$$

To implement MRAC, the control signal (28) is expressed in terms of realizable transfer functions. Substitution of (30) into (28) yields

$$U(z^{-1}) = \left[\frac{1}{1-\hat{K}^u(z^{-1})}\right]\left[\frac{\hat{K}\, B_m(z^{-1})}{z^{-1}A_m(z^{-1})}\right] R(z^{-1}) + \left[\frac{1}{1-\hat{K}^u(z^{-1})}\right]\left[\hat{K}^y(z^{-1})\right] Y(z^{-1}) . \tag{31}$$

In turn, substitution of (2), (16), (25), and (26) into (31) leads to the closed-loop transfer function

$$\frac{Y(z^{-1})}{R(z^{-1})} = \left[\frac{B_m(z^{-1})}{A_m(z^{-1})}\right] \left\{ \frac{\left(\frac{\hat{b}_1}{Q^2}\right) \frac{W(z^{-1})}{z^{-1}}}{1 + \left(\frac{\hat{b}_1}{Q^2}\right) \frac{W(z^{-1})}{z^{-1}}} \right\} . \tag{32}$$

The FP MRAC is thus characterized by an unity negative feedback system. Because of the zero at the origin, $\frac{W(z^{-1})}{z^{-1}}$ has an equal number (n) of poles and zeros. When the controller gains $\{\hat{K}, \hat{K}_i^u, \hat{K}_i^y\}$ converge, the dynamics of the process <u>and</u> reference model characterize the closed-loop MRAC.

The transfer function (32) indicates that as the control penalty $Q \to +\infty$, the closed-loop poles approach the process poles and the MRAC provides little control action. As the control penalty $Q \to 0$, the closed-loop poles cancel the process zeros and the additional zero at the origin, and the process output $Y(z^{-1})$ tracks the full-parallel reference model ② without delay.

The behavior of (32) can be visualized for finite values of the control penalty Q through the second-order process

$$W(z^{-1}) = \frac{b_1 z^{-1} + b_2 z^{-2}}{1 - a_1 z^{-1} - a_2 z^{-2}} \tag{33}$$

with the real poles[5] $0 < p_1 < 1$ and $0 < p_2 < 1$. The open-loop zero $z_o = -b_2/b_1$.

---

[5]From [11], the transfer function (33) models the cascade of a D/A and the analog process

$$G(s) = \frac{K}{(T_1 s+1)(T_2 s+1)} .$$

The real poles $p_1 = \exp[-T/T_1]$ and $p_2 = \exp[-T/T_2]$.

A control penalty is incorporated into the identification and input error methods to reduce the:

- o Magnitude of the control signal and thereby the effect of control signal saturation, and
- o Noise in the control signal,

while retaining reasonable closed-loop performance [19]. For the second-order process (33), the root-locus illustrates that a control penalty shifts the closed-loop poles into the complex interior of the unit circle and thus leads to an under-damped process output. The authors have found experimentally that increasing the control penalty reduces control signal noise and saturation by shifting the closed-loop poles into the highly oscillatory regions of the z-plane.

The action of the controller (31) in the absence of a control penalty (Q = 0) can be interpreted as follows. Since $[1-\hat{K}^u(z^{-1})] = \frac{1}{b_1} \frac{B_p(z^{-1})}{z^{-1}}$, the (n-1) controller poles cancel the process zeros and the feedback gains $\{K_i^y\}$ shift the n process poles to the origin, (n-1) of which are then cancelled by the controller zeros [19]. The closed-loop MRAC displays a deadbeat response. In terms of the frequency response of the closed-loop MRAC, the feedback loop acts as a high-pass filter when the process under control is a low-pass filter. The feedback filter amplifies measurement noise and deteriorates the quality of the control signal [19]. When the control signal saturates in the D/A converter, the large feedback gains lead to a marginally stable system. Inclusion of a control penalty reduces the feedback gains by the scale factor $\Omega(b_1,\hat{b}_1,Q)$ or π, increases the stability margin and attenuates noise in the feedback loop at the expense of deteriorating the performance of the closed-loop MRAC.

4.2 Root-Locus Interpretation of a SP Reference Model Adaptive Controller

A SP reference model in block ③ is intended to reduce the effect of control signal saturation and improve the quality of the control signal [20]. When utilized in conjunction with a control penalty, the actions of the controller and the performance of the closed-loop MRAC are markedly different from those of the FP reference model adaptive controller.

Since the SP reference model is specified by (27), the SP reference signal

$$X^R(z^{-1}) = \frac{B_m(z^{-1})}{z^{-1}} R(z^{-1}) + \frac{A^o(z^{-1})}{z^{-1}} Y(z^{-1}). \tag{34}$$

Substitution of (34) into (28) yields the control signal

$$\begin{aligned} U(z^{-1}) = & \left[\frac{1}{1-\hat{K}^u(z^{-1})}\right]\left[\hat{K}\,\frac{B_m(z^{-1})}{z^{-1}}\right] R(z^{-1}) \\ & + \left[\frac{1}{1-\hat{K}^u(z^{-1})}\right]\left[\hat{K}^y(z^{-1}) + \hat{K}\,\frac{A^o(z^{-1})}{z^{-1}}\right] Y(z^{-1}). \end{aligned} \tag{35}$$

When the controller gains converge, substitution of (2), (4), (16), (25), and (26) into (35) leads to the closed-loop transfer function

$$\frac{Y(z^{-1})}{R(z^{-1})} = B_m(z^{-1}) \left\{ \frac{\left(\frac{\hat{b}_1}{Q^2}\right) \frac{W(z^{-1})}{z^{-1}}}{1 + A_m(z^{-1})\left(\frac{\hat{b}_1}{Q^2}\right) \frac{W(z^{-1})}{z^{-1}}} \right\}. \tag{36}$$

The reference signal $R(z^{-1})$ is filtered to produce $F(z^{-1})$ which is the input to a feedback system with process output $Y(z^{-1})$. In parallel with the study of the FP reference model, interest focuses on the closed-loop transfer function

$$\frac{Y(z^{-1})}{F(z^{-1})} = \frac{\left(\frac{\hat{b}_1}{Q^2}\right) \frac{W(z^{-1})}{z^{-1}}}{1 + A_m(z^{-1}) \left(\frac{\hat{b}_1}{Q^2}\right) \frac{W(z^{-1})}{z^{-1}}} \quad . \tag{37}$$

In contrast to the closed-loop FP transfer function in (32) which has n poles and n zeros, the SP transfer function (37) has (n+m-1) poles and (n+m-1) zeros, m of which are located at the origin.

The transfer function (37) indicates that as the control penalty $Q \rightarrow +\infty$, n of the closed-loop poles approach the process poles and (m-1) poles approach the origin. Consequently little control effort is exerted. As $Q \rightarrow 0$, m of the closed-loop poles approach the poles of the reference model and (n-1) poles approach the process zeros. The closed-loop transfer function (36) replicates the transfer function of the FP reference model ② ; that is,

$$\frac{Y(z^{-1})}{R(z^{-1})} \longrightarrow \frac{B_m(z^{-1})}{A_m(z^{-1})} \quad . \tag{38}$$

The behavior of (37) can be exemplified for $0 < Q < +\infty$ through the second-order process (33). Block ② is specified by the second-order reference model

$$\frac{X(z^{-1})}{R(z^{-1})} = K_o z^{-1} \frac{(1 - z_o z^{-1})}{(1 - p_o z^{-1})^2} \tag{39}$$

with a critically damped response. The root-locus of (37) illustrates the difference between incorporating FP and SP reference models in the adaptive controller block ③ . The FP reference model forces the controller to drive the process to follow the reference model ② output without delay. The SP reference model requires the controller to shift the response characteristics of the process toward those of the reference model.

The action of the SP reference model adaptive controller (35) upon convergence can be explained further. When Q = 0, $[1-\hat{K}^u(z^{-1})] = \frac{1}{b_1}\frac{B_p(z^{-1})}{z^{-1}}$. The (n-1) controller poles cancel the process zeros and the (n+m) feedback gains $\{\hat{K}_i^y | i = 1,2,\ldots,n\}$ and $\{a_i^o | i = 1,2,\ldots,m\}$ shift m of the process poles to the reference model poles and (n-m) of the process poles to the origin which are then cancelled by the controller [19]. To avoid the aforementioned pitfalls of the FP reference model, it is essential that the order m of the reference model equal the order n of the process. When Q ≠ 0, (n-m) poles are placed at the origin. The controller now shifts (n+m-1) poles to the locations specified by (37), instead of simply cancelling the process zeros.

The advantage of the SP reference model arises from the fact that the feedback gains shift the closed-loop poles to the poles of the reference model rather than to the origin. Adaptive state-variable feedback designs accomplish this pole shifting. Since the feedback gains are smaller, the stability margin is larger when the control signal is limited by the D/A. Furthermore, the frequency response characteristics of the SP feedback loop are markedly different from those of the FP MRAC [19]. The SP reference model attenuates the higher frequencies. The amount of attenuation depends upon the difference (n-m) between the order n of the process and the order m of the SP reference model.

When a control penalty is introduced, the closed-loop poles approach the process poles. For the second-order process (33) the SP path of the closed-loop poles is through a less oscillatory region of the z-plane than the FP path. When the sampling period is a small fraction of the dominant process time-constants, this phenomenon generalizes to higher-order processes. In this case the process poles always lie closer to the poles of a properly designed reference model ② than the origin. When a control penalty is introduced to enhance the control signal, the performance of the closed-loop SP MRAC does not deteriorate to the extent of the FP MRAC. This behavioral comparison is magnified in their microcomputer implementation.

V. ADJUSTMENT MECHANISMS

5.1 Introduction

The single-input---single-output MRAC highlighted in Section III can be characterized by the equation error

$$e(j) = \sigma L(q^{-1})[\underline{\beta} - \underline{\hat{\beta}}(j-\ell)]^T \underline{\theta}(j-1) . \qquad (40)$$

In (40), the information vector $\underline{\theta}_{[2n\times 1]}$ contains n past input-output pairs, $\underline{\beta}_{[2n\times 1]}$ contains the unknown process parameters, $\underline{\hat{\beta}}_{[2n\times 1]}(j-\ell)$ contains estimates of these 2n unknown parameters at the $(j-\ell)$-th computational cycle, σ is a constant whose sign and range can be determined from process modeling, and $L(q^{-1})$ is a linear filter whose coefficients depend on the design and may be either constant or time-varying. The index ℓ specifies the number of sampling periods between successive parameter adjustments. When $\ell > 1$, MRAC designs [19] require that the linear filter $L(q^{-1}) = 1$. It is noted parenthetically that (2n+1) and 3n gains are adapted in the auxiliary fixed controller [7] and extended information vector [21] designs, respectively.

5.2 Output Error

For the output error design without enhancement, (40) reduces to (17) in which $\sigma = b_1$, $L(q^{-1}) = 1$, and the output error is prescribed by (13). When the SP reference model is introduced to enhance the MRAC design, (40) again reduces to (17), but the reference signal $x^R(j)$ is calculated according to (27) in both the controller (16) and output error equation (13). When state-variable filters [3-5] are incorporated, $L(q^{-1}) = \frac{D(q^{-1})}{A_m(q^{-1})}$ where

$$D(q^{-1}) = 1 + \sum_{i=1}^{m_1} d_i q^{-i}$$

is the state-variable filter

$$D(q^{-1})\ \underline{\theta}^f(j) = \underline{\theta}(j) \ . \tag{41}$$

The filtered information vector $\underline{\theta}^f(j-1)$ replaces $\underline{\theta}(j-1)$ in (40). The SP reference signal (27) is utilized in $\underline{\theta}(j-1)$ and in the controller (16) while the error calculation (13) processes the FP reference signal (4) and (12). It is interesting to note that if $D(q^{-1}) = A_m(q^{-1})$ in the state-variable filter, then $L(q^{-1}) = 1$ and (40) again reduces to (17) but $\underline{\theta}^f(j-1)$ replaces $\underline{\theta}(j-1)$. Furthermore, if the state-variable filter is applied only to the FP output error (4), (12) and (13), which is analogous to the FP identifier design of Landau [16], then (40) emerges directly.

5.3 Input Error

The input error design also leads to (40). Without enhancement, (40) reduces to (24) in which

$\sigma = \sigma(b_1, \hat{b}_1 0) = \dfrac{b_1}{\hat{b}_1}$. When a control penalty is introduced, $\sigma = \sigma(b_1, \hat{b}_1, Q)$, as defined in (24), and the parameter vector $\underline{\beta}$ is specified by (26). When the SP reference model (27) and state-variable filter (41) are incorporated, the foregoing summary of the output error design applies directly to the input error MRAC.

5.4 Identification Error

In the identification error method, $\sigma = 1$ in (40). The SP identifier outlined in Section III results in $L(q^{-1}) = 1$, with the elements of the parameter vector $\underline{\beta}$ being the coefficients of the process transfer function, and the information vector $\underline{\theta}(j-1)$ containing n past process inputs and outputs. Baradello [14] combines augmented error and state-variable filters to enhance the convergence of the augmented output of the SP identifier which, in turn, improves the performance of the inverted controller. In this case $L(q^{-1}) = \dfrac{D(q^{-1})}{R(q^{-1})}$, where $R(q^{-1})$ is the augmented error filter and $D(q^{-1})$ is the state-variable filter. The FP identifier designs of Landau [16] lead to $L(q^{-1}) = \dfrac{D(q^{-1})}{A_p(q^{-1})}$, where $D(q^{-1})$ processes the a posteriori FP identification error $e^o(j)$ according to

$$v(j) = D(q^{-1})e^o(j)$$

where

$$A_p(q^{-1})e^o(j) = [\underline{\beta} - \hat{\underline{\beta}}(j)]^T\underline{\theta}(j-1)$$

and hence

$$v(j) = \frac{D(q^{-1})}{A_p(q^{-1})}[\underline{\beta} - \hat{\underline{\beta}}(j)]^T\underline{\theta}(j-1) \ . \tag{42}$$

In (42), $A_p(q^{-1})$ is unknown and time-varying, and the information vector $\underline{\theta}(j-1)$ contains n past inputs and outputs of the a posteriori FP identifier. Recent developments in the application of extended information and parameter vectors [21] allows $L(q^{-1}) = 1$. These FP identifiers have not been incorporated in the MRAC designs summarized in this paper.

5.5 Stability

The design objective of MRAC adjustment mechanisms is to guarantee the asymptotic stability of the FP output error (4) and (13). As demonstrated in Section III, this output error can be related through a stable linear filter to an error signal satisfying (40). Consequently, if (40) is asymptotically stable, then the output error is also asymptotically stable.

Mechanisms for estimating the parameters $\hat{\underline{\beta}}(j-\ell)$ in (40) can be designed from several viewpoints. The adjustment of $\hat{\underline{\beta}}(j-\ell)$ can be specified to insure that there is a Lyapunov function which guarantees the asymptotic stability of e(j) in (40). This approach is followed by Ionescu and Monopoli [3], Narendra and Lin [5] and Neuman and Morris [7]. An augmented error signal is introduced to guarantee bounded signals; however, $L(q^{-1})$ must satisfy technical requirements. Landau [16], Suzuki and Takashima [4], Baradello [14] and Neuman and Baradello [2] utilize the hyperstability approach to design adjustment laws. In this approach, the equation error model (40) is decomposed into an equivalent autonomous negative feedback system with a linear block in the forward path and a nonlinear, time-varying block in the feedback path [22]. The linear block is characterized by the filter $L(q^{-1})$, which must be strictly circle positive real [23], and the adjustment laws evolve from the requirement that the inputs and outputs of the nonlinear block be positively correlated. Astrom, et al. [15] invoke the theory of stochastic convergence to show that $\hat{\underline{\beta}}(j) \to \underline{\beta}$ as $j \to \infty$. In each approach the linear filter $L(q^{-1})$ can be eliminated through proper choice of the error signal leading to (40) or the filter $D(q^{-1})$ or by extending the information and parameter vectors.

5.6 Adjustment Mechanisms

Gradient and least-squares adjustment laws emerge from stabilizing (40). The parameters are updated according to

$$\hat{\underline{\beta}}(j) = \hat{\underline{\beta}}(j-\ell) + \frac{1}{\lambda} \underline{P}(j-\ell)\underline{\theta}(j-1)s(j) \ . \tag{43}$$

In the Lyapunov designs highlighted in this paper, $\underline{P}$ is a constant matrix multiplied by the algebraic sign of σ defined in (40). The augmented error s(j) is updated according to [19]

$$s(j) = \frac{\Delta_e(q^{-1})e(j) + \Delta_s(q^{-1})s(j)}{1 + \frac{K_q}{\lambda} \underline{\theta}^T(j-1) \ \underline{P} \ \underline{\theta}(j-1)}$$

where

$$\Delta_e(q^{-1}) = \begin{cases} 1 & \text{for } \ell > 1 \\ \dfrac{D(q^{-1})}{L(q^{-1})} & \text{for } \ell = 1 \end{cases}$$

$$\Delta_s(q^{-1}) = \begin{cases} \mu q^{-\ell} & \text{for } \ell > 1 \\ \left[1 - (1-\mu q^{-1}) \dfrac{D(q^{-1})}{L(q^{-1})}\right] & \\ + \dfrac{K_q}{\lambda} [1-D(q^{-1})]\underline{\theta}^T(j-1)\underline{P\theta}(j-1) & \text{for } \ell = 1, \end{cases}$$

$0 < \mu^2 < 1$, $\lambda > 0$ and $K_q \geq \frac{|\sigma|}{2}$.

To guarantee bounded signals [3], $\underline{P}$ is multiplied by $\dfrac{K_K}{K_K + \underline{\theta}^T(j-1)\ \underline{P}\ \underline{\theta}(j-1)}$ in (43) and $\underline{\theta}^T(j-1)\ \underline{P}\ \underline{\theta}(j-1)$ is replaced by $[K_K + \underline{\theta}^T(j-1)\ \underline{P}\ \underline{\theta}(j-1)]$ in the filtering of the augmented error; $K_q > \frac{|\sigma|}{4}$ and $K_K > \frac{\lambda}{2K_q}$.

In the hyperstable designs [19],

$$s(j) = \frac{\Delta_e(q^{-1})e(j) + \Delta_s(q^{-1})s(j)}{1 + \frac{K_q}{\lambda}\underline{\theta}^T(j-1)\ \underline{P}(j-\ell)\ \underline{\theta}(j-1)}$$

in which $\mu = 0$ and $K_q = \sigma$. For $0 < \lambda < 1$, the adaptive gain matrix $\underline{P}(j-\ell)$ is updated according to $\underline{P}^{-1}(j) = \lambda\underline{P}^{-1}(j-\ell)+\gamma\underline{\theta}(j-1)\underline{\theta}^T(j-1)$ with $\underline{P}^{-1}(0) > 0$. $\gamma \leq \frac{2}{\alpha}$ is a constant relaxation gain, where

$$\alpha > \frac{1}{\sigma} \quad \text{for } \ell > 1$$

and

$$\sigma L(q^{-1}) > \frac{1}{\alpha} \quad \text{for } \ell = 1.$$

If $\lambda = 1$ and $\gamma = 0$ so that $\underline{P}(j) = P(j-\ell) = \underline{P}(0)$ and if $\underline{P}(0)$ is a diagonal matrix, then (43) prescribes a gradient adjustment mechanism [17]. The computational effort for gradient schemes is minimal because (43) are decoupled. This approach is followed by Ionescu and Monopoli [3], Suzuki and Takashima [4], Baradello [14], Neuman and Baradello [2] and Neuman and Morris [7]. At each computational cycle, gradient adjustment mechanisms minimize the one-step cost function

$$J[\hat{\underline{\beta}}(j-\ell)] = \frac{1}{2}\,e^2(j) \tag{44}$$

with respect to the adjustable controller parameters $\hat{\underline{\beta}}(j)$. The mechanism attempts to orthogonalize the parameter error vector $[\underline{\beta} - \hat{\underline{\beta}}(j-\ell)]$ with respect to the information vector $\underline{\theta}(j-1)$ so that

$$[\underline{\beta} - \hat{\underline{\beta}}(j-\ell)]^T \underline{\theta}(j-1) = 0 \quad \text{for all } j. \tag{45}$$

The performance of the closed-loop MRAC depends heavily on the ability of the adjustment mechanism to satisfy (45) when $\hat{\underline{\beta}}(j-\ell) \neq \underline{\beta}$. This occurs when the estimates are biased, as is often the case with gradient mechanisms, or when the process is undergoing large changes. To insure that (45) is satisfied, the constant adaptation gains $\underline{P}$ in (43) must be large thus making the mechanism sensitive to changes in the error signal. It is the authors' experience that large adaptation gains can amplify noise in the error signal and destroy the quality of the control signal.

When $\gamma \neq 0$, $\underline{P}(j)$ is time-varying and (43) leads to least-squares adjustment mechanisms which have been incorporated by Landau [21], Astrom, et al. [15], Boroson [24] and Morris [19]. Since (43) are coupled, the computational effort is much greater than for gradient mechanisms. Since the index ℓ in (43) can be selected by the engineer when $L(q^{-1}) = 1$, the parameters can be updated every ℓ computational cycles, while the control signal $u(j)$ can be calculated at each sampling instant to insure stability.

Least-squares adjustment mechanisms minimize the multi-step cost function

$$J[\hat{\underline{\beta}}(j)] = \gamma \sum_{i=0}^{j/\ell} \lambda^i e^2(j-i\ell) \; . \tag{46}$$ [6]

The mechanism attempts to orthogonalize the parameter error vector $[\underline{\beta} - \hat{\underline{\beta}}(j)]$ with respect to a set of $(1 + \frac{j}{\ell})$ weighted information vectors so that

[6] The error signal $e(j)$ is computed from the parameter estimates $\hat{\underline{\beta}}(j)$ that are available at the j-th computational cycle.

$$[\underline{\beta} - \hat{\underline{\beta}}(j)]^T \underline{\Phi}(j-1) \text{ DIAGONAL}[1\ \lambda \ldots \lambda^{j/\ell}] \underline{\Phi}^T(j-1) = 0 \tag{47}$$

where

$$\underline{\Phi}(j) = [\underline{\theta}(j)\ \underline{\theta}(j-\ell) \ldots\ldots\ldots \underline{\theta}(0)].$$

When (47) is satisfied, it is far more difficult for $\hat{\underline{\beta}}(j) \neq \underline{\beta}$. Even though the parameters may be biased, better estimates can be obtained than with gradient schemes. The number of past information vectors incorporated in (47) depends on the weights in (46), which in turn depend on the relaxation gain λ in (43). Improved estimates require smaller adaptation gains to satisfy (46). Thus, least-squares mechanisms are less sensitive to noise in the error signal and the quality of control signal is preserved. Although implementation of Lyapunov and hyperstable designs is similar, the Lyapunov algorithms appear to offer more flexibility in the choice of adaptation gains than do the hyperstable designs.

5.7 Experimental Evaluation

Both gradient and least-squares adjustment mechanism MRAC have been programmed and evaluated on a PDP 11/40 minicomputer. The process under control is simulated on an EAI TR-20 analog computer. The algorithms are written in Fortran IV and A/D and D/A converters provide the interface between the analog and digital computers. The minicomputer is utilized to implement and evaluate the MRAC which are summarized in Section III and incorporate the adjustment mechanisms reviewed in this section. In terms of adaptation speed [25] and quality of control signal, the gradient mechanisms perform adequately on processes with high open-loop gain.

In the more practical case in which the process gain is low and there is a real need for adaptive control, the performance of the least-squares mechanisms surpasses that of the gradient mechanisms.

To demonstrate the feasibility of practical implementation, both gradient and least-squares MRAC have been programmed on the MDS-800 Microcomputer Development System based on the Intel 8080A microcomputer [26]. The gradient algorithm for the identification error method incorporates an augmented error signal and state-variable filters. The flexible algorithm includes an operating system for communicating with an operator and a 24-bit software floating point package. The algorithm requires 886 bytes of program memory and 70 milliseconds to execute the computations for controlling second-order processes [7]. This is to be compared with 117 bytes of program memory and a 12 millisecond computational cycle for a PSD controller [27]. Considerable savings in memory space and computation time can be realized when the adaptive controller is tailored for a specific application.

The least-squares identification error MRAC is similarly implemented on the MDS-800. The algorithm requires 1919 bytes of program memory and 130 milliseconds of computation time for controlling second-order processes [28]. The authors have verified experimentally that, since the index ℓ in (40) can be selected by the engineer, the minimum sampling period for this implementation is reduced from 130 milliseconds (when $\ell = 1$) to 54 milliseconds (when $\ell = 5$). When the floating point accuracy and matrix operations are tailored to a specific application and hardware floating-point packages (such as the AM 9511 [29]) are utilized, the program memory space and computation times can be reduced phenomenally [28].

When $\lambda = 1$, the least squares criterion (46) is satisfied with all weights being equal. As $j \to \infty$, the number of information vectors included in (47) grows beyond bound. New information is being averaged with past data and the reaction to a changing process continually diminishes. This situation can be avoided by choosing $\lambda < 1$. As $\lambda \to 0$, however, the number of information vectors included in (47) approaches unity, the adaptive gain matrix $\underline{P}(j)$ becomes unbounded, and the adjustment mechanism becomes highly sensitive to measurement errors. It is possible to choose λ such that the reaction of the least-squares tuning mechanism to a changing process surpasses that of the gradient tuning mechanism while retaining the quality of the control signal. It remains to determine the trade-off between improved performance and increased computational effort and program memory requirements.

VI. CONCLUSIONS

In Sections III-V model reference adaptive controllers (MRAC) are reviewed in a common framework. This review is intended to enable the control engineer to design microcomputer MRAC. For practical applications these MRAC are subject to drifting process gains and parameters, a limited dynamic range for the control signal transmitted by the D/A, and inaccurate measurements and computations induced by the finite word-length of the microcomputer.

These practical constraints deteriorate the performance of the input and output error methods. Variations in the process gains result in the parameter σ in the filtered equation error (40) being both unknown and time-varying and this limits the stability of these methods. When the control signal saturates in the D/A, the input and output error methods are no longer characterized by (40) and continued application of this filtered equation error can cause erroneous action by the adjustment mechanism. The identification error method is immune from these problems.

Since the microcomputer accentuates measurement and computational errors, the FP reference model is impractical for controlling low-pass processes. The high-pass characteristics introduced by the FP reference model amplify high frequency disturbances, whereas the SP reference model does not. The SP reference model increases the stability margin of the closed-loop MRAC over that of the FP reference model.

Measurement errors also influence the adjustment mechanism. The high-gains required for gradient mechanisms to maintain reasonable tracking accuracy are sensitive to noise and can destroy the quality of the control signal. Least-squares mechanisms filter measurement noise over successive samples.

On the basis of this review, a practical and robust microcomputer MRAC would consist of an enhanced identification error design incorporating a SP reference model and least-squares adjustment mechanism. Synthesis and experimental verification of this analysis are detailed in [19].

REFERENCES

1. Martin-Sanchez, J. M., "A New Solution to Adaptive Control," Proceedings of the IEEE, 64, No. 8, pp. 1209-1218, Aug. 1976.
2. Neuman, C. P., and Baradello, C. S., "Microcomputer Implementation of Adaptive Controllers," Proceedings of the First International Symposium on Mini- and Microcomputers in Control, San Diego, Ca., January 8-9, 1979, pp. 57-62.
3. Ionescu, T., and Monopoli, R., "Discrete Model Reference Adaptive Control with an Augmented Error Signal," Automatica, 13, No. 5, pp. 507-517, September 1977.
4. Suzuki, T., and Takashima, S., "A Hyperstable Scheme for Discrete Model Reference Adaptive Control System," International Journal of Control, 28, No. 2, pp. 245-252, August 1978.
5. Narendra, K. S., and Lin, Y. H., "Design of Stable Model Reference Adaptive Controllers," Proceedings of the International Workshop on Applications of Adaptive Control, Yale Univ., New Haven, Ct., Aug. 23-25, 1979, pp. 15-32.

6. Johnson, Jr., C. R., and Tse, E., "Adaptive Implementation of One-Step-Ahead Optimal Control via Input Matching," IEEE Transactions on Automatic Control, AC-23, No. 5, pp. 865-872, October 1978.
7. Neuman, C. P., and Morris, R. L., "Mini and Micro-Computer Implementation of Adaptive Control Systems," Proceedings of the First International Symposium on Mini- and Micro-computers in Control, San Diego, Ca., January 8-9, 1979, pp. 63-68.
8. Kuo, B. C., "Digital Control Systems," Champaign, Illinois, SRL Publishing Co., 1977.
9. Oppenheim, A. V., and Schafer, R. W., "Digital Signal Processing," Englewood Cliffs, N. J., Prentice-Hall, Inc., 1975.
10. Neuman, C. P., and Schuhmacher, J. A., "DIGIT: A Computational Package for Digital System Design," Department of Electrical Engineering, Carnegie-Mellon University, Pittsburgh, Pa., October 1979.
11. Neuman, C. P., and Baradello, C. S., "Digital Transfer Functions for Microcomputer Control," IEEE Transactions on Systems, Man, and Cybernetics, SMC-9, No. 12, pp. 856-860, Dec. 1979.
12. Jury, E. I., "Theory and Application of the z-Transform Method," Huntington, N. Y., R. E. Krieger Publishing Co.,1973.
13. Peterson, W. C., and Nassar, A. H., "On the Synthesis of Optimum Linear Feedback Control Systems," Journal of the Franklin Institute, 306, No. 3, pp. 237-256, Sept. 1978.
14. Baradello, C. S., "Design of Adaptive Algorithms for Microcomputer Process Control," Ph.D. Dissertation, Department of Electrical Engineering, Carnegie-Mellon University, Pittsburgh, Pa., 1978.
15. Astrom, K. J., et al., "Theory and Applications of Self-Tuning Regulators," Automatica, 13, No. 5, pp. 457-476, Sept. 1977.
16. Landau, I. D., "Adaptive Control: The Model Reference Approach," New York, Marcel Dekker, 1979.

17. Mendel, J. M., "Discrete Techniques of Parameter Estimation: The Equation Error Formulation," New York, Marcel Dekker,1973.

18. Yuan, J. C., and Wonham, W. M., "Probing Signals for Model Reference Identification," IEEE Transactions on Automatic Control, AC-22, No. 4, pp. 530-538, August 1977.

19. Morris, R. L., "Design of Microcomputer Adaptive Controllers," Ph.D. Dissertation, Department of Mechanical Engineering, Carnegie-Mellon University, Pittsburgh, Pa., 1979.

20. Neuman, C. P., and Morris, R. L., "Design and Microcomputer Implementation of Adaptive Controllers," Proceedings of the First International Micro- and Mini-Computer Conference, Houston, Texas, November 14-16, 1979.

21. Landau, I. D., "Elimination of the Real Positivity Condition in the Design of Parallel MRAS," IEEE Transactions on Automatic Control, AC-23, No. 6, pp. 1015-1020, December 1978.

22. Landau, I. D., and Silveira, H. M., "A Stability Theorem with Applications to Adaptive Control," IEEE Transactions on Automatic Control, AC-24, No. 2, pp. 305-312, April 1979.

23. Siljak, D. D., "Algebraic Criteria for Positive Realness Relative to the Unit Circle," Journal of the Franklin Institute, 295, No. 6, pp. 469-476, April 1973.

24. Boroson, D., "Estimation and Identification in Linear Discrete-Time Systems: Deterministic Hyperstability and Stochastic Convergence," Ph.D. Dissertation, Department of Electrical Engineering and Computer Science, Princeton University, Princeton, N. J., 1977.

25. Neuman, C. P., and Baradello, C. S., "Experimental Evaluation of Adaptive Identifiers," Proceedings of the International Workshop on Applications of Adaptive Control, Yale University, New Haven, Ct., August 23-25, 1979, pp. 153-163.

26. "MDS-800 Intellec Microcomputer Development System: Hardware Reference Manual and User's Reference Manual," Santa Clara, Ca., Intel Corporation, 1975.

27. Auslander, D. M., Takahashi, Y., and Tomizuka, M., "Direct Digital Process Control Practice and Algorithms for Microprocessor Applications," Proceedings of the IEEE, 66, No. 2, pp. 199-208, February 1978.

28. Neuman, C. P., and Morris, R. L., "Microcomputer Requirements for Model Reference Adaptive Controllers," Proceedings of the Thirteenth Asilomar Conference on Circuits, Systems and Computers, Pacific Grove, Ca., November 5-7, 1979.

29. Parker, R. O. and Kroeger, J. H., "The AM9511 Arithmetic Processing Unit," Sunnyvale, Ca., Advanced Micro Devices, Inc., 1978.

DISCRETE TIME MODELLING OF THE TORQUE RESPONSE OF A SPARK-IGNITED FUEL-INJECTED ENGINE

H. G. Hopkins
R. H. Borcherts

Engineering and Research Staff
Ford Motor Company
Dearborn, Michigan

The dynamics of the torque output of a spark ignited engine are modelled using Landau's hyperstable model reference adaptive system. A set of low order difference equations were shown to be sufficient to model the dynamics resulting from perturbations of commands to throttle, spark advance and exhaust gas recirculation at a nominal operating point.

I. INTRODUCTION

Modelling forms the foundation upon which control strategies are built. For a spark ignited internal combustion engine, the required model relates engine actuator commands, throttle angle, spark advance, exhaust gas recirculation (EGR) and air fuel ratio

ISBN 0-12-514060-6

as control inputs to engine outputs, torque, engine speed (RPM) and emissions. Much has been done to establish static input/output relationships [1-4]. Cassidy, Athans and Lee [5] have gone a step further by constructing a linearized dynamic model from sinusoidal frequency response data. This paper describes preliminary results of discrete time modelling and identification of engine torque responses to perturbations from nominal commands to throttle, EGR and spark advance inputs. Air/fuel (A/F) ratio is assumed to be controlled to attempt to maintain stoichiometric conditions for combustion (no excess oxygen or fuel). We assume the engine model is composed of three single input subsystems. This multi-input single output system is first reduced to three single input single output (SISO) systems and then validated with multi-input experiments. Although this approach may yield a non-minimal representation, its great practical advantage is in allowing us to apply well developed SISO methods. Of these methods we select Landau's hyperstable model reference adaptive system [6] because of its adaptivity and robustness. Since the engine is a nonlinear system, it will be necessary to continually adapt its representation and to alter feedback controls accordingly. Landau's algorithm has been successfully applied on line to achieve this [7]. The algorithm's convergence to unbiased parameter estimates in the presence of zero mean Gaussian noise is used to advantage by being unaffected by the random nature of combustion events. Since it is expected that implementation of any control algorithm will be on a microprocessor, effects of limited word length upon output measurement and computational accuracy are of paramount importance. It has been shown that Landau's algorithm is well suited to such an environment [8].

II. DESCRIPTION OF A SPARK IGNITED ENGINE AND TEST FACILITY

Spark advance is the angular measure, generally in degrees, at which the spark plug fires before the piston reaches its highest point of travel (top dead center or TDC). Although the maximum power achieved for a given engine speed and throttle position is uniquely determined by spark advance, modifications to this value of spark are required to (1) provide smoothness of engine operation, (2) obtain minimum emissions and (3) prevent knock due to low octane rating of fuels.

EGR refers to the returning of exhaust gas to the intake manifold through a control valve. Its primary use is to reduce the emission of oxides of nitrogen (NOX) by lowering peak temperatures during combustion. EGR serves to decrease the amount of A/F charge and increase the heat capacity of the mixture entering the cylinder.

The throttle control has traditionally been mechanically linked to the accelerator pedal to demand greater or lesser power from the engine. This control is most often a butterfly valve in the throat of the carburetor regulating the flow of A/F mixture into the intake manifold.

Although the A/F ratio for combustion may be in a range of roughly 8/1 to 20/1, for best performance of a catalytic converter and good engine performance stoichiometric conditions should be maintained to within plus or minus .05 A/F ratios. Without becoming very involved in the details of fuel delivery systems, we note that deviations from this desired condition occur frequently during transient engine operation. Lean excursions (excess oxygen) most often exist during throttle opening and rich (excess fuel) during throttle closing. Fuel injection systems have been implemented to lessen these effects by decreasing the transport delay times for fuel delivery to a cylinder relative to delays in a carbureted engine. Feedback control may be implemented through the use of a sensor which indicates whether the mixture is rich or lean.

The test facility used is shown schematically in Figure 1. The engine is a 5.7L sequentially fuel injected V-8 equipped with a fast response solenoid actuated EGR valve and Jordan feedback controller used to actuate the throttle. The EGR valve, spark advance, fuel pulse width and throttle are all under the control of a computer with the ability to sample torque, engine speed (RPM), intake manifold pressure, fuel pulse width and output voltage of the A/F sensor. Supervision of the data collection task is done through a PDP-11, with a microprocessor implementation of spark advance and A/F mixture control.

Figure 2 shows changes in intake manifold absolute pressure (MAP) due to step commands to the throttle and EGR actuators,

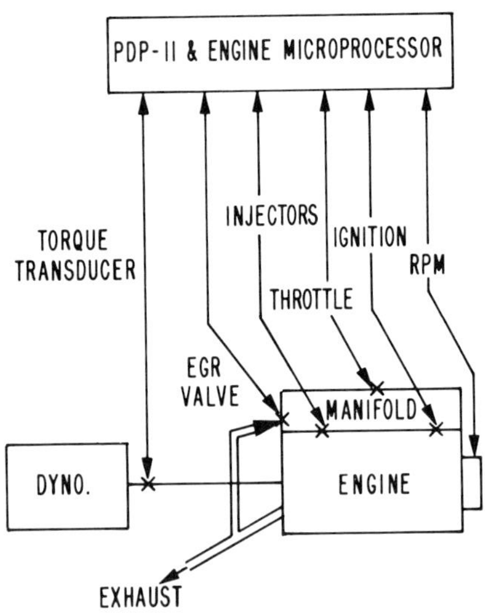

FIGURE 1. Schematic of test facility.

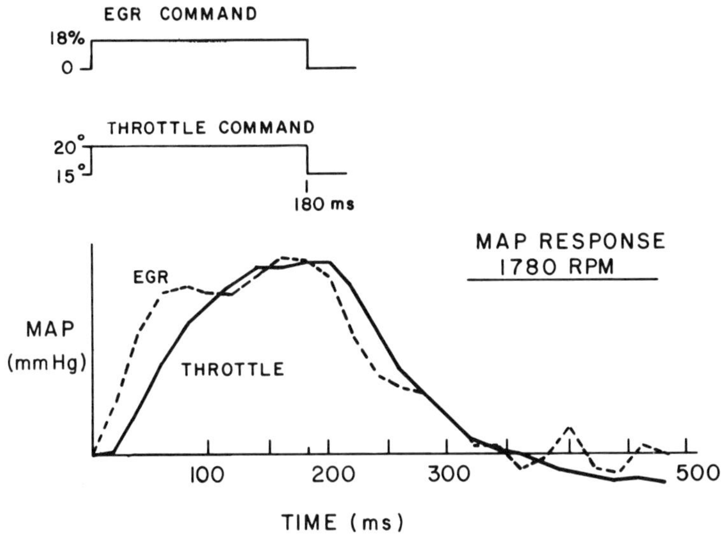

FIGURE 2. Intake manifold pressure response due to pulse changes in EGR and throttle.

respectively. Assuming changes in MAP track changes in mass flow through the intake manifold, the response times of mass flow due to these actuators is of the order of 200ms.

A Lebow torque transducer and instrumentation amplifier with a 500Hz bandwidth were used to sense torque. However, a bandwidth of 20Hz was shown experimentally to be sufficient to preserve the dominant features to the torque response. With this bandwidth we chose to fulfill the Nyquist limit by a sample interval of 20ms.

Since it is the intent here to describe the dominant features, we chose to average several torque responses to the same commanded input. An average of ten proved sufficient.

III. METHODOLOGY

Schematically, the assumed structure for the engine is shown in Figure 3. Define the total change in torque due to changes of commands as y(k). Let the contribution of the jth subsystem be expressed as $y_j(k)$, which implies that y(k) = y1(k) + y2(k) + y3(k). Let the jth command be denoted as $u_j(k)$. The jth subsystem can then be expressed as

$$x_j(k+1) = A_j x_j(k) + B_j u_j(k)$$

$$y_j(k) = C_j x_j(k)$$

The system order and structure of $[A,B,C]_j$ found experimentally to adequately represent the engine are given in Appendix A. For the jth subsystem the identification problem reduces to the determination of a_{ij} and c_{ij}.

Landau's algorithm is now briefly outlined. Implicit in the following is that the algorithm will be applied to each subsystem and hence subscripts indicating this are dropped. An equivalent representation of a subsystem is

$$y_k = p^T w_k$$

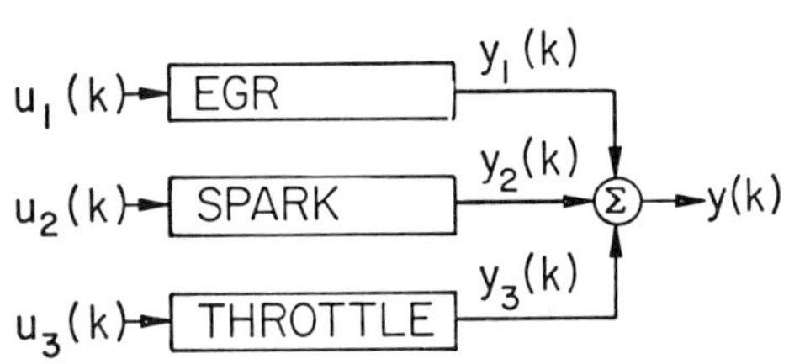

FIGURE 3. Assumed input-output structure for engine.

where the unknown subsystem parameters are grouped as

$$p^T = [-a_{n-1},\dots,-a_0,c_m,\dots,c_0]$$

$$w_k^T = [y_{k-1},\dots,y_{k-n},u_{k+m-n},\dots,u_{k-n}].$$

Define the following notation for parameter estimates:

$$\hat{p}_k^T = [-\hat{a}_{n-1},\dots,-\hat{a}_0,\hat{c}_m,\dots,\hat{c}_0]$$

output estimate and input vector:

$$\hat{w}_k^T = [\hat{y}_{k-1},\dots,\hat{y}_{k-n},u_{k+m-n},\dots,u_{k-n}]$$

output error:

$$e_k = y_k - \hat{p}_{k+1}^T\hat{w}_k$$

"A priori" error is formed using the current output measurement and an estimate using the current parameter estimate. This is combined with a weighted sum of previous output errors to form a filtered output error:

$$v_k = y_k - \hat{p}_k^T\hat{w}_k + \sum_{i=1}^{n} g_i e_{k-i}$$

where g_i are chosen such that

$$\frac{z^n + \sum_{i=1}^{n} g_i z^{n-i}}{z^n + \sum_{i=1}^{n} a_i z^{n-i}}$$

is discrete strictly positive real. The estimate vector is perturbed in a direction prescribed by $F_k\hat{w}_k$ with a step size of

$$v_k/(1 + \hat{w}_k^T F_k \hat{w}_k)$$

$$\hat{p}_{k+1} = \hat{p}_k + \frac{F_k \hat{w}_k}{1 + \hat{w}_k^T F_k w_k} v_k$$

The direction of perturbation responds to further input/output information by altering F_k such that

$$F_{k+1} = F_k - \frac{F_k \hat{w}_k w_k^T F_k}{1 + \hat{w}_k^T F_k \hat{w}_k}$$

Since a_i are unknown the choice of g_i presents difficulty. There are two approaches to remedy this. The estimate vector can be augmented to also estimate g_i such that the discrete strictly positive real criterion is met. However, this solution introduces an undesirable increase in computational complexity if the

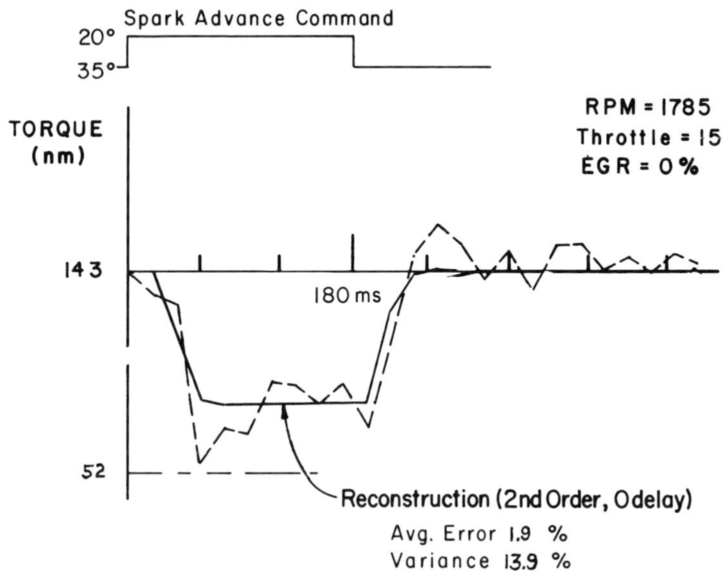

FIGURE 4. Torque response due to a 15° decrease in spark advance.

algorithm is to be implemented on-line. Another solution is to acquire and use some knowledge of the unknown parameters. Off-line this knowledge is easily obtained by guessing g and trying it. The choice of g could be constant or time varying, $g_i = a_i$ for example. Both choices result in locally stable identifiers. It was found that g_i equal to zero resulted in convergence of the identifier for the system considered here.

Initial conditions for the algorithm are the following:

1. $F = 1000 * I$
2. $\hat{p} = 0$
3. $\hat{w} = 0$

The last condition is applied since the model is for changes of torque about a nominal operating point. As in [8], the algorithm will be judged to have converged when $w^T Fw$ is less than 0.5. With output measurements scaled for a maximum of 1, convergence to final estimates occurs within 15 samples in all cases.

IV. RESULTS

The presentation of results will be done in two stages. The first establishes the models of the subsystems and the second validates the proposed structure using multi-input experiments.

Torque responses due to independent commands to spark, EGR and throttle are presented in Figures 4, 5 and 6, respectively. The response due to a 15 degree change in spark for a duration of 180ms with 18% EGR and 15 degree throttle opening is shown in Figure 4. We note that there is no noticeable delay and that a second order model describes the torque dynamics with an average error of less than 2% and a variance of 14%.

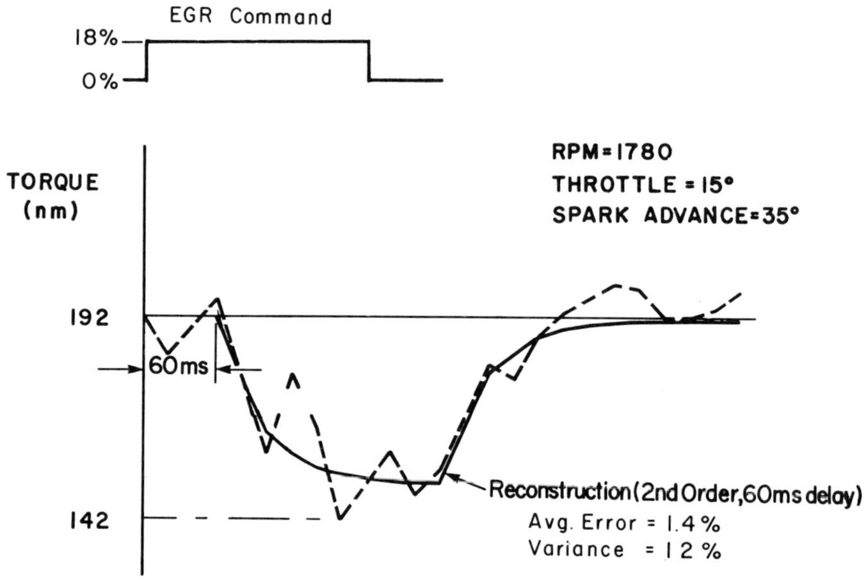

FIGURE 5. Torque response due to a 18% increase in exhaust gas recirculation.

The response due to opening the EGR valve, shown in Figure 5, has a delay of three sample intervals or 60ms. Upon inclusion of this delay, the dominant dynamics can be modelled by a second order system with an average error of less than 2% and a variance of 12%.

Torque response due to a 180ms pulsed opening of the throttle is shown in Figure 6. As with the EGR subsystem there is a 60ms delay. However, in order to have an error comparable with those of the other subsystems fourth order dynamics were required to describe it. Note that the reconstruction based on a fourth order response begins to include the higher frequency components in

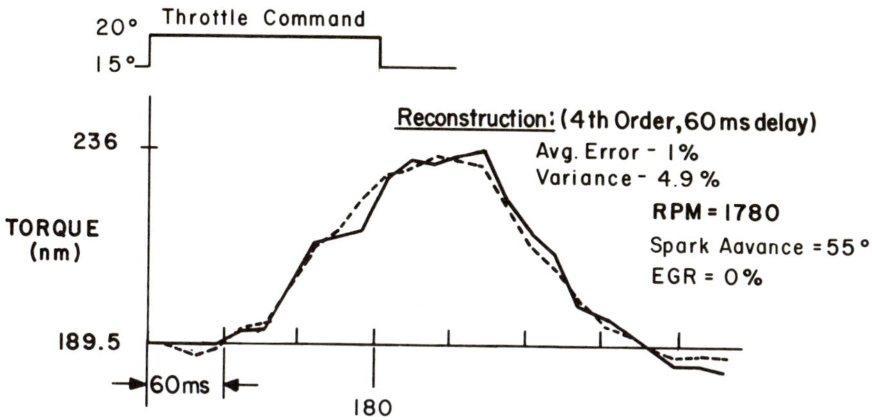

FIGURE 6. Torque response due to a 5° increase in throttle opening.

addition to dominant dynamics. For this reason the average error between experimental and reconstructed data is less than 0.1% with a variance of 4.9%.

The resulting models from the above experiments are summarized in Appendix B. These experiments show that the dominant dynamics of the subsystems can be identified by Landau's algorithm. To validate the assumed multi-input system structure we turn to Figures 7 and 8.

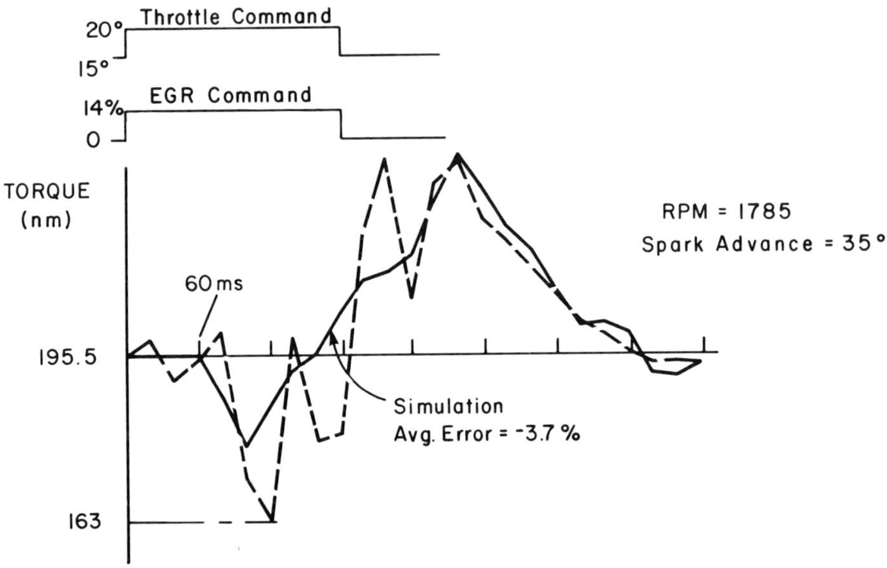

FIGURE 7. Torque response due to a throttle and EGR increases. Simulation based upon models shown in Figures 5 and 6.

Note in Figure 7 both throttle and EGR are opened for 180ms simultaneously. Although there exists a high frequency component in the torque response, the dominant dynamics are modelled by adding contributions of the throttle and EGR subsystems. The average error incurred in doing this is -3.7%.

Figure 8 depicts results of simultaneous changes of throttle, spark and EGR. Throttle is commanded to open 5 degrees, EGR to close and spark advance to be retarded by 15 degrees. Note first, the dynamics used to represent the closing of the EGR are used for its opening. Secondly, the dynamics used to represent the change in spark is over a significantly different range than was used to generate Figure 5, namely 45 to 25 degrees instead of

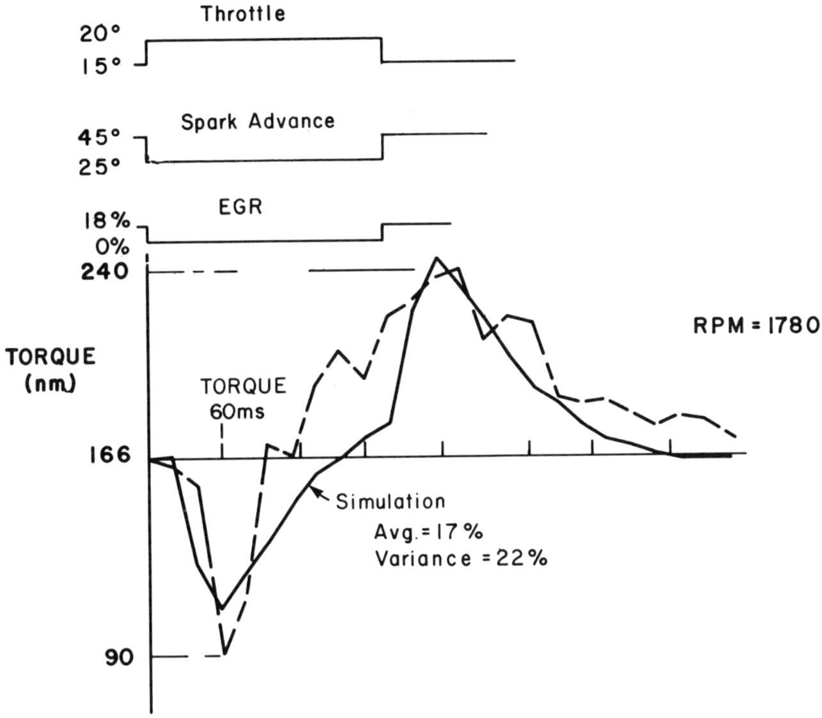

FIGURE 8. Torque response due to throttle increase and EGR and spark advance decreases. Simulation based upon models shown in Figures 4, 5 and 6.

35 to 20 degrees. Nevertheless the dominant dynamics appear to be well modelled with an average error of 17% and a variance of 22%. To this extent the assumption of system structure is valid.

An important feature shown in Figure 8 is the well-known need to coordinate EGR and spark advance controls. The command to open throttle in a driving situation indicates the desire for an increase in torque (power), but because of the faster response of the spark subsystem, initially there is a torque decrease. Currently this torque decrease is lessened by commanding spark to change in proportion to change in the measured amount of EGR. However, knowledge of the engine system dynamics and the application of modern control theory could potentially yield a superior method of coordinating engine controls.

V. CONCLUSION

In this paper we have shown the feasibility of applying a linear system identification technique to represent the dynamics of a spark ignited engine at a nominal operating point. In effect, we have linearized and at the same time identified the engine's dominant dynamics. Assuming regions of validity for models generated at operating points over the entire torque plane can be obtained, we have established an experimental procedure upon which to base control strategies generated by linear quadratic theory.

ACKNOWLEDGMENTS

The authors wish to acknowledge Dr. Neal Laurance, Ford Motor Company, and Professor Petar Kokotovic, University of Illinois, for their support and encouragement during the course of this investigation. Also, special mention is deserved for Mr. Keith Hampton, University of Michigan, for his experimental talents and technical expertise.

REFERENCES

1. Prahbakar, R., Citron, S. J., and Goodson, R. E., "Optimization of Automobile Engine Fuel Economy and Emissions," ASME 75 WA/Aut-19, Dec. 1975.
2. Rishavy, E. A., Hamilton, S. C., Ayers, J. A., and Keane, M.A., "Engine Control Optimization for Best Fuel Economy with Emission Constraints," SAE 770075, Feb. 1977.
3. Auiler, J. E., Zbrozek, J. D., and Blumberg, P. N., "Optimization of an Automotive Engine Calibrations for Better Fuel Economy-Methods and Applications," SAE 770076, Feb. 1977.
4. Cassidy, J. F., Jr., "A Computerized On Line Approach to Calculating Optimum Engine Calibrations," SAE 770078, Feb. 1977.
5. Cassidy, J. F., Jr., Athans, M., and Lee, W., "On the Design of Electronic Automotive Engine Controls Using Linear Quadratic Control Theory," IEEE Conference on Decision and Control, San Diego, Ca., Jan. 1979.
6. Landau, I. D., "An Asymptotic Unbiased Recursive Identifier for Linear Systems," Proc. IEEE Conference on Decision and Control, Nov. 1974.
7. Landau, I. D., "A Survey of Model Reference Adaptive Techniques - Theory and Applications," Automatica, 10, 368, 1974.
8. Hopkins, H. G., "A Feasibility Study of Real Time Process Identification Using a Microprocessor," M. S. Thesis, University of Illinois, March 1977.

APPENDIX A

SUBSYSTEM 1 - EGR RESPONSE MODEL

$$A_1 = \begin{bmatrix} 0 & 1 & 0 & 0 & 0 \\ a_{11} & a_{01} & 1 & 0 & 0 \\ 0 & 0 & 0 & 1 & 0 \\ 0 & 0 & 0 & 0 & 1 \\ 0 & 0 & 0 & 0 & 0 \end{bmatrix}$$

$$B_1^T = [0 \quad 0 \quad 0 \quad 0 \quad 1]$$

$$C_1 = [c_{01} \quad c_{11} \quad 0 \quad 0 \quad 0]$$

SUBSYSTEM 2 - SPARK RESPONSE MODEL

$$A_2 = \begin{bmatrix} 0 & 1 \\ a_{12} & a_{02} \end{bmatrix}$$

$$B_2^T = [0 \quad 1]$$

$$C_2 = [c_{02} \quad c_{12}]$$

SUBSYSTEM 3 - THROTTLE RESPONSE MODEL

$$A_3 = \begin{bmatrix} 0 & 1 & 0 & 0 & 0 & 0 & 0 \\ 0 & 0 & 1 & 0 & 0 & 0 & 0 \\ 0 & 0 & 0 & 1 & 0 & 0 & 0 \\ a_{33} & a_{23} & a_{13} & a_{03} & 1 & 0 & 0 \\ 0 & 0 & 0 & 0 & 0 & 1 & 0 \\ 0 & 0 & 0 & 0 & 0 & 0 & 0 \\ 0 & 0 & 0 & 0 & 0 & 0 & 0 \end{bmatrix}$$

$$B_3^T = [\,0 \quad 0 \quad 0 \quad 0 \quad 0 \quad 0 \quad 1]$$

$$C_3 = [c_{03} \quad c_{13} \quad c_{23} \quad c_{33} \quad 0 \quad 0 \quad 0]$$

APPENDIX B

SUBSYSTEM 1 - EGR RESPONSE MODEL

$$A_1 = \begin{bmatrix} 0 & 1 & 0 & 0 & 0 \\ .22 & .15 & 1 & 0 & 0 \\ 0 & 0 & 0 & 1 & 0 \\ 0 & 0 & 0 & 0 & 1 \\ 0 & 0 & 0 & 0 & 0 \end{bmatrix}$$

$$B_1^T = [0 \quad 0 \quad 0 \quad 0 \quad 1]$$

$$C_1 = [-.678 \quad -.906 \quad 0 \quad 0 \quad 0]$$

SUBSYSTEM 2 - SPARK RESPONSE MODEL

$$A_2 = \begin{bmatrix} 0 & 1 \\ -.1 & .402 \end{bmatrix}$$

$$B_2^T = [0 \quad 1]$$

$$C_2 = [2.98 \quad -.0473]$$

SUBSYSTEM 3 - THROTTLE RESPONSE MODEL

$$A_3 = \begin{bmatrix} 0 & 1 & 0 & 0 & 0 & 0 & 0 \\ 0 & 0 & 1 & 0 & 0 & 0 & 0 \\ 0 & 0 & 0 & 1 & 0 & 0 & 0 \\ -.44 & .59 & -.124 & .696 & 1 & 0 & 0 \\ 0 & 0 & 0 & 0 & 0 & 1 & 0 \\ 0 & 0 & 0 & 0 & 0 & 0 & 1 \\ 0 & 0 & 0 & 0 & 0 & 0 & 0 \end{bmatrix}$$

$$B_3^T = [0 \quad 0 \quad 0 \quad 0 \quad 0 \quad 0 \quad 1]$$

$$C_3^T = [.151 \quad 3.11 \quad -.545 \quad 1.01 \quad 0 \quad 0 \quad 0]$$

APPLICATION OF ADAPTIVE SYSTEMS IN PROCESS CONTROL

H. Unbehauen
Chr. Schmid

Department of Electrical Engineering
Ruhr-University Bochum
Germany

This paper gives a review of applications of adaptive systems in process control. After a statement of problems in process control, the structures and design methods for adaptive systems are briefly discussed. From the field of applications in the cement industry, the steel and metallurgical industry, the chemical industry, the paper industry and some miscellaneous areas of process industry, the most interesting developments in adaptive control during the last five years are discussed. The discussion shows that only relatively few applications have been reported from the broad field of process industries. It is pointed out that the reason for this situation is due to the complicated dynamic behaviour of these industrial processes as well as the conservative management of many industrial processes. Thus it is recommended that more experimental work should be done in connection with proven theoretical methods. These methods, however,

ISBN 0-12-514060-6

must provide a systematic design procedure, which can be understood and also implemented by industrial engineers.

I. INTRODUCTION

A detailed technical review of adaptive control in process industries covering the entire range of applications and methods is clearly beyond the aims of this paper. This paper deals with applications of adaptive control in the cement industry, the steel and metallurgical industry, the chemical industry, the paper industry and some miscellaneous areas of process industry. The review does not in any way claim to be complete, but it tries to discuss the most interesting developments published during the last five years. Thus this paper is directly connected to the authors' report of 1975 [1]. There are a number of survey papers on adaptive control - partly in book form - dealing with general aspects [2-4] as well as surveys of more specific fields of adaptive systems [5-9]. However, only very few papers are directly concerned with industrial applications [10]. With the rapid appearance of very cheap microprocessors during the last five years new possibilities for the industrial application of adaptive systems have been opened up [11]. Therefore, adaptive control has become very attractive again during the last few years.

Given the limits of this paper, it would seem most useful to concentrate attention on the following broad questions:

1) Why is adaptive control necessary in the process industry?
2) Which adaptive design methods fulfil the requirements for application in the process industry?
3) What is the actual status of adaptive control in the various areas of process industry?

4) What conclusions can be drawn from the actual situation to the further development of adaptive control in process industry?

The following sections try to give answers to these questions.

II. PROBLEMS OF ADAPTIVE CONTROL IN THE PROCESS INDUSTRIES

During the past 30 years it has often been predicted that adaptive control would become a common sophistication in process industries. This control technique claims to compensate for changes in process parameters attaining or holding a defined, usually optimal, set of system conditions given a certain initial indeterminateness or varying working conditions of a time-varying system. This can normally be attained by adaptation of parameters, system structures or signals using the information received during the adaptation process. High costs of development, implementation and operation of such control systems must be justified economically if the control performance is affected significantly by such changes or if the adaptive control system serves as a substitute for the time-consuming process of plant experiments, such as extensive measurements, modelling, evaluation, controller design and controller tuning.

The number of proposed adaptive control algorithms is very large relative to the number of actual applications in process industry. Successful applications are mostly based on particular solutions, in which variations of process parameters are known and used directly to adapt a controller improving performance significantly. From the present state of application, however, it can be concluded that progress in the industrial application of adaptive control systems appears only if some methods become standard and thus generally applicable. From the domain of theory we have an arsenal of methods, but there is a lack of experience of the efficiency of the proposed methods as well as of the existing possibilities of converting methods into practice. The innovation process of introducing adaptive control as a standard of technique is time-consuming, despite the acceleration of new hardware techniques, so more emphasis should be put on the clarification of essential prerequisites, such as practicability, limits of applicability, assumptions for design and influence of real restrictions. From the practical point of view this requirement can be rephrased by the following questions, which should be answered before deciding on the implementation of adaptive control:

- What are the variations in process dynamics and are they significant?
- What degree of improvement can be expected using adaptive methods and are the costs justified?
- Are alternative solutions a better means to solve the problem?

From the methodological point of view additional requirements must be fulfilled by a suitable adaptive control and design method:

- guaranteed stability or a maximal stability region,
- a small amount of a priori knowledge about the process,
- robustness in the case of disturbances,
- general applicability to different processes,
- high speed of the adaptive algorithm to compensate for parameter variations,
- a small amount of computer time and capacity.

A lot of adaptive structures and methods violate these requirements. Only a small set of such structures are really suitable. The right selection is a critical task.

III. ADAPTIVE STRUCTURES AND DESIGN METHODS

General applicability of adaptive methods would have to become a reality if these systems are to be designed in a systematic way. This applies both to theory and practice. It would allow one to generalise results and would make applications simpler and more commonplace. A first step in the systematic design is the characterizing of the control system features, such as the basic structure and the principle of design. The mode and principle of action are basic features, which can be described in a basic structure, thus a rough description of the structure is the first step in the design of an adaptive control system.

A. Basic Structures

For the design of adaptive control systems today three main basic control system structures are relevant. The first structure, the parallel reference model structure (Fig. 1), allows the identification of plant dynamics and the adaptation of the controller parameters to achieve a given model behaviour for the closed loop.

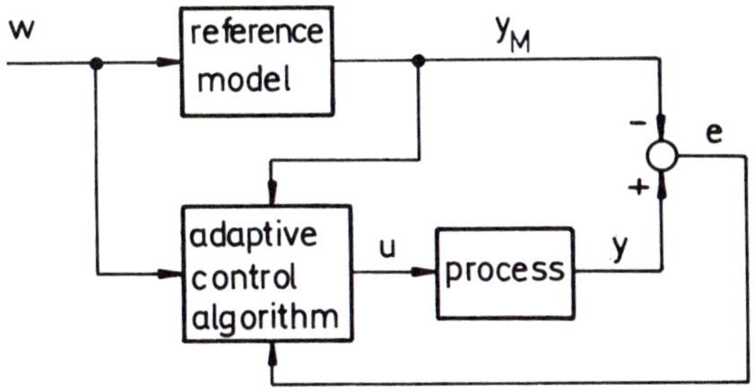

FIGURE 1. Model reference adaptive control system with a parallel reference model.

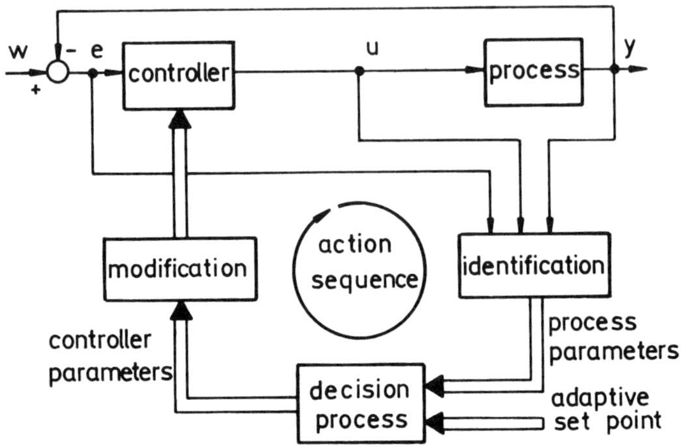

FIGURE 2. Adaptive control system structure with closed loop adaptation.

The second structure, the closed loop structure, is based on a classical control loop (Fig. 2). Changes of parameters or other disturbances are detected by an identification block. Depending on the chosen criterion (adaptive set point) the controller parameters are obtained by a decision process and the controller

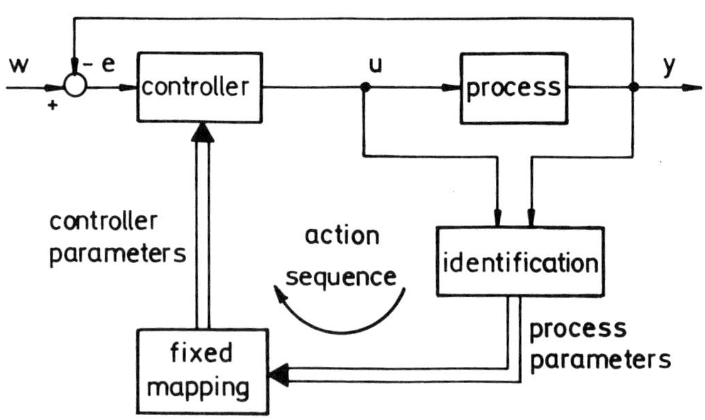

FIGURE 3. Adaptive control system structure with open loop adaptation.

is corrected accordingly. In this case, the adaptive system is realized by a second closed loop control system, since the effect of controller modification is fed back to the decision process by the basic control loop and the identification process. Thus an adaptive error forces the adaptation process. In the third basic structure, the open loop structure, the decision process is reduced to a fixed mapping of the process parameters to the controller parameters. The original decision process is already realized in the design phase of the adaptive control system. This type of structure is widespread and in vogue today, since it allows one to tune a wide range of controllers using a manifold of popular on-line process identification methods. The assumption for the faultless operation of systems with this structure is a good knowledge of the actual process dynamics. Thus the identification block has to satisfy high requirements of precision and speed. Only powerful identification methods can be applied in this situation and the requirements for good control performance are met only in the case of robust and reliable on-line identification algorithms. All systems designed using the so-called certainty equivalence principle [9] show this structure, since the uncertainties of estimated process parameters are not taken into account in the controller design procedure.

B. Principles of Design

The advances in design of adaptive systems are oriented according to their principles. These principles fix the theoretical concept of design, which is divided into two phases. In the first phase the main problem is to find out the adaptive law for the controller parameters after some pre-experimental work. The result is an adaptive control algorithm. Finally in the second phase the free parameters are chosen and the realization of the algorithm is taken into account. Four different design principles must be considered [12].

In the continuous design principle all signals are treated as continuous signals in the first phase. Then in the second phase the designed system is prepared for digital signal processing by discretization. In the discrete principle the process with its continuous input and output signals is treated as a sampled data system. This allows one to apply the theory of discrete systems. The result of the first phase is a system of nonlinear difference equations describing the entire control structure. Thereafter in the second phase numerical specification of the free design parameters follow. Much emphasis should be put on the discrete principle, since this allows a simpler design of a larger class of systems, e.g. for deadtime processes and for disturbances inside the process and disturbances in the measured process variables. In addition various on-line parameter estimation methods which have been developed during the last years, can be applied simply.

Adaptive control systems show a highly nonlinear behaviour. Therefore the guaranteed stability of the whole system may be a general aim of the design procedure. Using linear design methods global stability is never guaranteed. For high performance systems, therefore, methods from nonlinear theory must be used. In the stability design principle the global stability of the whole system is the basis of the design procedure. Therefore in the first phase the aim is to find an algorithm which guarantees stability in the whole parameter and signal space. For that purpose the direct method of Lyapunov or the hyperstability theory of Popov is frequently used.

For the stochastic design principle in the first phase the influence of unknown stochastic disturbances on the control system is already considered. This principle has to be applied in the case of large disturbances. Not only an adaption to the process but also to the kind of disturbances can be performed.

IV. ADAPTIVE CONTROL SYSTEM IN THE CEMENT INDUSTRY [13-22]

The process for cement production includes the following three subsystems:

a) Raw material blending: The raw material consists of the following four most important oxides: CaO, SiO_2, Al_2O_3, Fe_2O_3. The raw materials of different composition are contained in the feeder tanks, from which they are fed into the raw mill by a conveyor belt. Afterwards the ground mix of raw materials (rawmeal) is placed into the silo.

b) Rotary kiln: The homogenized silo content is fed to the kiln where as a result of the sinter process a very hard clinker is produced.

c) Cement grinding mill: The closed loop cement ball grinding mill works very similarly to the raw mill. The input material flow consists of the clinker usually mixed with gypsum and slag. The fresh material flow, together with the reflux, is fed into the ball grinding mill. The mill output is brought by an elevator to the separator where the final product (cement) is separated from the reflux (breeze).

Different adaptive control schemes have been applied to raw material blending and to cement grinding mills (Table I), which will be described briefly.

TABLE I. Practical applications of adaptive systems in process control.

field	references	PROCESS					ADAPTATION		
		plant	behaviour	output variables	time response (time constants)	manipulated variables	method	adapted variables	status
CEMENT INDUSTRY	[13] to [14]	cement raw material blending	a) stochastic varying amounts of the constituent oxides b) static non-linear characteristic with extremum	a) control of chemical composition of raw-meal (with minimum variance) b) control of maximum quantity of rawmeal (gradient evaluation)	time constants in minutes	weight feeder rates	modified self-tuning minimum variance regulator (MIMO-system)	set point of the weight feeder (moduli values)	pilot experiments 1976/78
	[16] [17]	cement grinding mills	a) stochastic variations in the quality of cement b) static non-linear characteristic with extremum	a) control of fineness (Blain value) of the final product (cement) b) control of maximum quantity of final product	time constants in minutes	a) speed of separator spreading plate b) mill input feed rate	a) selftuning minimum-variance control of fineness b) selftuning extremum regulator of quantity	parameters of the self-tuning regulator	pilot experiments 1976/78
	[18]			breeze flow control	$T_1=T_2$=3-6min T_t=4min	fresh input material feed	parameter adaptation by feed forward control	controller parameters	in operation 1977/78
	[19]			a) mill loading b) power consumption	T_1=3.3min T_2=1.1min T_t=2.5min	a) fresh input material b) set point of the cascaded inner loop	parameter adaptation by sensitivity models using a gradient method		pilot experiments 1976/78

S T E E L I N D U S T R Y	[24] [25]	plate mill	-statistical properties -nonlinear dynamics	-rolling torque -roll-separating force -power -final thickness -final shape -final width (final pass schedule)	very fast (seconds)	-screw down -load roll gauge -speed	step by step adaptation of model equations for optimal reference values (and optimal pass schedule)	adapted model equations	in operation 1973
	[24]	hot strip mill		-roll gap -thickness -final temperature -roll force -strip tension					in operation 1973/75
	[26]	tandem cold reduction mill		-roll gap -thickness -strip speed -roll force -strip tension					in operation 1975
	[27]	hot strip coiling	maintaining coil temperature constant is very difficult	coiling temperature	very fast	spray banks	matching model equations for controlling the error from strip to strip	model parameters	in operation 1975
	[28]		varying radius of coil	speed control	very fast	armature voltage	feed-forward parameter adaptation	controller paramaters	in operation 1979
	[29]	LD-converter	incomplete knowledge of dynamic behaviour of the converter and content of input	-decarburization rate -oxygen content in the slag	long dead times (∿20sec)	-oxygen flow rate (stop point of blowing) -lance height above the bath surface	predicitve algorithm using a model and Wiener filter ("learning algorithm")	model parameters	in operation
	[30]			-carbon content -metal temperature			self-adaptive model technique (self-tuning predictor)		in operation 1971
	[31]			-decarburization -carbon content of the steel			predictive model adaptation using the technique of hyperstable model reference systems		in operation 1975

TABLE I. (continued)

field	references	PROCESS					ADAPTATION		
		plant	behaviour	output variables	time response (time constants)	manipulated variables	method	adapted variables	status
METALLURGICAL INDUSTRY	[32]	blast furnace	time-varying, nonlinear characteristics of reduction process	quality and quantity of hot iron	very slow (hours)	-blast temperature -blast humidity	self-adaptive model technique (self-tuning predictor)	model-parameters	proposed for application 1978
	[33]	ore crusher	stochastic variation of input material quality	maximum quantity of crushed ore	long dead time (minutes)	mill input feed rate	self-tuning minimum variance regulator	parameters of the self-tuning regulator	in operation 1974
	[34]	sinter-plant	time-varying nonlinear behaviour, large measurement noise and input disturbances	permeability of sinter mixture (sintering rate)	PT_1T_t-process (T_t=20s, T_1=30s)	water addition (water valve position)	fuzzy-set theory	levels of input and output fuzzy sets	in operation 1976
	[35]	bulk flotation in nickel recovery	-nonlinear behaviour -statistical properties depending on ore minerals	-pH-value (varying between 3.5 and 5) -metal value of concentrates	slow	reagent	model adaptation to the process using regression analysis for prediction	model parameters	in operation 1975/76
						feed	optimum seaking using evolutionary operation principle	adaptation of pH-value	in operation 1975/76
	[36]	wet grinding mill in Ni-ore concentrator plant	-nonlinear behaviour -statistical changes in grindability	grinding noise (as a measure for mill power)	slow dead time	-feed rate of fine ore -lumps -water	tuning of parameters of a normal PI-controller with changing process conditions	controller parameters	in operation 1975

	[37]	titanium dioxide kiln	statistical properties	composition of process outlet	very slow (hours)	fuel rate	self-tuning regulator	parameters of the regulator	in operation 1977
CHEMICAL INDUSTRY	[38]	double-effect evaporator	a)deterministic disturbances -feed flow rate -feed concentration -feed enthalpy b)dynamics depending on feed rates	a)first effect hold up b)second effect hold up c)second effect concentration	1/2 - 1 hour	a)steam flow b)first effect bottoms c)second effect bottoms	model reference adaptive control Lyapunov-design	state-feedback controller with feedforward	pilot experiments 1972/73
	[39]	packed bed tubular reactor	nonlinear behaviour, high sensitivity to load changes (wall temperature)	extent of a)n-butane b)propane c)hydrogene	∿10min	feedrate of a)n-butane b)hydrogen	model reference adaptive control Lyapunov-design	state feedback controller	pilot experiments 1977
	[40]	distillation column	a)determinstic disturbances of feed flow rate b)dynamics depending on direction and magnitude of setpoint and feed flow rate changes	top product composition	T_t=1min	reflux flow-rate	self-tuning regulation(compared with PI-controller)	parameters of the self-tuning regulator	pilot experiments 1977
	[42] [43]			top product: a)concentration b)flow rate		a)reflux rate b)heating power	dual control algorithm (cautions control algorithm)	parameters of the algorithm	pilot experiments 1977
	[44]	cracking furnace	a)drift b)limited rate of cooling or heating	temperature	∿100sec	fuel flow	variable control period	control period	pilot experiments 1977/78
	[45]	neutralization of pH-value	nonlinear titration curve	pH-value	not significant	acid/bases-flow	PD-controller with deadband characteristic	width and gain of deadband	in operation 1973

TABLE I. (continued)

field	references	PROCESS					ADAPTATION		
		plant	behaviour	output variables	time response (time constants)	manipulated variables	method	adapted variables	status
PAPER INDUSTRY	[46]	paper machine	stochastic disturbances with unknown characteristics	a) moisture b) basis weight	large known time delays	a) steam pressure b) thick stack flow	MIMO-self-tuning regulator	parameters of the regulator	in operation 1972
	[47] [48]			moisture		steam pressure	self-tuning regulator with feedforward		in operation 1973
	[49]	continuous digester	a) nonlinear time varying b) stochastic disturbances	Kappa-number (quality of the pulp)	slow	steam pressure	self-tuning regulator		in operation 1973
	[50]	glasshouse heating systems	disturbances a) wind b) temperature c) radiation	glasshouse temperature	∿1/2h	heating pipe temperature	least-squares-like gradient identification of process gain-factor and adaptation of a PI-controller	gain of a master PI-controller	pilot experiments 1977
	[51]	refrigerant compressor test	nonlinear and unknown characteristics	a) evaporating pressure b) condensing pressure c) temperature at compressor suction inlet	∿20min	a) expansion b) flow rate of cooling water c) electrical power of the heater	nonlinear state and parameter estimation using Kalman-filter technique. Optimum control using quadratic performance index	all parameters of feedback law	experiments 1977

MISCELLANEOUS AREAS OF PROCESS INDUSTRY	[52]	air heater	nonlinear behaviour (air flow dependant)	air-temperature		mixing valve position	recursive least squares parameters estimation	parameters of a deadbeat-controller	pilot experiments 1978
	[53]	condenser cooling process	nonlinear static characteristics	effective power		r.p.m.of condenser pump	extremum control using Hammerstein model	r.p.m. of condenser pump	pilot experiments 1976
	[54]	epitaxy reactor	time-variant dynamics	a) thickness b) sensitivity of epitaxy layers		a) dropping gas flow b) duration of reaction	parameter identification by Kalman filtering	inverse model parameters	in operation 1974
	[55]	vaporization reactor	unknown process dynamics	rate of vaporization		electrical vaporization heating	recursive least squares parameter estimation	parameters of a deadbeat controller	in operating 1976
	[56]	heat exchanger	load dependent time-variant dynamics	outlet temperature	2-4min	heat flow rate	comparison of 3 methods: a) adaptation of a PI-controller using test signals b) adaptation of compensating controller by direct identification c) parallel model technique using hyperstability theory	controller parameters	pilot experiments 1976

A. Adaptive Control of Raw Material Blending

As the main goal in cement production is to produce cement of a given quality at minimum cost, two control tasks should be fulfilled:

- control of the maximum quantity of rawmeal.
- control of the chemical composition of rawmeal.

Both control loops are coupled. However when a good quantity control system is applied, e.g. by using an autonomous adaptive extremum controller, then the quality control can be investigated separately. For the raw mill especially the control of chemical composition is extremely important, while the control of the input feed rate and fineness is not so interesting. The quality characteristics (Ca, Si, Al-moduli) must be kept on given values with minimum variance. This is necessary because the components of the raw materials and their physical characteristics are very variable. For this purpose a self-tuning algorithm for adaptive control of the quality was developed [13-16].

The technological scheme of the process is shown in Fig. 4. In the feeder tanks are raw materials of different composition. The weigh feeders are controlled by the computer. The rubble is fed into the mill by a conveyor belt. Afterwards the ground mix of raw materials (rawmeal) is placed in the silo. Before the silo the meal is sampled and the sample is fed into an X-ray Fluorescence Analyser (FRA). This provides analysed oxide values, which are used as process-output signals for the computer. According to the self-tuning algorithm for MIMO-systems the computer adjusts directly the scale factors of the weigh feeders on the basis of the moduli values computed from the measured-oxides of the mill output and the quantity of the ground rawmeal fed to the silo. By the applied self-tuning control strategy the minimization of the variance of the output signals (oxide moduli) from varying reference values is assured.

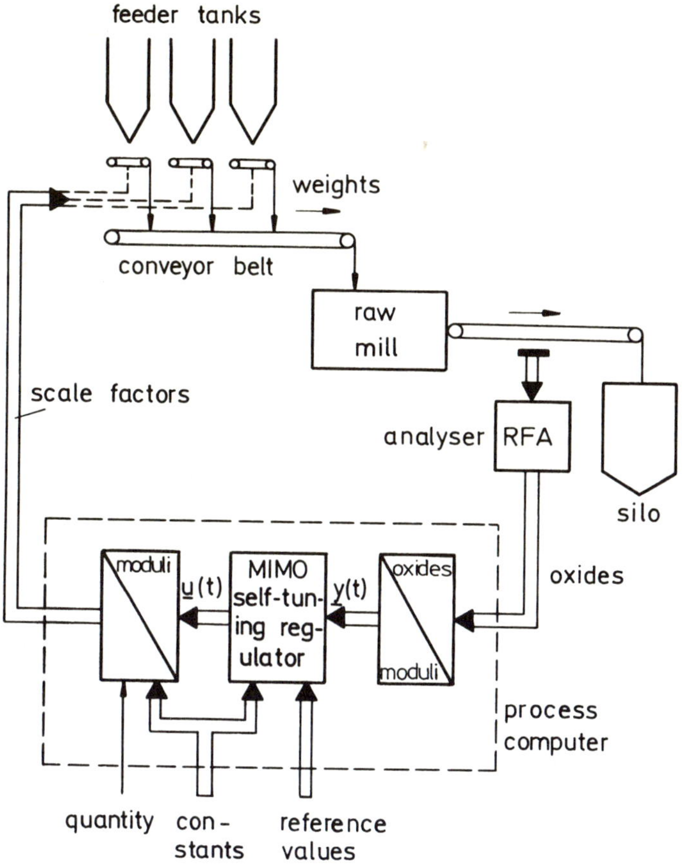

FIGURE 4. Block scheme of the adaptive control of raw material blending.

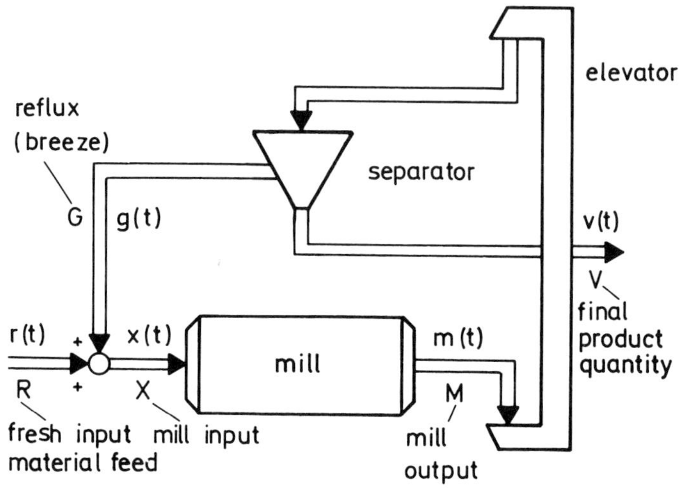

FIGURE 5. Sketch of the mill (lower-case letters refer to time-varying signals; capital letters to steady-state values).

B. Adaptive Control of Cement Grinding Mills

A sketch showing the principle of a closed circuit mill is given in Fig. 5. The basic operation is very similar to that of the raw mill. The steady-state relationship (described by capital letters)

$$V = R \tag{1}$$

$$M = X = R + G = V + G \tag{2}$$

$$U = X = M \tag{3}$$

can be obtained directly from this figure. The static characteristic of the mill has an extremum (maximum) $V = V(X)$ for the input material flow as is shown in Fig. 6. The purpose of an

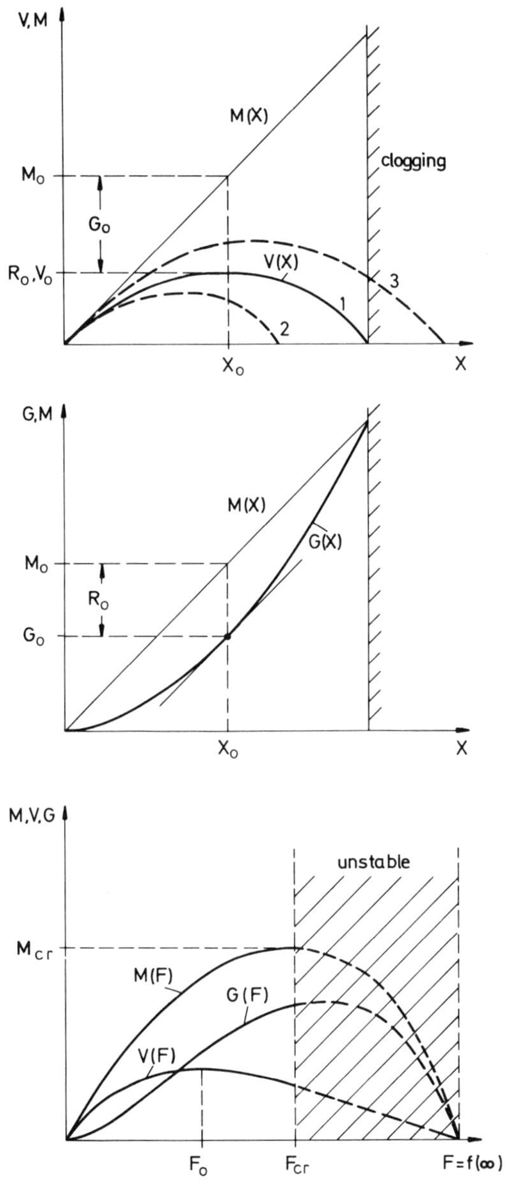

FIGURE 6. Nonlinear characteristics of a closed circuit mill.

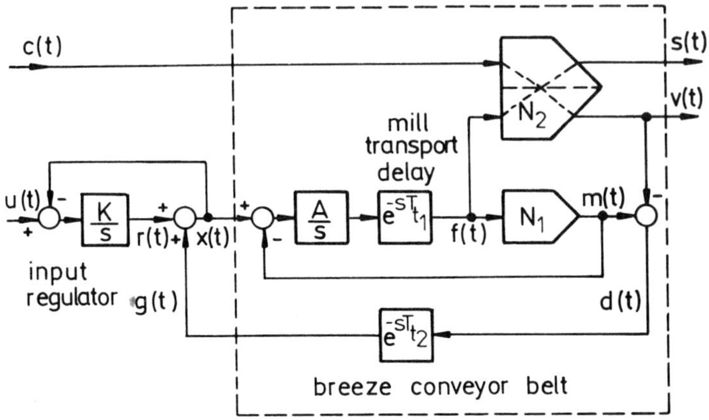

FIGURE 7. Block diagram of the closed circuit cement mill and separator system.

optimal control may be to seek the optimum quantity X_o under varying process conditions and to ensure the optimum values M_o and V_o by this means.

The mill has a second nonlinear characteristic with respect to the degree of filling f(t) or F (see Fig. 7) showing also extremum features, which is also illustrated in Fig. 6. The area to the right of F_{cr} (critical degree of filling) cannot be reached.

From the consideration on the material balances of

R = V and M = V

two alternative control concepts seem to be reasonable and are usually applied in the cement industry:

a) $X = R + G = \text{const}$ (4)

b) $G = \text{const.}$ (5)

In addition to these control concepts several models for the dynamic properties of the closed circuit mill have been proposed during the last few years. Based on different models, also different strategies for adaptive control, systems have been developed which are briefly described in the following.

1. SELF-TUNING EXTREMUM MILL CONTROL [17]. For the control concept X = const it seems to be reasonable to ensure a follow-up control between x(t) and m(t). The specially developed dynamic model for the mill, including the separator, transport delay of the mill, the breeze conveyor belt and the elevator together with the control loop for the input of fresh material feed is shown in Fig. 7. The integrator with the transfer function A/s describes the essential part of the mill dynamics because it guarantees that the system can get to its steady state only if the conditions $m(\infty) = M = X = x(\infty)$ and thus $r(\infty) = R = V = v(\infty)$ hold. The internal feedback loop with input x(t) and output m(t) represents, therefore, the conservation of matter. The nonlinearity N_1 is in the inner loop and the transfer function $e^{-sT_{t_1}}$ represents the transport delay of the mill. This loop includes the degree of filling f(t) from which the finished product can be obtained via N_2.

The recycling flow of breeze can be written as

$$d(t) = m(t) - v(t) \tag{6}$$

which after the dead time T_{t_2} arrives as $g(t) = d(t-T_{t_2})$ at the input of the mill, where

$$x(t) = r(t) + g(t) \tag{7}$$

holds. The input control loop with the reference value u(t) as input and output signal x(t) is a servo control system for which $x(\infty) = u(\infty)$ in the steady state. The input regulator ensures the constancy of x(t) by changing r(t) when the recycling flow of breeze g(t) acts as disturbing variable.

The block diagram described above represents the basis of a new adaptive method for the optimal control of the closed-circuit ball-grinding mill. For this purpose a multivariable adaptive system based on a modified self-tuning regulator algorithm was realized at a hierarchically higher control level, the operation of which can be considered autonomous. There are two separated adaptive loops as can be seen from Fig. 8. The Blain value s(t) of the cement as a measure of quality is adjusted to a given

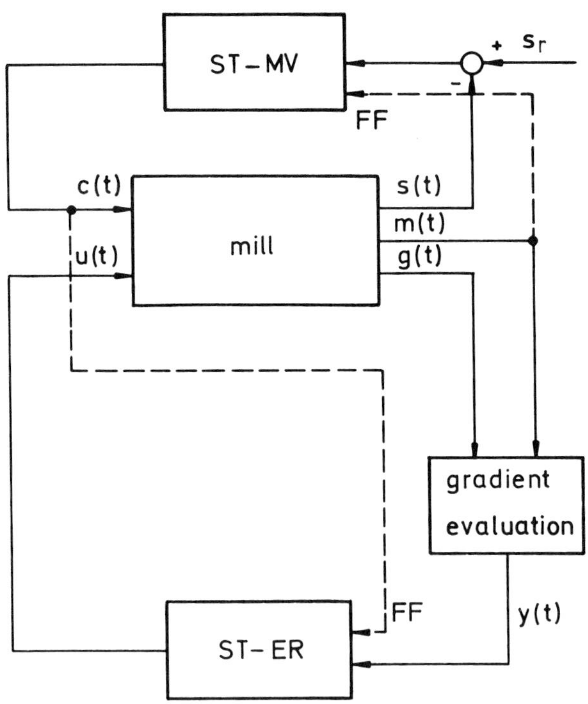

FIGURE 8. Scheme of the MIMO-self-tuning extremum mill control.

reference value s_r with minimum variance by self-tuning control (ST - MV) of the speed c(t) of the separator spreading plate. Since the characteristics of the separator strongly depend on its load, a feedforward disturbance compensation (FF) has to be introduced into the self-tuning algorithm. As the quantity of the rawmeal v(t) cannot be measured in the investigated process a method was elaborated which uses the available values of the recycling flow of breeze g(t) and the power intake of the elevator proportional to m(t). So much fresh material has to be given to the closed loop mill as to ensure the reaching of the maximum of the nonlinearity V(X), i.e. the maximum quantity of the cement production of a given fineness. This condition is fulfilled if dG/dX is kept equal to unity. For this purpose a modified self-tuning extremum regulator (ST-ER) algorithm was used to generate the control law u(t). Also here a feedforward disturbance compensation (FF) is necessary because changes of c(t) (speed of the separator spreading plate) strongly influence this loop too.

2. ADAPTIVE BREEZE FLOW CONTROL [18]. Here the control concept G = const was used. The dynamic behaviour of the feed-forward path of the mill is described by the transfer function (PT_2T_t-behaviour)

$$G_{M_o}(s) = \frac{L\{m(t)\}}{L\{x(t)\}} = \frac{e^{-sT_t}}{(1-sT_M)^2} \quad . \tag{8}$$

In the closed-circuit cement grinding system the dynamic behaviour of the separator can be neglected against that of the mill. Thus the steady-state operation of the separator gives

$$g(t) = km(t) \text{ or } G = kM$$

$$v(t) = (1-k)m(t) \text{ or } V = (1-k)M$$

with $k = k(m) < 1$.

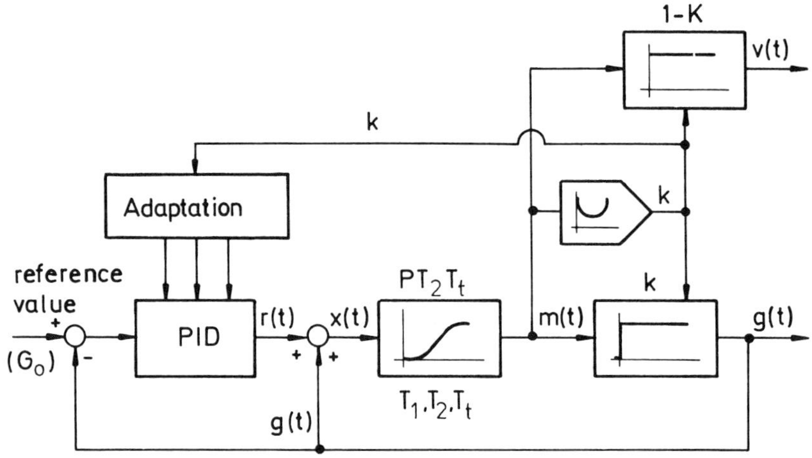

FIGURE 9. Adaptation of breeze flow rate controller.

The closed circuit cement grinding mill thus is described for k = const by the transfer function

$$G_{M_c}(s) = \frac{L\{g(t)\}}{L\{r(t)\}} = \frac{kG_{M_o}(s)}{1 - kG_{M_o}(s)} \tag{9}$$

according to Fig. 9. However, as the separator setting k is a nonlinear function, the plant behaviour depends strongly on the mill output flow rate $m(t)$. As k is directly connected to the recirculation rate μ of the mill

$$\mu = \frac{X}{V} = \frac{X}{R} = \frac{G+R}{R} = \frac{G+V}{V} = \frac{1}{1-k} , \tag{10}$$

changes of μ (usually $1.5 \leq \mu \leq 5$) influence strongly the control behaviour of the plant. According to changes of k or μ the parameters of the PID-controller are adapted in order to keep the control action of the breeze flow rate constant.

3. ADAPTIVE CONTROL FOR CEMENT MILLS USING SENSITIVITY MODELS [19]. The applied control concept here was M = X = R + G = const. The model of the feed-forward path of the mill is described again by Eq. (8). For the separator the nonlinear characteristic g = g(m) which directly corresponds to G = G(x) of Fig. 6 is assumed. In addition to the previous model the delay of the breeze conveyor belt is considered. Thus we obtain the model structure shown in the block diagram of Fig. 10. The control scheme consists of two cascaded controllers. The inner control loop has a PI-controller for the mill loading (measured by a microphone) by acting on the fresh input material feed while the outer control loop has a PID-controller for the power consumption of the elevator system (as measured by the mill output flow rate m(t)). All five controller parameters were made adaptive so as to minimize the integral criteria of the two error signals. Thereby a direct parameter adaptation was applied, using a sensitivity model based on the gradient method. This sensitivity model consists of a linear

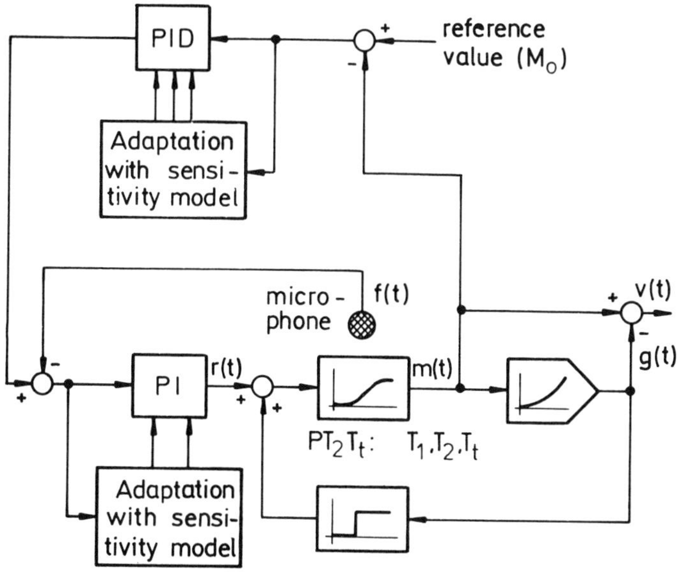

FIGURE 10. Adaptation of the control loops.

model of the plant and the controller. Therefore the nonlinearity of the separator had to be linearized for the sensitivity model. The proportional coefficients of the gradient algorithm are normalized and automatically adapted to the behaviour of the mill.

V. ADAPTIVE CONTROL IN THE STEEL AND METALLURGICAL INDUSTRIES [24-37]

Many processes in the steel and metallurgical industries are nonlinear and nonstationary. Therefore, the application of classical control methods has not led to the desired results. More than ten years ago the first proposals were made for improving these processes by application of adaptive control systems. However, the processes in the steel and metallurgical industries are characterized by a large variety of very different properties. Very often the specific target variables cannot be directly measured, or at least not without considerable time delay. In some parts of these multivariable processes both continuous and discontinuous nonlinear processes with small reaction times are coupled. Furthermore the disturbances are usually distributed through the entire complex process. On the other hand there are also very slow processes (e.g. blast furnaces) where no continuous measurements of the target variables are possible.

Thus in the steel and metallurgical industries there exist processes with very great differences concerning the degree of automation. Where direct or at least good indirect measurements of the target variables are available, process computers have been already widely introduced. In these cases adaptive methods have been successfully applied, as Table I indicates. From this table it becomes obvious that the adaptive methods applied in the steel industry are mainly based on highly sophisticated process models.

Assuming these process models, measurements are used directly for a step-by-step adaptation of the parameters of the model equations for steel mills [24-26] and also for updating predictive models of LD-converters [29-31].

In metallurgical plants the self-tuning regulator has been given serious consideration. This technique has been applied with dramatic success, e.g. to ore crushing plants [33]. The main advantage of this technique is that it identifies a process from operating data without the need for lengthy experiments.

The main success of adaptive control has been in various applications to steel mills. Therefore, in the following these applications will be discussed briefly.

A. Applications in the Steel Industry

Metal rolling mill systems appear to represent the most successful applications in the iron and steel industry of sophisticated process control systems. Adaptive control modes are successfully used on a broad basis. While the later discussion uses a hot strip mill as a specific example, it applies generally to plate mills and tandem cold reduction mills.

The control modes include feedback, feedforward, predictive and adaptive control [23]. Feedback control is used for speed, roll position and automatic gauge control. Feedforward control is applied in several of the process model calculations, including roll force and temperature. These process models are extremely important for the operation of the plant. The order in which model programs run, iteration and adaption of the model permit accurate performance with relatively straightforward algorithms.

Predictive control with bar-to-bar feedback is applied for head end thickness and temperature using the above mentioned process models. Predictive control is also used for in-bar control of finishing and coiling temperatures because of significant transport lags between points of control and points of measurement.

Thermal and dimensional conditions in the mill change with the pace and duration of rolling. Rolling characteristics may vary from bar to bar even for the same stated steel grade code. The predictive control must adapt to this changeable environment for accurate control of the physical properties of the steel strip. The control system must be and is adaptive, implying three criteria:

1) Definition of optimum operating conditions, e.g. product dimensions and temperatures.
2) Comparison of actual with desired values of controlled variables, e.g. roll force and temperature.
3) Adjustment of system parameters for correction of the actual performance towards that desired. The system makes automatic on-line adjustments to the process model to represent more accurately the parameters related to the product chemistry, mill condition, and other environmental factors.

Having determined the reduction distribution along the mill stands, the reference values for conventional control of roll gap, thickness after stand, strip speed, roll force and tension are calculated for each stand by means of mathematical process models. These models consist of roll force, roll gap, rolling torque, forward slip and material hardness models and are constantly updated by means of adaptation constants during rolling in order to achieve a more optimal set-up calculation for the mill. For the model adaptation regional adaptation is preferred because it is almost impossible to employ reasonable techniques to a complex equation that represents the global range of a process. Thus the actually measured values are inserted in the mathematical partial models and new preliminary adaptation constants calculated [26]. By

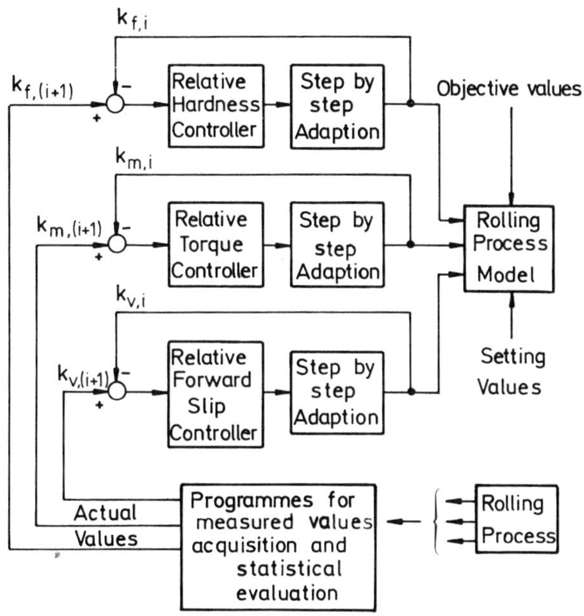

FIGURE 11. Adaptation of mathematical process models (k_f, k_m, k_v adaptation constants for material hardness, spindle torque and forward slip).

comparing the old adaptation constants k_i and these preliminary adaptation constants k_{i+1} the old adaptation constants are adapted. Using the newly calculated adaptation constants new rolling parameters are calculated and issued to the mill according to Fig. 11.

Constant recalculation and resetting results in diminishing differences between the calculated and actual mill parameters. Once a predetermined number of adaptations has been made a regression analysis of the successively calculated adaptation

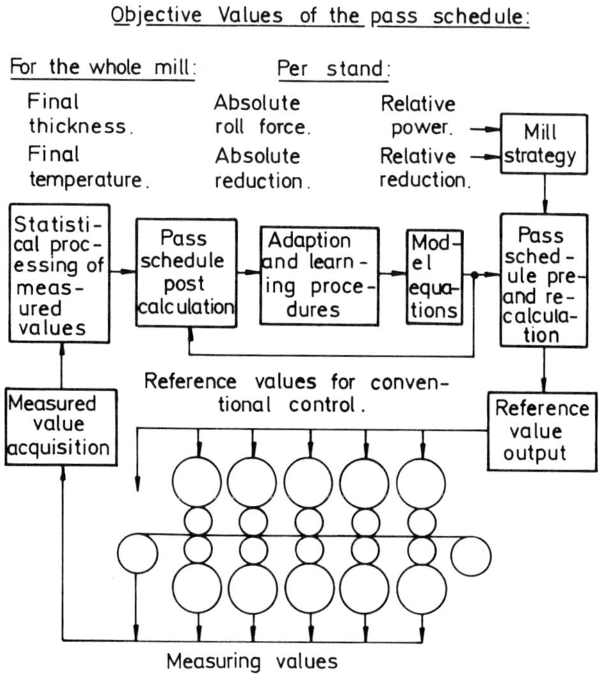

FIGURE 12. Process control of a continuous strip mill.

constants is made to adapt the material constants. Fig. 12 represents the process control in the form applied for hot strip mills and cold strip tandem mills. The manipulated and output variables of the process recorded for the individual strip segments by the measured value acquisition are presented to the pass schedule post-calculation via the statistical measured value processing system. In using the model equations the pass schedule post-calculation derives from the measured values the parameter vector of the adaptation constants for each stand. The adaptation then updates the model equations as described before. Thus the pass schedule pre- and re-calculation have each adapted model equations available. To calculate the required optimum pass schedule further specification of the objective values of the pass and related technological factors are required.

Objective values are the final values of thickness and exit temperature. The related technological factors include, for example, the power distribution through the stands and the technologically prescribed limit values.

If the optimum pass schedule is fixed with regard to these technological factors, then the process manipulated variables are derived. The "reference value directive" makes available these process manipulated variables at the correct instants to the subordinate single variable controllers. From these considerations it is obvious that the adaptive multivariable control of steel rolling mills has the function of so matching the model equations that they correctly represent the relationships between the output and manipulated variables.

VI. ADAPTIVE CONTROL IN THE CHEMICAL INDUSTRY [38-45]

Applications of adaptive control schemes in chemical plants are very rare. Nevertheless some work is in progress and aimed at the design and implementation of simple and robust adaptive controllers to replace conventional ones. Investigations are being carried out on some pilot plants like distillation columns, evaporators or reactors according to Table I. Robust algorithms capable of self-tuning or adaptation of parameters are desirable, particularly if large excursions in the plant operation occur. Model reference adaptive control (MRAC) systems are used to adjust the inputs and the parameters of a plant so that its outputs track those of a reference model [38,39]. This is usually based on the lowest model order that adequately represents the process dynamics. MRAC systems are insensitive to inherent modelling errors, they do not require the time-consuming calculations for process identification and optimization and offer a feasible approach for tuning multivariable control systems. Difficulties in controlling chemical processes are often associated with variable time constants and time delays and process nonlinearities. Any digital algorithm used therefore must be capable not only of achieving indefinite steady state regulation but must be able to deal satisfactorily with any sudden large disturbance in the set point or measured plant output. This and load disturbance rejection are problems which can be solved by self-tuning regulators [40,41].

As a representative application of adaptive control let us consider the following example of a distillation column.

A. Adaptive Control of Distillation Columns

Distillation columns can be regarded as typical chemical plants to demonstrate the feasibility of adaptive control techniques. A schematic and very simplified diagram of a (binary) distillation column is shown in Fig. 13. The mixture of (two) components, fed into the column, is split into a light component (top product) and into heavy components (bottom product). The principal control inputs are the internal reflux rate $\dot{M}_R$ and the heating power, which is proportional to the steam flow rate $\dot{M}_S$. These inputs control the concentration c_T and the exit flow rate $\dot{M}_T$ of the top product. The additional valves V_2 and V_4 are for level control in the condenser and reboiler, which do not figure in the whole control scheme as their function is only local. The main control objective is to hold the top product concentration c_T and flow rate $\dot{M}_T$ on constant values despite feed flow rate disturbances $\Delta\dot{M}_F$. From time to time it is necessary to change the throughput of the column to meet changing demands. In order to minimize the disturbances

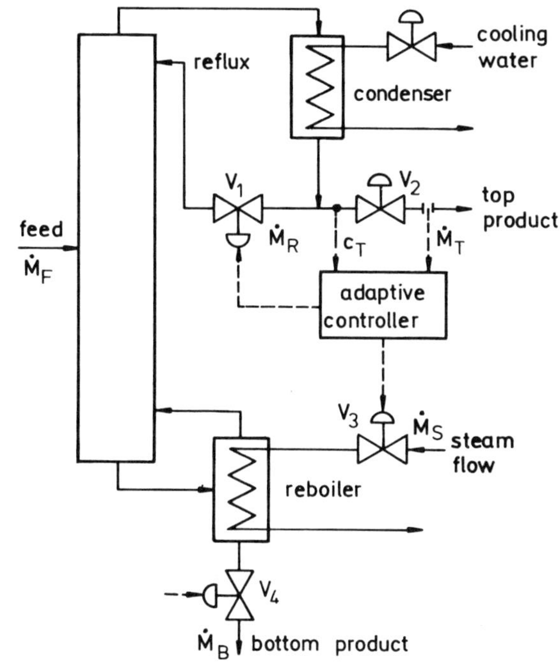

FIGURE 13. Schematic diagram of a distillation column and the control scheme.

to the column conditions resulting from these necessary changes, a supervisory control scheme (not indicated in Fig. 13) applies the changes in set point to the feed flow controllers. Since the column behaviour depends on the feed rates, feedforward control of c_T is an accepted control scheme, but it requires measurement of disturbances, which may not be feasible or not always possible. As the operating conditions change, the controller parameters should be adapted since the column behaviour is nonlinear.

Gain and time constants of the plant are strongly dependent upon the feed rate $\dot{M}_F$ and also upon the direction and magnitude of $\Delta\dot{M}_F$. In addition, stochastic fluctuations influence the process variables. So a self-tuning controller [40,41] or a dual control strategy [42,43] may be appropriate components in such interactive plants.

VII. ADAPTIVE CONTROL IN THE PAPER INDUSTRY [46-49]

Paper machines have been controlled digitally for many years. Digital control of base weight and moisture content is an example where adaptive control techniques are applied. The control objectives are to maintain the best possible uniform quality, under the production rate chosen and with the given grade of paper. Further, waste is to be minimized. The specific aims are to keep the base weight and moisture as close as possible to a chosen setpoint and the mean moisture content as high as possible; the deviations should have minimum statistical variance. Further, the loss of fiber should be minimized in the case of paper breaks.

Design of an advanced control system based on modern control theory certainly requires more knowledge about the paper process than is necessary in most cases for direct application of some conventional control schemes. A very simplified diagram of a paper machine is shown in Fig. 14. It can conveniently be divided into several subprocesses, headbox, wet end, press section, dryer section and drive system. The thick stock from the machine chest enters through the thick stock valve and is mixed with white water from the wire pit in a pump, the tubes and the pressurized headbox. The thin stock consistency level and the flow rate in the headbox are controlled by this mixing pump. On the wire the fibers are

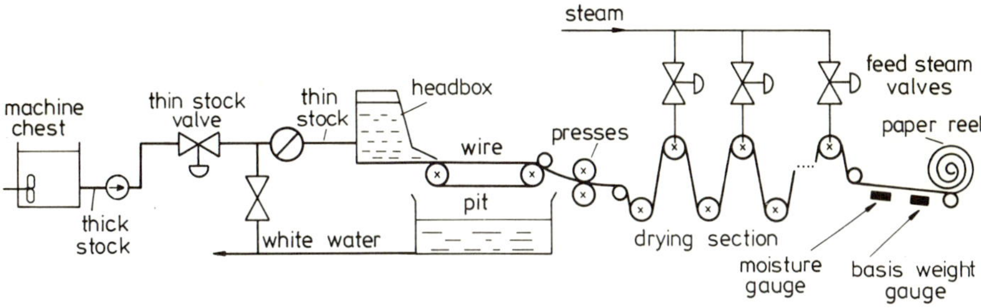

FIGURE 14. Schematic diagram of a paper machine.

separated from the water and a paper sheet is formed. After the wire, the machine has a pickup and presses. In the drying section the moisture content is reduced by steam groups. The steam pressure in each group can be controlled separately by steam feed valves.

The object of control is to operate the thick stock flow valve and also the steam pressure in the drying section in such a way that the base weight and the moisture deviation of the manufactured paper from predetermined set points are minimized. For the purpose of control it is of major interest to understand the dynamics which describe how fluctuations in concentration propagate through the system. The processes involved, mixing and separation of water fiber mixture, are very complicated and a detailed understanding of the mixing and separation phenomena is not available. So there is no possibility of determining the characteristics of the disturbances in the system.

In this case minimum variance control strategies are appropriate, which in addition adapt to the process dynamics. This can be successfully performed by means of self-tuning controllers as shown in Table I.

VIII. ADAPTIVE CONTROL OF MISCELLANEOUS AREAS OF PROCESS INDUSTRY [50-56]

There is also interest in the application of adaptive systems in other areas of process control. One remarkable application is the MIMO-adaptive control of a refrigerant compressor test plant (Fig. 15) with 3 inputs and 3 outputs [51]. Since in refrigerant compressor performance tests the static characteristics must be examined in a wide range as accurately and quickly as possible, a very fast control system is necessary, which shows stable and optimal behaviour for the whole range of operating conditions. Thus the control objective is to change the process from one set of operating conditions to another in an optimal manner. This was unsuccessfully accomplished using single-variable PID-control with fixed parameters. Because of the interaction of the process

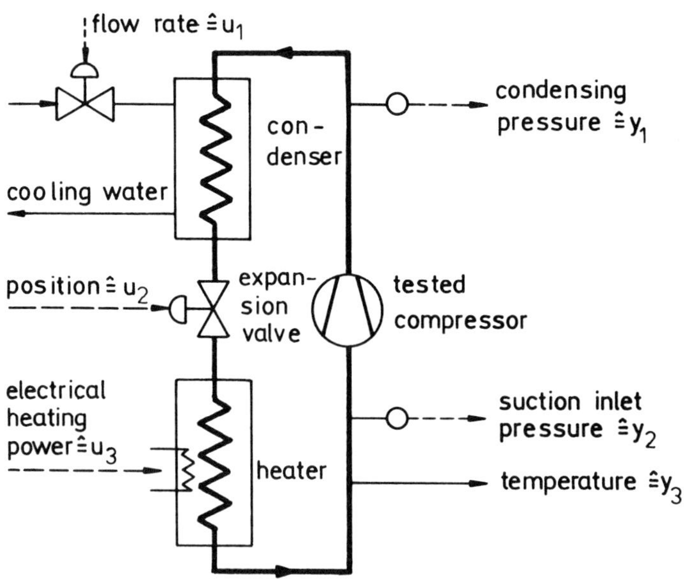

FIGURE 15. Schematic diagram of a refrigerant test plant.

variables and the unknown nonlinear behaviour, PID-control shows low performance and instabilities. In an adaptive solution a decomposed discrete state space model is used to estimate both states and parameters by Kalman filters. Based on the estimated result the optimal control is determined using a quadratic performance index. In spite of a high computing time (compared with PID-control) a satisfactory control behaviour could be obtained over a wide range of operating conditions.

Another interesting example of the successful application of adaptive control has been in pilot experiments on a heat-exchanger [56]. This system is characterized by a strongly load-dependent time-variant dynamical behaviour. Three different adaptive control systems have been designed and compared, as can be seen from Table I. As a result of this study it has been shown that the method using a parallel model of the closed loop combined with hyperstability theory gave an excellent adaptive control action and, at the same time, needed the smallest effort for realization.

These two applications and others (see Table I), demonstrate the practicability of adaptive control methods which are based on modern control theory.

IX. DISCUSSION OF THE ACTUAL SITUATION

While extensive use of many adaptive control techniques has been made in other areas, e.g. aircraft and space control systems, electrical drives and signal processing, during the last years only relatively few applications have been reported from the broad field of process industries, as we have seen from the previous sections. During the last 20 years a large number of reports have been published on the application of identification and estimation techniques in process industries. However, only very few reports deal with the adaptive control schemes of the type considered here and especially with the continuous use of such control systems for more than a short pilot-scale experimental period.

The reason for this situation cannot be explained only by the relatively complicated dynamic behaviour of these industrial processes under external and internal disturbances, drifts, parameter changes etc. Often the main reason is blamed on the conservative management of many processes, who will not hazard their very expensive plants by applying modern control methods which are unproven and still "in leading reins". In addition among the production and instrumentation staff there is an absence of sufficiently trained personnel who understand these control strategies. In practice most of the industrial processes are regulated using two-point or cascaded PI-control with fixed parameters. Sometimes a derivative term is added to improve stability. These regulators are simple to implement and to tune, with low sensitivity to small process disturbances, and in the vast majority of cases perform excellently. However, at odd times, there remain some loops which cannot be controlled satisfactorily. These problems arise mostly from the extreme nonlinear behaviour of the process, from large and time-varying transport delays and from strong interactions between process variables. So the controller has to be retuned whenever the operating point is changed. This is also the case in many starting phases of process industry plants.

Although there has recently been some experimental work on the implementation of modern adaptive techniques in nearly all areas of process industry plants, in laboratories as well as in some factories there is still no great interest to accept these techniques. Despite the advantages of more sophisticated algorithms they will not be adopted unless it can be clearly demonstrated, that they are far superior to classical control in those few loops and they do not introduce other problems into an interactive process industry plant. Optimum long term steady state regulation, good set point following properties, and the need to ensure that any control action applied and the effect it produces on the plant does not in any way endanger plant safety, are the main general conflicting requirements. In order to fulfil the desired dual role of long term steady state regulation and large disturbance control, an adaptive controller must be both robust and flexible in its control actions. What the real behaviour of such an adaptive controller in extreme situations might be, is difficult to predict from simulations or experiments on small scale processes.

The application of adaptive control in most fields of process industry is to this date in a phase of discovery. It is necessary to identify problems in which adaptive control theory can offer real advantages for process control. The hitherto known applications should not conceal the fact that the financial expense often did not yield the desired commercial benefit, since often a presumably inadequate knowledge of the processes limited successful application.

From this brief discussion of the actual situation of adaptive control in process industries it becomes obvious that research engineers as well as industrial engineers must cooperate closely in the further developments of this field. The research engineers should try in the future not only to develop other new adaptive methods, but to investigate common features of various methods that already exist. This would also allow the transfer of knowledge and results to other methods and applications. As only a few such trials have been conducted, e.g. [12], much remains to be done in the future.

The research engineer should suggest for practical application only those methods which fulfil the following conditions:

- possibility of a systematic design of a guaranteed stable and efficient adaptive controller using only little a priori knowledge;

- the expense of realization by mini or micro computers must be low, especially with regard to the number of numerical operations during one sampling period and the memory capacity;
- general applicability of the method to a variety of processes.

According to these conditions the self-tuning regulator principle and the hyperstable design principle seem to be the most promising groups of adaptive systems. Furthermore for practical application, rules for a systematic design must be provided. It would be desirable for the industrial engineer to have program packages for a complete interactive design procedure which can be understood easily. As only few such possibilities are available up to now [57], much effort has still to be made.

On the other side the industrial engineer must provide more possibilities for applications of adaptive methods. However, for this he must fulfil the following conditions:

- a good possibility of describing dynamical processes properly;
- a better education in modern control and thus more confidence in adaptive control systems;
- intensive cooperation with the research engineer during the design and realization phase.

In spite of all this, more experimental work should be done crystallizing a group of generally applicable adaptive controllers from theory and thus arousing a large scope and incentive for application.

X. CONCLUSION

While we can agree that adaptive control methods have worked, reports on applications are thin and process performance data from such systems even thinner. Until these voids in the technical field are filled, industrial engineers are likely to continue to use time-proven classical methods understood and valued from proven performance. Therefore, more interaction between industry and research institutes is necessary. Industry should initiate orientation to provide examples of real world process control systems and problems. The academics should avoid structuring hypothetical problems to justify minor extensions to basic theory, and instead study real world problems with the intention of developing the extensions needed to apply basic theory to these problems. What is needed is interactive communication on a regular basis, and not merely transmittal of papers. However, engineers in industry must also learn to describe clearly their control problems in a way which may be accepted by research engineers.

REFERENCES

1. Unbehauen, H. and Schmid, Chr., "Status and Industrial Application of Adaptive Control Systems," Automatic Control Theory and Application 3, No. 1, pp. 1-12, 1975.
2. Asher, R. B., Andrisani, D. and Dorato, P., "Bibliography on Adaptive Control Systems," Proceed. IEEE 64, No. 8, pp. 1226-40 1976.
3. Zypkin, J., "Adaption und Lernen in Kybernetischen Systemen," Oldenbourg Verlag, München, 1970.
4. Zypkin, J., "Adaptive System Theory Today and Tomorrow," Proceed. IFAC-Symposium Ischia (Italy), pp. 47-67, 1973.
5. Landau, I. D., "Model Reference Adaptive Systems - a Survey," ASME Trans. J. of Dynamic Systems, Measurement and Control 94, pp. 119-132, 1972.
6. Hang, C. C. and Parks, P. C., "Comparative Studies of Model Reference Adaptive Control Systems," IEEE Trans. Aut. Control 18, pp. 419-428, 1973.

7. Lindorff, D. and Carroll, R., "Survey of Adaptive Control Using Lyapunov Design," Int. J. Control 18, pp. 897-914, 1973.
8. Lüders, G. and Narendra, K., "Lyapunov Functions for Quadratic Differential Equations with Applications to Adaptive Control," IEEE Trans. Aut. Control 17, pp. 798-801, 1972.
9. Wittenmark, B., "Stochastic Adaptive Control Methods - a Survey," Int. J. Control 21, No. 5, pp. 705-730, 1975.
10. Landau, I. D. and Unbehauen, H., "The Development of Adaptive Systems in Germany and France," ASME Trans. J. of Dynamic Systems, Measurement and Control 96, pp. 405-413, 1974.
11. Huguenin, F., "Entwicklung von Regelalgorithmen für Mikrorechner," Elektroniker, No. 11, pp. 11-20, 1978.
12. Schmid, Chr., "Ein Beitrag zur Realisierung Adaptiver Regelungssysteme," Ph.D. Dissertation, Ruhr-Universität, Bochum, 1979.
13. Keviczky, L., Kolostori, J. and Hilger, M., "On Simultaneous Optimal Control of a Raw Material Blending and Ball Grinding Mill," Proceed. IFAC-Symposium, Johannesburg, pp. 143-158, 1976.
14. Keviczky, L., Hetthessy, J., Hilger, M., and Kolostori, J., "Self-Tuning Adaptive Control of Cement Raw Material Blending," Automatica 14, pp. 525-32, 1978.
15. Hilger, M., Keviczky, L., Kolostori, J.,and Szijj, F., "Moderne Regelungsmethoden und die bei Ihrer Anwendung Erworbenen Erfahrungen," Preprints Fachtagung, "Regelungstechnik in Zementwerken," Bielefeld (Germany), Sept. 1978.
16. Csaki, K., Keviczky, L., Hetthessy, J., Hilger, M. and Kolostori, J., "Simultaneous Adaptive Control of Chemical Composition, Fineness and Maximum Quantity of Ground Materials at a Closed Circuit Ball Mill," Proceed. IFAC-Congress, Helsinki, pp. 453-460, 1978.

17. Keviczky, L., and Vajk, I., "A Self-Tuning Extremum Regulator," Proceed. 12th Asilomar Conference on Circuits, Systems and Computers, pp. 69-72, 1978.

18. Heinrich, K. und Kinske, H., "Durchsatzoptimierung von Sichterumlaufmühlen," Preprints Fachtagung "Regelungstechnik in Zementwerken," Bielefeld (Germany), Sept. 1978.

19. Kunze, E. und Salaba, M., "Praktische Erprobung Eines Adaptiven Regelungsverfahrens an Einer Zementmahlanlage," Preprints Fachtagung "Regelungstechnik in Zementwerken," Bielefeld (Germany),Sept. 1978.

20. Schulz, R., "Prozeßregelung Einer Kugelmühle," Preprints Fachtagung "Regelungstechnik in Zementwerken," Bielefeld (Germany), Sept. 1978.

21. Schulze, H., "Erfahrungen mit Einem Systemprogramm für Adaption und Regelung in Zementwerken," IFAC/IFIP-Symposium, Zürich, pp. 436-450, 1974.

22. Zhivogladov, V. and Mirkin, B., "Adaptation and Direct Digital Control of the Cement Industry Technological Process with Application of Distributed Check," IFAC/IFIP-Symposium, Zürich, pp. 451/462, 1974.

23. Müller, W. E., "Automation in Metal Processing," Proceed. IFAC-Symposium, Johannesburg, pp. 15-30, 1976.

24. Seyfried, H. W. and Stöle, D., "Application of Adaptive Control in Rolling Mill Area, Especially for Plate Mills," Proceed. IFAC-Congress, Boston, paper 46.1, 1974.

25. Stelzer, R., Heideprim, J. and Griese, F. W., "Adaptivalgorithmen für die Prozeßmodelle Einer Rechnergesteuerten Grobblechstraße,"Stahl und Eisen 93, Nr. 23, S. 1100-1105, 1973.

26. Alberts, R. and Joubert, A., "Automation of Iscor's Five Stand Tandem Cold Mill," Proceed. IFAC-Symposium, Johannesburg, pp. 597-605, 1976.

27. Ono, M., Kurokawa, T., Hirao, F., Takeda, E., and Nakano, K., "The Computer Control System of Hot Strip Coiling Temperature," Proceed. IFAC-Congress, Helsinki, pp. 159-166, 1978.

28. Dittmar, E., "Mikrocomputer-Einsatz in der Automatisierung," Vogel-Verlag, Würzburg, 1979.

29. Voll, H., Danby, P. and Ramelot, D., "Improvement in Oxygen Furnace Controllability," Prepr. IFAC/IFIP-Symposium, Zürich, pp. 99-111, 1974.

30. Yasui, T., "Dynamic Control of LD Converter with Digital Computer," Proc. IFAC/IFIP-Symposium, Zürich, pp. 112-121, 1974.

31. Landau, I. D. and Muller, L., "A New Method for Carbon Control in Basic Oxygen Furnace," Proceed. IFAC-Symposium, Johannesburg, pp. 257-261, 1976.

32. Wellstead, P. E., Montagner, G. and Gale, S., "A Self-Adaptive Model for the Prediction of Blast Furnace Hot Metal Production and Quality," Control System Centre Report UMIST, 1978.

33. Borisson, U. and Syding, R., "Self-Tuning Control of an Ore Crusher," IFAC-Symposium on Stochastic Control, Budapest, pp. 491-499, 1974.

34. Carter, G. and Rutherford, D., "A Heuristic Adaptive Controller for a Sinter Plant," IFAC-Symposium, Johannesburg, pp. 315-324, 1976.

35. Mattila, O., and Pentillä, H., "Experience of Computer Automation at Kotalahti Concentrator," Proceed. IFAC-Symposium, Johannesburg, pp. 461-473, 1976.

36. Uronen, P., Tarvainen, M. and Aurasmaa, H., "Audiometric Control System of Wet Semiautogenous and Autogenous Mills," Proceed. IFAC-Symposium, Johannesburg, pp. 215-233, 1976.

37. Dumont, G. A. and Bélanger, P. R., "Self-Tuning Control of a Titanium Dioxide Kiln," Trans. IEEE AC-23, No. 4, pp. 532-538, 1978.

38. Oliver, W. K., Seborg, D. E., and Fisher, D. G., "Model Reference Adaptive Control Based on Liapunov's Direct Method, Part I and II," Chem. Eng. Commun, pp. 125-140, 1973.

39. Tremblay, J. P., and Wright, J. D., "Multivariable Model Reference Adaptive Control of a Pilot Scale Packed Bed Tubular Reactor," IFAC/IFIP-Symposium Den Haag, paper M2-3, pp. 513-525, 1977.

40. Sastry, V. A., Seborg, D. E., and Wood, R. K., "Self-Tuning Regulator Applied to a Binary Distillation Column," Automatica 13, No. 4, pp. 417-424, 1977.

41. Morris, A. J., Fenton, T. P. and Nazer, Y., "Application of Self-Tuning Regulators to the Control of Chemical Processes," IFAC/IFIP-Symposium Den Haag, paper M1-3, pp. 447-455, 1977.

42. Badr, O., Rey, D., and Ladet, P., "Adaptive Dual Control for a Multivariable Distillation Process," IFAC/IFIP-Symposium, Den Haag, paper A1-3, pp. 141-147, 1977.

43. Badr, O., Rey, D., and Ladet, P., "On the Dual-Adaptive Control and its Practical Applications," Automatica 15, pp. 91-96 1979.

44. D'Hulster, F. M. and van Cauwenberghe, A. R., "An Adaptive Controller for Slow Processes with Variable Control Periods," IFAC-Congress Helsinki, pp. 461-467, 1978.

45. Shinsky, F. G., "Adaptive pH Controller Monitors Nonlinear Process," Control Engineering, No. 2, pp. 57-59, 1974.

46. Cegrell, T. and Hedquist, T., "Successful Adaptive Control of Paper-Machines," Preprint IFAC/IFIP-Symposium Den Haag, paper PV-2, pp. 485-491, 1973.

47. Borisson, U., "Self-Tuning Regulators - Industrial Application and Multivariable Theory," Report 5713, Lund Institute of Technology, Lund, 1975.

48. Aström, K. J., Ljung, L., Borisson, U., and Wittenmark, B., "Theory and Applications of Adaptive Regulators Based on Recursive Parameter Estimation," IFAC-Congress Boston, paper 50.1, 1974.

49. Cegrell, T., and Hedquist, T., "A New Approach to Continuous Digester Control," Proc. IFAC/IFIP-Symposium, Zürich, pp. 300-311, 1974.

50. Udink ten Cate, A. J., and van de Vooren, J., "Digital Adaptive Control of a Glasshouse Heating System," IFAC/IFIP-Symposium Den Haag, paper M2-2, 1977.

51. Fujii, S., Fujimoto, H., Shibata, A., Shimada, S., and Hajiri, "Application of an Adaptive Control in a Refrigerant Compressor Test," IFAC/IFIP-Symposium Den Haag, paper M2-5, 1977.

52. Kurz, H., Isermann, R., and Schumann, R., "Development, Comparison and Application of Various Parameter-Adaptive Digital Control Algorithms," Proc. IFAC-Congress Helsinki, pp. 443-452, 1978.

53. Bamberger, W., "Adaptive On-Line-Optimierung des Statischen Verhaltens Dynamisch Träger Prozesse mit dem Programmpaket 'OLIOPT'," VDI-Bericht 276, pp. 131-140, 1977.

54. Rabut, C., "Self Adaptive Numerical Control of an Epitaxi Reactor Using Kalman-Bucy Filtering," IFAC-Congress Boston, paper 52.4, 1974.

55. Kapfer, W., "On-Line Identifikation Einer Aufdampfanlage Mittels ProzeBrechner," VDI-Bericht 276, pp. 101-107, 1977.

56. Unbehauen, H. and Schmid, Chr., "Adaptive Regelung Eines Wärmetauschers mit dem ProzeBrechner," Proceed. 21, Internat. Wiss. Koll. TH Ilmenau, pp. 109-117, 1976.

57. Schmid, Chr., "CAD of Adaptive Systems," Proceed. IFAC-Symposium on Computer Aided Design of Control Systems, Zürich, 1979.